T0211371

LONDON MATHEMATICAL SOCIETY LECTURE NOTE SERIES

Managing Editor: Professor M. Reid, Mathematics Institute,
University of Warwick, Coventry CV4 7AL, United Kingdom

The titles below are available from booksellers, or from Cambridge University Press at
http://www.cambridge.org/mathematics

London Mathematical Society Lecture Note Series: 428

Geometry in a Fréchet Context

A Projective Limit Approach

C. T. J. DODSON
University of Manchester, UK

GEORGE GALANIS
Hellenic Naval Academy, Piraeus, Greece

EFSTATHIOS VASSILIOU
University of Athens, Greece

CAMBRIDGE
UNIVERSITY PRESS

CAMBRIDGE
UNIVERSITY PRESS

Shaftesbury Road, Cambridge CB2 8EA, United Kingdom

One Liberty Plaza, 20th Floor, New York, NY 10006, USA

477 Williamstown Road, Port Melbourne, VIC 3207, Australia

314–321, 3rd Floor, Plot 3, Splendor Forum, Jasola District Centre, New Delhi – 110025, India

103 Penang Road, #05–06/07, Visioncrest Commercial, Singapore 238467

Cambridge University Press is part of Cambridge University Press & Assessment, a department of the University of Cambridge.

We share the University's mission to contribute to society through the pursuit of education, learning and research at the highest international levels of excellence.

www.cambridge.org
Information on this title: www.cambridge.org/9781316601952

First published 2016

A catalogue record for this publication is available from the British Library

ISBN   978-1-316-60195-2   Paperback

Contents

Preface

The aim of the authors is to lay down the foundations of the projective systems of various geometrical structures modelled on Banach spaces, eventually leading to homologous structures in the framework of Fréchet differential geometry, by overcoming some of the inherent deficiencies of Fréchet spaces. We elaborate this brief description in the sequel.

Banach spaces, combining a metric topology (subordinate to a norm), and a linear space structure (for representing derivatives as linear approximations to functions in order to do calculus), provide a very convenient setting for many problems in functional analysis, which we need for handling calculus on function spaces, usually infinite dimensional. They are a relatively gentle extension from experience on finite dimensional spaces, since many topological properties of spaces and groups of linear maps, as well as many of the existence and uniqueness theorems for solutions of differential equations carry over to the infinite dimensional case.

Manifolds and fibre bundles modelled on Banach spaces arise from the synthesis of differential geometry and functional analysis, thus leading to important examples of global analysis. Indeed, many spaces of (differentiable) maps between appropriate manifolds admit the structure of Banach manifolds (see, for instance, J. Eells [Eel66, § 6]).

On the other hand, as mentioned also in [Eel66], Riemannian manifolds, represented as rigid maps on infinite dimensional function spaces, arise as configuration spaces of dynamical systems, with metrics interpreted as kinetic energy. Much of the calculus of variations and Morse theory is concerned with a function space in differential geometry— the Euler-Lagrange operator of a variational problem is interpreted as a gradient vector field, with integral curves the paths of steepest ascent. Some eigenvalue problems in integral and differential equations are interpretable via Lagrangian multipliers, involving infinite dimensional function spaces from differential geometry—such as focal point theory and geometric consequences of the inverse function theorem in infinite dimensions.

However, in a number of situations that have significance in global analysis and physics, for example, physical field theory, Banach space representations break down. A first step forward is achieved by weakening the topological requirements: Instead of a norm, a family of seminorms is considered. This leads to Fréchet spaces, which do have a linear structure and their topology is defined through a sequence of seminorms.

Although Fréchet spaces seem to be very close to Banach spaces, a number of critical deficiencies emerge in their framework. For instance, despite the progress in particular cases, they lack a general solvability theory of differential equations, even the linear ones; also, the space of continuous linear morphisms between Fréchet spaces does not remain in the category, and the space of linear isomorphisms does not admit a reasonable Lie group structure.

The situation becomes much more complicated when we consider manifolds modelled on Fréchet spaces. Fundamental tools such as the exponential map of a Fréchet-Lie group may not exist. Additional complications become particularly noticeable when we try to collect Fréchet spaces together to form bundles (over manifolds modelled on atlases of Fréchet spaces), in order to develop geometrical operators like covariant derivatives and curvature to act on sections of bundles. The structure group of such bundles, being the general linear group of a Fréchet space, is not a Lie group—even worse, it does not have a natural topological structure. Parallel translations do not necessarily exist because of the inherent difficulties in solving differential equations within this framework, and so on.

This has relevance to real problems. The space of smooth functions $C^\infty(I, \mathbb{R})$, where I is a compact interval of $\mathbb{R}$, is a Fréchet space. The space $C^\infty(M, V)$, of smooth sections of a vector bundle V over a compact smooth Riemannian manifold M with covariant derivative ∇, is a Fréchet space. The C^∞ Riemannian metrics on a fixed closed finite-dimensional orientable manifold has a Fréchet model space. Fréchet spaces of sections arise naturally as configurations of a physical field. Then the moduli space, consisting of inequivalent configurations of the physical field, is the quotient of the infinite-dimensional configuration space $\mathcal{X}$ by the appropriate symmetry gauge group. Typically, $\mathcal{X}$ is modelled on a Fréchet space of smooth sections of a vector bundle over a closed manifold.

Despite their apparent differences, the categories of Banach and Fréchet spaces are connected through projective limits. Indeed, the limiting real product space $\mathbb{R}^\infty = \lim_{n \to \infty} \mathbb{R}^n$ is the simplest example of this situation. Taking notice of how $\mathbb{R}^\infty$ arises from $\mathbb{R}^n$, this approach extends to arbitrary Fréchet spaces, since always they can be represented by a countable sequence of Banach spaces in a somewhat similar manner. Although careful concentration to the above example is salutary, (bringing to mind the story of the mathematician drafted to work on a strategic radar project some 70 years ago, who when told of the context said "but I only know Ohms Law!" and the response came, "you only need to know Ohms Law, but you must know it very, very well"), it should be emphasized that the mere properties of $\mathbb{R}^\infty$ do not answer all the questions and problems referring to the more complicated geometrical structures mentioned above.

The approach adopted is designed to investigate, in a systematic way, the extent to which the shortcomings of the Fréchet context can be worked round by viewing, under sufficient conditions, geometrical objects and properties in this context as limits of sequences of their Banach counterparts, thus exploiting the well developed geometrical tools of the latter. In this respect, we propose, among other generalizations, the replacement of certain pathological structures and spaces such as the structural group of a Fréchet bundle, various spaces of linear maps, frame bundles, connections on principal and vector bundles etc., by appropriate entities, susceptible to the limit process. This extends many classical results to our framework and, to a certain degree, bypasses its drawbacks.

Apart from the problem of solving differential equations, much of our work is motivated also by the need to endow infinite-dimensional Lie groups with an exponential map [a fact characterizing–axiomatically– the category of (infinite-dimensional) *regular* Lie groups]; the differential and vector bundle structure of the set of infinite jets of sections of a Banach vector bundle (compare with the differential structure described in [Tak79]); the need to put in a wider perspective particular cases of projective limits of manifolds and Lie groups appearing in physics (see e.g. [AM99], [AI92], [AL94], [Bae93]) or in various groups of diffeomorphisms (e.g. [Les67], [Omo70]).

For the convenience of the reader, we give an outline of the presen-

tation, referring for more details to the table of contents and the introduction to each chapter.

Chapter 1 introduces the basic notions and results on Banach manifolds
and bundles, with special emphasis on their geometry. Since there is not
a systematic treatment of the general theory of connections on Banach
principal and vector bundles (apart from numerous papers, with some
very fundamental ones among them), occasionally we include extra details on specific topics, according to the needs of subsequent chapters.
With a few exceptions, there are not proofs in this chapter and the
reader is guided to the literature for details. This is to keep the notes
within a reasonable size; however, the subsequent chapters are essentially
self-contained.

Chapter 2 contains a brief account of the structure of Fréchet spaces and
the differentiability method applied therein. From various possible differentiability methods we have chosen to apply that of J.A. Leslie [Les67],
[Les68], a particular case of Gâteaux differentiation which fits well to the
structure of locally convex spaces, without recourse to other topologies.
Among the main features of this chapter we mention the representation
of a Fréchet space by a projective limit of Banach spaces, and that of
some particular spaces of continuous linear maps by projective limits of
Banach functional spaces, a fact not true for arbitrary spaces of linear
maps. An application of the same representation is proposed for studying differential equations in Fréchet spaces, including also comments on
other approaches to the same subject. Projective limit representations
of various geometrical structures constitute one of the main tools of our
approach.

Chapter 3 is dealing with the smooth structure, under appropriate conditions, of Fréchet manifolds arising as projective limits of Banach manifolds, as well as with topics related to their tangent bundles. The case of
Fréchet-Lie groups represented by projective limits of Banach-Lie groups
is also studied in detail, because of their fundamental role in the structure of Fréchet principal bundles. Such groups admit an exponential
map, an important property not yet established for arbitrary Fréchet-
Lie groups.

Chapter 4 is devoted to the study of projective systems of Banach principal bundles and their connections. The latter are handled by their
connection forms, global and local ones. It is worthy of note that any
Fréchet principal bundle, with structure group one of those alluded to
in Chapter 3, is always representable as a projective limit of Banach

principal bundles, while any connection on the former bundle is an appropriate projective limit of connections in the factor bundles of the limit. Here, related (or conjugate) connections, already treated in Chapter 1, provide an indispensable tool in the approach to connections in the Fréchet framework. We further note that the holonomy groups of the limit bundle do not necessarily coincide with the projective limits of the holonomy groups of the factor bundles. This is supported by an example after the study of flat bundles.

Chapter 5 is concerned with projective limits of Banach vector bundles. If the fibre type of a limit bundle is the Fréchet space $\mathbb{F}$, the structure of the vector bundle is fully determined by a particular group (denoted by $\mathcal{H}_0(\mathbb{F})$ and described in § 5.1), which replaces the pathological general linear group $GL(\mathbb{F})$ of $\mathbb{F}$, thus providing the limit with the structure of a Fréchet vector bundle. The study of connections on vector bundles of the present type is deferred until Chapter 7.

Chapter 6 contains a collection of examples of Fréchet bundles realized as projective limits of Banach ones. Among them, we cite in particular the bundle $J^\infty(E)$ of infinite jets of sections of a Banach vector bundle E. This is a non trivial example of a Fréchet vector bundle, essentially motivating the conditions required to define the structure of an arbitrary vector bundle in the setting of Chapter 5. On the other hand, the generalized bundle of frames of a Fréchet vector bundle is an important example of a principal bundle with structure group the aforementioned group $\mathcal{H}_0(\mathbb{F})$.

Chapter 7 aims at the study of connections on Fréchet vector bundles the latter being in the sense of Chapter 5. The relevant notions of parallel displacement along a curve and the holonomy group are also examined. Both can be defined, despite the inherent difficulties of solving equations in Fréchet spaces, by reducing the equations involved to their counterparts in the factor Banach bundles.

Chapter 8 is mainly focused on the vector bundle structure of the second order tangent bundle of a Banach manifold. Such a structure is always defined once we choose a linear connection on the base manifold, thus a natural question is to investigate the dependence of the vector bundle structure on the choice of the connection. The answer relies on the possibility to characterize the second order differentials as vector bundle morphisms, which is affirmative if the connections involved are properly related (conjugate). The remaining part of the chapter is essentially an application of our methods to the second order Fréchet tangent bundle and the corresponding (generalized) frame bundle.

We conclude with a series of open problems or suggestions for further applications, within the general framework of our approach to Fréchet geometry, eventually leading to certain topics not covered here.

These notes are addressed to researchers and graduate students of mathematics and physics with an interest in infinite-dimensional geometry, especially that of Banach and Fréchet manifolds and bundles. Since we have in mind a wide audience, with possibly different backgrounds and interests, we have paid particular attention to the details of the exposition so that it is as far as possible self-contained. However, a familiarity with the rudiments of the geometry of manifolds and bundles (at least of finite dimensions) is desirable if not necessary.

It is a pleasure to acknowledge our happy collaboration, started over ten years ago by discussing some questions of common research interest and resulting in a number of joint papers. The writing of these notes is the outcome of this enjoyable activity. Finally, we are very grateful to an extremely diligent reviewer who provided many valuable comments and suggestions on an earlier draft, we have benefited much from this in the final form of the monograph.

Manchester – Piraeus – Athens,
February 2015

1

Banach manifolds and bundles

The geometry of Banach manifolds and bundles has been greatly developed since the 1960s and now there are many papers and a number of books covering a great variety of related topics. Here we intend to fix our notation and give a brief account of the basic results which will be used in the main part of the present work. Occasionally, some topics are dealt with in more detail. These refer to subjects either not easily found in the literature or their methods have a particular interest and cover explicit needs of the exposition.

1.1 Banach manifolds

The main references for this section are [AMR88], [AR67], [Bou67], and [Lan99], where the reader may find the necessary details.

1.1.1 Ordinary derivatives in Banach spaces

Let $\mathbb{E}$ and $\mathbb{F}$ be two Banach spaces. We denote by $\mathcal{L}(\mathbb{E}, \mathbb{F})$ the (Banach) space of continuous linear maps between $\mathbb{E}$ and $\mathbb{F}$. In particular, we set $\mathcal{L}(\mathbb{E}) := \mathcal{L}(\mathbb{E}, \mathbb{E})$, which is a Banach algebra. On the other hand, $\mathcal{L}is(\mathbb{E}, \mathbb{F})$ denotes the (open) set of invertible elements (viz. linear isomorphisms) of $\mathcal{L}(\mathbb{E}, \mathbb{F})$, while $\mathcal{L}is(\mathbb{E}) := \mathcal{L}is(\mathbb{E}, \mathbb{E})$. The latter space, viewed as a group under the composition of automorphisms, is denoted by $\mathrm{GL}(\mathbb{E})$ and is called the **general linear group** of $\mathbb{E}$.

A map $f : U \to \mathbb{F}$ ($U \subseteq \mathbb{E}$ open) is called **differentiable at** x if there exists a map $Df(x) \in \mathcal{L}(\mathbb{E}, \mathbb{F})$, the (Fréchet) **derivative of** f **at** x, such that

$$\lim_{h \to 0} \frac{\|f(x+h) - f(x) - [Df(x)](h)\|}{h} = 0 \qquad (h \neq 0).$$

The (*total*) **derivative**, or **differential**, of f is $Df \colon U \to \mathcal{L}(\mathbb{E}, \mathbb{F})$. If Df is continuous, then we say that f is of class C^1. Inductively, we set

$$D^k f = D(D^{k-1} f) \colon U \longrightarrow \mathcal{L}^k(\mathbb{E}, \mathbb{F}) \equiv \mathcal{L}(\mathbb{E}, \mathcal{L}^{k-1}(\mathbb{E}, \mathbb{F})),$$

if the latter derivative exists. The map f will be called **smooth**, or (of class) C^∞, if the derivatives D^k exist for every k and are continuous.

For an excellent treatment of the differential calculus in Banach spaces we refer also to [Car67(a)].

1.1.2 Smooth structures

A **Banach manifold** M is a smooth manifold whose differential structure is determined by local charts of the form $(U, \phi, \mathbb{B})$, where the **ambient space** or **model** $\mathbb{B}$ is a Banach space. If all the charts have the same model $\mathbb{B}$ (a fact ensured in the case of a connected manifold), we say that B is **modelled on** $\mathbb{B}$ or it is a $\mathbb{B}$-**manifold**. If there is no ambiguity about the model, the charts will be simply denoted by (U, ϕ). The (maximal) **atlas** inducing the differential structure is denoted by $\mathcal{A}$.

A Banach space $\mathbb{B}$ is a Banach manifold whose differential structure is determined by the global chart $(\mathbb{B}, \mathrm{id}_{\mathbb{B}})$.

For the sake of simplicity, unless otherwise stated, differentiability is assumed to be of class C^∞, a synonym of smoothness. Usually, a Banach manifold is assumed to be *Hausdorff*, equipped with smooth *partitions of unity*.

1.1.3 Smooth maps

A map $f \colon M \to N$ is said to be **smooth at** $x \in M$ if there are charts (U, ϕ) and (V, ψ) of M and N, respectively, such that $x \in U$, $f(U) \subseteq V$, and the **local representation** or **representative** of f, with respect to the previous charts,

(1.1.1) $\psi \circ f \circ \phi^{-1} \colon \phi(U) \longrightarrow \psi(V)$

is smooth at $f(x)$ in the sense of ordinary differentiability in Banach spaces. Short-hand notations for (1.1.1) are f_{VU} or $f_{\psi\phi}$. We also write $f_{\beta\alpha}$ for the local representation of f with respect to the charts (U_α, ϕ_α) and (U_β, ϕ_β), with $f(U_\alpha) \subseteq U_\beta$.

1.1.4 The tangent space

A **smooth curve at** $x \in M$ is a smooth map $\alpha \colon J \to M$ with $\alpha(0) = x$, where J is an open interval of $\mathbb{R}$ containing 0. Two curves α and β at x are called **equivalent** or **tangent** if there is a chart (U, ϕ) at x such that

$$(1.1.2) \qquad (\phi \circ \alpha)'(0) = (\phi \circ \beta)'(0)$$

Here we have that

$$(1.1.3) \qquad (\phi \circ \alpha)'(t) = [D(\phi \circ \alpha)(t)](1),$$

for every $t \in J$ such that $\alpha(t) \in U$. Clearly, (1.1.2) is equivalent to

$$(1.1.2') \qquad D(\phi \circ \alpha)(0) = D(\phi \circ \beta)(0).$$

The equivalence classes of curves as above are denoted by $[(\alpha, x)]$ (or $[\alpha, x]$ for complicated expressions of curves) and are called **tangent vectors at** x. The set of all tangent vectors at x is the **tangent space at** x, denoted by $T_x M$.

Considering any chart $(U, \phi) \equiv (U, \phi, \mathbb{B})$ at x, we check that $T_x M$ is in a bijective correspondence with $\mathbb{B}$ by means of the map

$$(1.1.4) \qquad \overline{\phi} \colon T_x M \longrightarrow \mathbb{B} \colon [(\alpha, x)] \mapsto (\phi \circ \alpha)'(0).$$

Therefore, $T_x M$ becomes a Banach space and $\overline{\phi}$ a continuous linear isomorphism. The Banach structure of $T_x M$ is independent of the choice of the chart containing x. This is an immediate consequence of the following fact: If $(U, \phi, \mathbb{B})$ and $(U, \psi, \mathbb{B}')$ are two charts at x, then the following diagram is commutative:

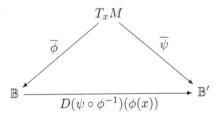

Considering a Banach space $\mathbb{B}$ as a smooth manifold, the tangent space $T_b \mathbb{B}$, for every $b \in \mathbb{B}$, is identified with $\mathbb{B}$ by means of $\overline{\mathrm{id}_{\mathbb{B}}}$ (see § 1.1.2). In particular, $T_t \mathbb{R}$ is an 1-dimensional vector space, with the natural basis

$$(1.1.5) \qquad \left. \frac{d}{dt} \right|_t := \overline{\mathrm{id}_{\mathbb{R}}}^{-1}(1).$$

1.1.5 The tangent bundle

As usual, the **tangent bundle** of a (Banach) manifold M is determined by the triple (TM, M, τ_M), where

$$TM := \bigcup_{x \in M}^{\cdot} T_x M \equiv \bigsqcup_{x \in M} T_x M$$

(disjoint union) is the **total space** and $\tau_M \colon TM \to M$ the **projection** of the tangent bundle, with $\tau_M([(\alpha, x)]) := x$.

The total space TM is a Banach manifold, whose structure is induced as follows: Given a local chart $(U, \phi) \equiv (U, \phi, \mathbb{B})$, we define the map

$$(1.1.6) \qquad\qquad \Phi \colon \pi^{-1}(U) \longrightarrow \phi(U) \times \mathbb{B}$$

by setting

$$(1.1.7) \qquad \Phi(u) := \big(\tau_M(u), \overline{\phi}(u)\big) = \big(x, (\phi \circ \alpha)'(0)\big),$$

if $u = [(\alpha, x)] \in T_x M$ and $x \in U$. Then the collection of all pairs $(\pi^{-1}(U), \Phi)$, obtained by running (U, ϕ) in the maximal atlas of M, determines a smooth atlas on TM, whose maximal counterpart induces the desired smooth structure on TM.

1.1.6 The differential of a smooth map

The tangent spaces and the tangent bundle provide the appropriate framework for the development of a differential calculus on manifolds. Precisely: if $f \colon M \to N$ is a smooth map between two Banach manifolds, then the **differential** or **tangent map of** f **at** x is the map

$$(1.1.8) \qquad\qquad T_x f \colon T_x M \longrightarrow T_{f(x)} N,$$

given by

$$(1.1.9) \qquad\qquad T_x f([(\alpha, x)]) := [(f \circ \alpha, f(x))].$$

This is a well-defined continuous linear map, independent of the choice of the representatives of the tangent vectors.

In various computations, the differential $T_x f$ is handled by using local charts and the derivative of the corresponding local representation (1.1.1) of f. More precisely, if $(U, \phi, \mathbb{E})$ and $(V, \psi, \mathbb{F})$ are local charts of M and N, respectively, such that $x \in U$ and $f(U) \subseteq V$ (as ensured by the smoothness of f at x), then one proves that the next diagram is commutative.

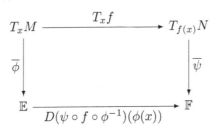

In particular, differentiating the map ϕ of a chart $(U, \phi, \mathbb{B})$, we obtain the following commutative diagram:

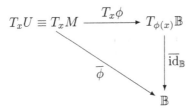

Frequently, omitting $\overline{\mathrm{id}}_\mathbb{B}$, we simply write

$$(1.1.10) \qquad \overline{\phi} \equiv T_x\phi.$$

1.1.7 Velocity vectors

Let $\alpha\colon J \to M$ be a smooth curve. The **tangent** or **velocity vector at** $\alpha(t)$ (or, simply, at t) is the vector

$$(1.1.11) \qquad \dot{\alpha}(t) := T_t\alpha\Big(\frac{d}{dt}\Big|_t\Big) \in T_{\alpha(t)}M.$$

In particular, if α passes through x, i.e. $\alpha(0) = x$, then

$$(1.1.12) \qquad \dot{\alpha}(t) = [(\alpha, x)].$$

If the curve has a more complicated form, e.g. $f \circ \alpha$, then the corresponding velocity vector is denoted by $(f \circ \alpha)^{\bullet}(t)$ instead of $\widetilde{(f \circ \alpha)}(t)$.

1.1.8 The tangent map

Let $f\colon M \to N$ be a smooth map. The **tangent map** or **(total) differential** of f is obtained by gluing together the differentials $T_x f$, for all $x \in M$; that is,

$$(1.1.13) \qquad Tf\colon TM \longrightarrow TN\colon Tf\big|_{T_x M} = T_x f.$$

The following diagram is also commutative:

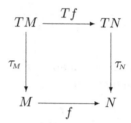

Note: For the differentials of maps on manifolds we prefer to use the *functorial* T instead of d, the latter been reserved for the exterior differential of differential forms.

1.1.9 Vector fields

A **vector field** on M is a **section** of the tangent bundle; that is, a map of the form $X\colon M \to TM$ such that $\pi \circ X = \mathrm{id}_M$. The set of *smooth* vector fields on M is denoted by $\mathcal{X}(M)$. The bracket of vector fields determines the structure of a Lie algebra on $\mathcal{X}(M)$.

A vector field X induces a derivation of the algebra of smooth functions on M by $X(f)(x) = T_x f(X_x)$, for every $x \in M$. For the correspondence between vector fields and derivations of smooth functions or Banach space valued maps on M see, for instance, [AMR88].

Given a chart $(U, \phi, \mathbb{B})$ of M and the corresponding chart $(\pi^{-1}(U), \Phi)$ of the tangent bundle (see §1.1.5), the local representation of $X \in \mathcal{X}(M)$, with respect to the previous charts, is the map $\Phi \circ X \circ \phi^{-1}$ (see §1.1.3) shown also in the following diagram:

Then, the *(local) **principal part*** of X (with respect to the above representation) is the map

(1.1.14) $\qquad X_\phi := \mathrm{pr}_2 \circ \Phi \circ X \circ \phi^{-1} \colon \phi(U) \longrightarrow \mathbb{B}.$

If we consider an indexed chart $(U_\alpha, \phi_\alpha, \mathbb{B})$, then we set

(1.1.14') $\qquad\qquad\qquad X_\alpha := X_{\phi_\alpha}.$

1.1.10 Related vector fields

Let $f \colon M \to N$ be a smooth map. Two vector fields $X \in \mathcal{X}(M)$ and $Y \in \mathcal{X}(N)$ are f-***related***, if $Tf \circ X = Y \circ f$. Equivalently,

$$T_x f(X_x) = Y_{f(x)}, \qquad x \in M.$$

1.1.11 Integral curves

A smooth curve $\alpha \colon J_\alpha \to M$ (J_α: open interval containing 0) such that $\alpha(0) = x$ and

$$X(\alpha(t)) = \dot{\alpha}(t), \qquad t \in J_\alpha$$

is called an ***integral curve*** of $X \in \mathcal{X}(M)$ with ***initial condition*** x. Locally, the problem of finding α reduces to the determination of a smooth curve $\beta \colon J_\beta \to \mathbb{B}$ such that $\beta(0) = \phi(x)$ and

(1.1.15) $\qquad\qquad \beta'(t) = X_\phi(\beta(t)), \qquad t \in J_\beta$

[recall also (1.1.14)]. The theory of differential equations in Banach spaces ensures the existence and uniqueness of such a β. Thus $\alpha = \phi^{-1} \circ \beta$ is an integral curve of X with initial condition $\alpha(0) = x$.

If M is a Hausdorff manifold, then there is a unique integral curve α with $\alpha(0) = x$, defined on a maximal interval of $\mathbb{R}$ containing 0.

1.2 Banach-Lie groups

Beside the references given in the begining of § 1.1, here we add [Bou72] and [Mai62].

1.2.1 Basic notations

A **Banach-Lie group** G is a Banach manifold with a compatible group structure, i.e. the **multiplication** or **product**

$$\boldsymbol{\gamma} \colon G \times G \longrightarrow G \colon (x,y) \mapsto \boldsymbol{\gamma}(x,y) := xy \equiv x \cdot y,$$

and the **inversion**

$$\boldsymbol{\alpha} \colon G \longrightarrow G \colon x \mapsto \boldsymbol{\alpha}(x) := x^{-1}$$

are smooth maps. $\boldsymbol{\gamma}$ comes from the Greek word γινόμενο meaning product. Observe the use of the bold typeface $\boldsymbol{\gamma}$ to distinguish the product from the normal γ usually denoting a curve. $\boldsymbol{\alpha}$ (bold typeface, again) is the first letter of αντιστροφή, the Greek word for inversion. The unit (element) of G is denoted by e.

The **left translation** by $g \in G$ is the diffeomorphism

$$\lambda_g \colon G \longrightarrow G \colon x \mapsto \lambda_g(x) := gx.$$

Similarly, the **right translation** by $g \in G$ is

$$\rho_g \colon G \longrightarrow G \colon x \mapsto \rho_g(x) := xg.$$

The differentials of $\boldsymbol{\gamma}$ and $\boldsymbol{\alpha}$, in terms of the translations, are given, respectively, by

(1.2.1) $$T_{(x,y)}\boldsymbol{\gamma}(u,v) = T_x\rho_y(u) + T_y\lambda_x(v),$$

(1.2.2) $$T_x\boldsymbol{\alpha}(u) = -T_e\lambda_{x^{-1}} \circ T_x\rho_{x^{-1}}(u) = -T_x(\lambda_{x^{-1}} \circ \rho_{x^{-1}})(u),$$

for every $x, y \in G$ and every $u \in T_xG$, $v \in T_yG$.

In the following subsections G will denote a Banach-Lie group.

1.2.2 Invariant vector fields

A vector field $X \in \mathcal{X}(G)$ is said to be **left invariant** if it is λ_g-related with itself, for every $g \in G$; that is,

$$T\lambda_g \circ X = X \circ \lambda_g, \qquad g \in G;$$

equivalently,

$$T_e\lambda_g(X_e) = X_g, \qquad g \in G.$$

The set of all left invariant vector fields on G forms a Lie subalgebra of $\mathcal{X}(G)$, denoted by $\mathcal{L}(G)$ and called the **Lie algebra of** G.

$\mathcal{L}(G)$ is in bijective correspondence with $T_e G$ by means of the linear isomorphism

(1.2.3) $\qquad\qquad \boldsymbol{h} \colon \mathcal{L}(G) \ni X \longmapsto X_e \in T_e G$

whose inverse is given by

(1.2.4) $\qquad\qquad \boldsymbol{h}^{-1}(v) = X^v; \qquad v \in T_e G,$

where $X^v \in \mathcal{L}(G)$ is defined by

(1.2.5) $\qquad\qquad X^v(x) = T_e L_x(v), \qquad x \in G.$

Therefore, $T_e G$ becomes a Lie algebra by setting (same symbol of bracket !)

$$[u, v] := \boldsymbol{h}\left(\left[\boldsymbol{h}^{-1}(u), \boldsymbol{h}^{-1}(v) \right] \right).$$

Equivalently, if $u = X_e$ and $v = Y_e$, for $X, Y \in \mathcal{L}(G)$, then

$$[X_e, Y_e] = [X, Y]_e,$$

under the appropriate interpretation of the bracket in each side.

For convenience, sometimes, we shall denote by $\mathfrak{g}$ the Lie algebra $T_e G$ with the previous structure. As is the custom, we shall denote the Lie algebra of G by $\mathfrak{g}$ and $\mathcal{L}(G)$ interchangeably, as a result of the identification (1.2.3).

1.2.3 The exponential map

The **exponential map** of G is the map

$$\exp \equiv \exp_G \colon T_e G \longmapsto G \colon v \mapsto \exp(v) := \alpha(1),$$

where α is the integral curve of $X = \boldsymbol{h}^{-1}(v) \in \mathcal{L}(G)$ with initial condition $\alpha(0) = e$. Recall that the left invariant vector fields are *complete*, thus the domain of α is $\mathbb{R}$.

1.2.4 The adjoint representation

The **adjoint representation** of G is the map $\mathrm{Ad} \colon G \to \mathrm{Aut}(\mathfrak{g})$, with

$$\mathrm{Ad}(g) := T_e(\rho_{g^{-1}} \circ \lambda_g) = T_e(\lambda_g \circ \rho_{g^{-1}}).$$

It is a smooth map whose differential at $e \in G$,

$$T_e \mathrm{Ad}(g) \colon T_e G \equiv \mathfrak{g} \longrightarrow \mathcal{L}(\mathfrak{g}),$$

is given by

$$\left(T_e\mathrm{Ad}(g)(X)\right)(Y) = [X,Y]; \qquad X,Y \in \mathfrak{g}.$$

1.2.5 Lie algebra-valued differential forms

Let B be a Banach manifold and let G be a Banach-Lie group with Lie algebra $\mathfrak{g}$.

Heuristically, a $\mathfrak{g}$-**valued differential form of degree** k ($\mathfrak{g}$-**valued** k-**form**, for short) on B is a smooth map ω assigning a k-alternating (antisymmetric) map $\omega_x \in \mathcal{A}_k(T_xB, \mathfrak{g})$ to each $x \in B$. Formally, ω is a smooth section of the vector bundle of k-alternating maps

$$A_k(TB, \mathfrak{g}) := \bigcup_{x \in B} \mathcal{A}_k(T_xB, \mathfrak{g}),$$

described in detail in §1.4.4(e) (see also § 1.4.1). The set of $\mathfrak{g}$-valued k-forms on B is denoted by $\Lambda^k(B, \mathfrak{g})$.

Important examples of $\mathfrak{g}$-valued forms are the Maurer-Cartan forms on a Lie group defined below, and the Maurer-Cartan differentials defined in the next subsection. More specifically, the **left Maurer-Cartan** (or **left canonical**) **form** on G is the 1-form $\omega^l \in \Lambda^1(G, \mathfrak{g})$ given by

$$\omega^l_g(v) := T_g\lambda_{g^{-1}}(v); \qquad g \in G, v \in T_gG.$$

Analogously, the **right Maurer-Cartan form** on G is the differential form $\omega^r \in \Lambda^1(G, \mathfrak{g})$ defined by

$$\omega^r_g(v) := T_g\rho_{g^{-1}}(v); \qquad g \in G, v \in T_gG.$$

The form ω^l is **left invariant**, i.e. $\lambda^*\omega^l = \omega^l$. Likewise, ω^r is **right invariant**, i.e. $\rho^*\omega^r = \omega^r$. The two forms satisfy the respective **Maurer-Cartan equations**:

$$d\omega^l = -\frac{1}{2}[\omega^l, \omega^l] = -\omega^l \wedge \omega^l,$$

$$d\omega^r = \frac{1}{2}[\omega^r, \omega^r] = \omega^r \wedge \omega^r.$$

For the exterior product, the bracket and the exterior differentiation of $\mathfrak{g}$-valued forms, we refer to the general theory of [Bou71, § 8.3], [Car67(b)] and [Nab00, § 4.2].

1.2.6 The Maurer-Cartan differentials

These differentials will be encountered in the study of local connection forms (see §1.7.2 below).

Let B be a Banach manifold, G a Banach Lie group and $f: B \to G$ a smooth map. Then the **left Maurer-Cartan differential** of f is the differential form $D^l f \equiv f^{-1} df \in \Lambda^1(G, \mathfrak{g})$ defined by

$$\left(D^l f\right)_x(v) \equiv \left(f^{-1} df\right)_x(v) := \left(T_{f(x)} \lambda_{f(x)^{-1}} \circ T_x f\right); \quad x \in B, v \in T_x B.$$

Analogously, the **right Maurer-Cartan differential** of f is the differential form $D^r f \equiv df.f^{-1} \in \Lambda^1(G, \mathfrak{g})$ defined by

$$\left(D^r f\right)_x(v) \equiv \left(df.f^{-1}\right)_x(v) := \left(T_{f(x)} \rho_{f(x)^{-1}} \circ T_x f\right); \quad x \in B, v \in T_x B.$$

It is immediate that

$$(1.2.6) \qquad f^{-1} df = f^* \omega^l \quad \text{and} \quad df.f^{-1} = f^* \omega^r,$$

where ω^l and ω^r are the Maurer-Cartan forms of G defined in §1.2.5.

Equations (1.2.6) justify our terminology. Other terms in use are **left** and **right differentials of** f (N. Bourbaki [Bou72, Ch. III, §3.17]), **logarithmic derivatives** (A. Kriegl and P. Michor [KM97, Ch. VIII, §38.1]), or **multiplicative differentials** (S.G. Kreĭn and N.I. Yatskin, [KJ80, Ch. I, §3]). Another legitimate term is **total left/right differentials** since $D^l f = D^r f = Tf$, for $G = (\mathbb{E}, +)$ and any smooth map $f: B \to \mathbb{E}$ (see the terminology of §1.1.8).

We list below a few properties of the Maurer-Cartan differentials, referring for details to the aforementioned sources.

$$D^r f^{-1} = -D^l f,$$
$$D^l f = \mathrm{Ad}\left(f^{-1}\right) D^r f,$$
$$D^r f^{-1} = -\mathrm{Ad}\left(f^{-1}\right) D^r f,$$
$$D^r(f \cdot h) = D^r f + .\mathrm{Ad}(f) D^r h,$$
$$D^l(f \cdot h) = D^l h + \mathrm{Ad}\left(h^{-1}\right) D^l f,$$

for all smooth maps $f, h: B \to G$. We recall that $f^{-1}: B \to G$ is given by $f^{-1}(x) := f(x)^{-1}$, for every $x \in B$. On the other hand, $\mathrm{Ad}(f) D^r h$ is the 1-form given by

$$\left(\mathrm{Ad}(f) D^r h\right)_x(v) = \mathrm{Ad}(f(x))\left((D^r h)_x(v)\right); \quad x \in B, v \in T_x B.$$

Analogously for the other expressions involving the adjoint representation.

We also have:

$$D^r f = 0 = D^l f \quad \Leftrightarrow \quad f \text{ locally constant,}$$
$$D^r f = D^r h \quad \Leftrightarrow \quad h = fC, \quad C \text{ locally constant,}$$
$$D^l f = D^l h \quad \Leftrightarrow \quad h = Cf, \quad C \text{ locally constant.}$$

Of particular interest are the equations with Maurer-Cartan differentials. For instance, let us consider the equation

$$(1.2.7) \qquad\qquad D^r x = \theta, \quad \text{where} \quad \theta \in \Lambda^1(M, \mathfrak{g}).$$

Let $(x_0, g_0) \in M \times G$. Then:

Equation (1.2.7) admits a unique solution $f \colon U \to G$ (U: open neighborhood of x_0) such that $f(x_0) = g_0$, if and only if $d\theta = \frac{1}{2}[\theta, \theta]$.

In this case θ is called **integrable**. If M is *simply connected*, then there exist global solutions.

Lifting (1.2.7) to the universal cover $\widetilde{M}$ of M, we obtain the equation

$$(1.2.8) \qquad\qquad D^r z = \widetilde{\pi}^* \theta,$$

where $\widetilde{\pi} \colon \widetilde{M} \to M$ is the natural projection. We fix an arbitrary $\widetilde{x}_0 \in \widetilde{M}$. If θ is integrable, there is a global solution $F_\theta \colon \widetilde{M} \to G$ of (1.2.8), called **fundamental**, with initial condition $F_\theta(x_0) = e$.

The **monodromy homomorphism** of (1.2.7) is the homomorphism

$$(1.2.9) \qquad \theta^\# \colon \pi_1(M) \longrightarrow G \colon [\gamma] \mapsto \theta^\#([\gamma]) := F_\theta(\widetilde{x}_0 \cdot [\gamma]),$$

where $\pi_1(M) \equiv \pi_1(M, \widetilde{x}_0)$ is the fundamental group of M.

Analogous results hold for equations with the left Maurer-Cartan differential, under the integrability condition $d\theta = -\frac{1}{2}[\theta, \theta]$.

1.3 Smooth actions

1.3.1 Definitions

A Banach-Lie group G **acts** (**from the right**) on a Banach manifold M if there is a smooth map $\delta \colon M \times G \longrightarrow M$, called **action**, such that:

(**A.1**) $\qquad\qquad\qquad \delta(x, e) = x,$

(**A.2**) $\qquad\qquad \delta(\delta(x, g_1), g_2) = \delta(x, g_1 g_2),$

for every $x \in M$ and $g_1, g_2 \in G$. For convenience, we write $x \cdot g$ or xg instead of $\delta(x, g)$.

An action δ will be called **effective** if

$$xg = x \ \forall \ x \in M \quad \Leftrightarrow \quad g = e.$$

The action will be called **free** if

$$xg = x, \text{ for some } x \in M, \quad \Rightarrow \quad g = e.$$

Finally, δ is called **transitive** if

$$\forall \ (x, y) \in M \times M \quad \Rightarrow \quad \exists \, g \in G : y = xg.$$

If $g \in G$ in the previous definition is uniquely determined, then the action is called **freely transitive**.

For a $g \in G$, the partial map

$$\delta_g \colon M \longrightarrow M : x \mapsto \delta_g(x) := x \cdot g$$

is a diffeomorphism. As a matter of fact, a smooth action is equivalently defined by requiring (A.2) and δ_g to be a diffeomorphism. The map δ_g is also denoted by R_g (especially in the case of principal bundles, §1.6) and called the **right translation of** M **by** g. In the case of the (obvious) action of a Lie group on itself, $R_g = \rho_g$, according to the notations of §1.2.1.

1.3.2 Fundamental (Killing) vector fields

Let $\delta \colon M \times G \to M$ be a smooth action. Given a left invariant vector field $X \in \mathcal{L}(G)$, we set

$$X^*(x) := T_e \delta_x(X_e); \qquad x \in M,$$

where, as usual, the partial map $\delta_x \colon G \to M$ is given by $\delta_x(g) = \delta(x, g)$, for every $g \in G$. Since $X^*(x) \in T_x M$, it follows that $X^* \colon M \to TM$ is a smooth vector field of M. It is called the **fundamental** (or **Killing**) **vector field** on M corresponding to $X \in \mathcal{L}(G)$, with respect to the action δ. Obviously, the vector fields X and X^* are δ_x-related, for every $x \in M$ (see §1.1.10).

The integral curve β of X^*, with initial condition $x \in M$, is given by

$$\beta(t) = x \cdot \alpha_X(t) = x \cdot \exp(tX); \qquad t \in \mathbb{R},$$

where α_X is the integral curve of X with initial condition e. Therefore, X^* is a complete vector field.

The map

$$^* \colon \mathcal{L}(G) \ni X \longmapsto X^* \in \mathcal{X}(M)$$

is a morphism of Lie algebras. If the action is effective, then the map
* is injective. Moreover, if the action is free, then, for every $X \in \mathcal{L}(G)$
with $X_e \neq 0$, it follows that $X_x^* \neq 0$, for every $x \in M$.

1.4 Banach vector bundles

We mainly follow [Bou67], [Lan99] and [AR67].

1.4.1 The structure of a Banach vector bundle

Let E, B be smooth manifolds, $\pi \colon E \to B$ a smooth map, and $\mathbb{E}$ a
Banach space. We also assume that $\{U_\alpha \subset B \,|\, \alpha \in I\}$ is an open cover
of B and, for each $\alpha \in I$, there is a diffeomorphism $\tau_\alpha \colon \pi^{-1}(U) \to U \times \mathbb{E}$
satisfying the following conditions:

(VB. 1) The diagram

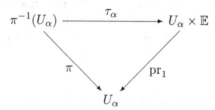

is commutative and the restriction of τ_α to the **fibre (over x)** $E_x = \pi^{-1}(x)$,

$$(1.4.1) \qquad \tau_{\alpha,x} \colon E_x \longrightarrow \{x\} \times \mathbb{E} \cong \mathbb{E}$$

is a bijection for every $x \in U_\alpha$.

(VB. 2) For two pairs (U_α, τ_α) and (U_β, τ_β), with $U_{\alpha\beta} = U_\alpha \cap U_\beta \neq \emptyset$,
the map

$$\tau_{\alpha,x} \circ \tau_{\beta,x}^{-1} \colon \mathbb{E} \longrightarrow \mathbb{E}$$

is an isomorphism of Banach spaces, for every $x \in U_{\alpha\beta}$.

(VB. 3) For (U_α, τ_α) and (U_β, τ_β) as above, the map

$$(1.4.2) \qquad T_{\alpha\beta} \colon U_{\alpha\beta} \ni x \longmapsto T_{\alpha\beta}(x) := \tau_{\alpha,x} \circ \tau_{\beta,x}^{-1} \in \mathcal{L}(\mathbb{E})$$

is smooth. As a matter of fact, $T_{\alpha\beta}(x) \in \mathrm{GL}(\mathbb{E})$.

We shall use the following terminology: (U_α, τ_α) is called a **trivializa-
tion of E** with **trivializing map** τ_α. The collection $\mathcal{C} = \{(U_\alpha, \tau_\alpha)\}_{\alpha \in I}$
is a **trivializing cover of E**.

Two trivializing covers of E are said to be **equivalent** if their union satisfies conditions (VB. 2) and (VB. 3). An equivalence class of trivializing covers determines the structure of a **Banach vector bundle** of **fibre type** $\mathbb{E}$, with **total space** E, **projection** π, and **base (space)** B.

▶ A vector bundle, as above, will be denoted by $\ell = (E, B, \pi)$. We refer to it either by ℓ or E if there is no ambiguity about its elements.

Because of (1.4.1), each fibre E_x admits the structure of a Banach space isomorphic to $\mathbb{E}$, and

(1.4.3) $$\tau_\alpha(u) = (x, \tau_{\alpha,x}(u)), \qquad u \in E_x.$$

Clearly, $\tau_{\alpha,x} = \mathrm{pr}_2 \circ \tau|_{\pi^{-1}(x)}$.

It is often useful to assume that the open sets U_α of the trivializing cover are the range of charts (U_α, ϕ_α) of the base B (this can always be intersecting a trivializing cover of E with the atlas of the smooth structure of B). Then, in analogy to the local structure of the tangent bundle of a manifold defined by (1.1.6) and (1.1.7), we may consider the map

(1.4.3′) $$\Phi_\alpha := (\phi_\alpha \times \mathrm{id}_\mathbb{E}) \circ \tau_\alpha \colon \pi^{-1}(U_\alpha) \to \phi_\alpha(U_\alpha) \times \mathbb{E},$$

and the commutative diagram on the next page. The triple $(U_\alpha, \phi_\alpha, \Phi_\alpha)$ is called a **vector bundle chart**, or **vb-chart** for short, and (U_α, Φ_α), or simply Φ_α, still a **trivialization** of E. As a matter of fact, a vector bundle structure is completely determined by a (maximal) **atlas** of compatible vector bundle charts. Here the compatibility is expressed by means of an isomorphism of local vector bundles (for details see also [AR67]).

$$
\begin{array}{ccc}
\pi^{-1}(U_\alpha) & \xrightarrow{\ \Phi_\alpha\ } & \phi_\alpha(U_\alpha) \times \mathbb{E} \\
\downarrow{\scriptstyle \pi} & & \downarrow{\scriptstyle \mathrm{pr}_1} \\
U_\alpha & \xrightarrow[\ \phi_\alpha\ ]{} & \phi_\alpha(U_\alpha)
\end{array}
$$

A vector bundle chart $(U_\alpha, \phi_\alpha, \Phi_\alpha)$ induces an isomorphism of Banach spaces $\Phi_{\alpha,x} \colon E_x \to \mathbb{E}$ such that the following analog of (1.4.3)

(1.4.4) $$\Phi_\alpha(u) = (\phi_\alpha, \Phi_{\alpha,x}(u)), \qquad u \in E_x.$$

holds true. Obviously,

(1.4.5) $$\tau_{\alpha,x} = \Phi_{\alpha,x}, \qquad x \in U_\alpha.$$

1.4.2 Transition maps

The maps $T_{\alpha\beta}$ defined by (1.4.2) are the **transition maps** or **functions** of the bundle $\ell = (E, B, \pi)$. They satisfy the **cocycle condition**

(1.4.6) $\quad T_{\alpha\beta}(x) = T_{\alpha\gamma}(x) \circ T_{\gamma\beta}(x), \qquad x \in U_{\alpha\beta\gamma} := U_\alpha \cap U_\beta \cap U_\alpha.$

It follows that

(1.4.7) $$T_{\alpha\alpha}(x) = \mathrm{id}_{\mathbb{E}} \quad \text{and} \quad T_{\beta\alpha}(x) = T_{\alpha\beta}(x)^{-1},$$

for every $x \in U_\alpha$ and $x \in U_{\alpha\beta}$, respectively.

The collection $\{T_{\alpha\beta}\}$ is the **cocycle** of ℓ (with respect to the trivializing cover $\{(U_\alpha, \Phi_\alpha)\}_{\alpha,\beta \in I}$). More precisely, in the formal language of cohomology theory, $\{T_{\alpha\beta}\}$ is a *1-cocycle*, [DP97].

Given an open cover $\mathcal{C} = \{U_\alpha \mid \alpha \in I\}$ of a smooth manifold B, and a collection of smooth maps $\{T_{\alpha\beta} \colon U_{\alpha\beta} \to \mathcal{L}(\mathbb{E})\}$, with $\mathrm{Im}(T_{\alpha\beta}) \subset \mathrm{GL}(\mathbb{E})$, and satisfying the cocycle condition (1.4.6), there exists a Banach vector bundle $\ell = (E, B, \pi)$ with transition maps $\{T_{\alpha\beta}\}$. More precisely, E is obtained by quotienting the set

$$\bigcup_{\alpha \in I} (\{\alpha\} \times U_\alpha \times \mathbb{E})$$

by the equivalence relation

$$(\beta, y, k) \sim (\alpha, x, h) \quad \Leftrightarrow \quad y = x, \; k = T_{\beta\alpha}(x)(h).$$

Then $\pi \colon [(\alpha, x, h)] \mapsto x$, and the trivializing maps $\tau_\alpha \colon \pi^{-1}(U_\alpha) \to U_\alpha \times \mathbb{E}$ are given by $\tau_\alpha([(\gamma, z, m)]) := (z, T_{\alpha\gamma}(z)(m))$. The bundle ℓ is *unique up to isomorphism* (see the application in the next subsection).

It is often convenient to connect the transition maps $T_{\alpha\beta}$ with vector bundle charts: If $(U_\alpha, \phi_\alpha, \Phi_\alpha)$ and $(U_\beta, \phi_\beta, \Phi_\beta)$ are two intersecting vb-charts, we define the smooth maps

(1.4.8) $$G_{\alpha\beta} \colon \phi_\beta(U_{\alpha\beta}) \longrightarrow \mathrm{GL}(\mathbb{E}) \subset \mathcal{L}(\mathbb{E}),$$

by setting

(1.4.9) $$G_{\alpha\beta}(\phi_\beta(x)) := \Phi_{\alpha,x} \circ \Phi_{\beta,x}^{-1}, \qquad x \in U_{\alpha\beta}.$$

As a result,

(1.4.10) $$T_{\alpha\beta}(x) = G_{\alpha\beta}(\phi_\beta(x)), \qquad x \in U_{\alpha\beta}.$$

1.4.3 Morphisms of vector bundles

Let $\ell_1 = (E_1, B_1, \pi_1)$ and $\ell_2 = (E_2, B_2, \pi_2)$ be two Banach vector bundles, of corresponding fibre types $\mathbb{E}_1$ and $\mathbb{E}_2$. A ***vector bundle morphism***, or ***vb-morphism*** for short, between ℓ_1 and ℓ_2 is a pair of smooth maps (f, h), with $f \colon E_1 \to E_2$ and $h \colon B_1 \to B_2$ satisfying the following properties:

(**VBM. 1**) The next diagram is commutative

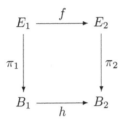

and the restriction of f to the fibres, namely

$$f_x := f|_{E_{1,x}} \colon E_{1,x} \longrightarrow E_{2,h(x)},$$

is a continuous linear map, for every $x \in B$.

(**VBM. 2**) For each $x_0 \in B$, there are trivializations

$$\tau_1 \colon \pi_1^{-1}(U_1) \longrightarrow U_1 \times \mathbb{E}_1 \quad \text{and} \quad \tau_2 \colon \pi_2^{-1}(U_2) \longrightarrow U_2 \times \mathbb{E}_2$$

with $x_0 \in U_1$, $h(U_1) \subseteq U_2$, and such that the map

$$U_1 \ni x \longmapsto \tau_{2,h(x)} \circ f_x \circ \tau_{1,x}^{-1} \in \mathcal{L}(\mathbb{E}_1, \mathbb{E}_2)$$

is smooth. Using the corresponding vb-charts (U_1, ϕ_1, Φ_1), (U_2, ϕ_2, Φ_2), condition (VBM. 2) is equivalent to the smoothness of

$$\phi_1(U_1) \ni \phi_1(x) \longmapsto \Phi_{2,h(x)} \circ f_x \circ \Phi_{1,x}^{-1} \in \mathcal{L}(\mathbb{E}_1, \mathbb{E}_2).$$

A vb-morphism, as above, will be also denoted by $(f, h) \colon \ell_1 \to \ell_2$. The composition of two vb-morphisms, as well as the notion of a ***vb-isomorphism*** are defined in the obvious way.

We shall mostly deal with vector bundles over the same base B and vb-morphism between them with $h = \mathrm{id}_B$. In this case we obtain the category $\mathcal{VB}_B$, in particular $\mathcal{VB}_B(\mathbb{E})$ if the bundles have the same fibre type $\mathbb{E}$. A morphism between bundles in the latter category will be also denoted by $f \colon E_1 \to E_2$.

As an application, we outline the following relationship between cocycles and vb-isomorphisms: Assume that $\ell = (E, B, \pi)$ and $\ell' = (E', B, \pi')$

are Banach vector bundles of the same fibre type $\mathbb{E}$. By appropriate restrictions, we may take trivializations of ℓ and ℓ' over the same open cover $\mathcal{C} = \{U_\alpha\}_{\alpha \in I}$ of B. Denote by $\{T_{\alpha\beta}\}$ and $\{T'_{\alpha\beta}\}$ the respective cocycles over $\mathcal{C}$. Then:

> There exists a vb-isomorphism (f, id_B) of ℓ onto ℓ' if and only if the cocycles $\{T_{\alpha\beta}\}$ and $\{T'_{\alpha\beta}\}$ are **cohomologous**.

The latter term means that there are smooth maps $h_\alpha \colon U_\alpha \to \mathcal{L}is(\mathbb{E})$ such that

$$T'_{\alpha\beta}(x) = h_\alpha(x) \circ T_{\alpha\beta}(x) \circ h_\beta(x)^{-1}; \qquad x \in U_{\alpha\beta},$$

for all indices $\alpha, \beta \in I$.

Indeed, if there is a vb-isomorphism (f, id_B), then we define h_α by setting $h_\alpha(x) = \tau'_{\alpha,x} \circ f_x \circ \tau_{\alpha,x}^{-1}$. The smoothness of h_α is ensured by (VBM 2).

Conversely, assume that the cocycles are cohomologous. We define the maps $f_\alpha \colon E_{U_\alpha} \to E'_{U_\alpha}$ with

$$f_\alpha(u) := \left((\tau'_{\alpha,x})^{-1} \circ h_\alpha \circ \tau_{\alpha,x} \right)(u),$$

for every $u \in E_{U_\alpha}$ with $\pi(u) = x$. It is smooth because

$$f_\alpha = (\tau'_{\alpha,x})^{-1} \circ \left(\pi, ev \circ (h_\alpha \circ \pi, \mathrm{pr}_2 \circ \tau_\alpha) \right),$$

where $\mathrm{pr}_2 \colon U_\alpha \times \mathbb{E} \to \mathbb{E}$ is the projection to the second factor and

$$ev \colon \mathcal{L}is(\mathbb{E}) \times \mathbb{E} \longrightarrow \mathbb{E} \colon (f, u) \mapsto f(u)$$

is the *evaluation map*. The assumption implies that the collection $\{f_\alpha\}$ determines a smooth bijection $f \colon E \to E'$. It remains to see that (f, id_B) is a vb-morphism by verifying conditions (VBM. 1)–(VBM. 2). The first is obviously satisfied. For the second condition, observe that the map $U_\alpha \ni x \mapsto \tau_{\alpha,x} \circ f_x \circ (\tau'_{\alpha,x})^{-1}$, whose smoothness is required, is precisely h_α, for every $\alpha \in I$. By the same token we prove that (f^{-1}, id_B) is also a vb-morphism, thus (f, id_B) is a vb-isomorphism.

The previous arguments justify the uniqueness—up to isomorphism—of the vector bundle E constructed from a cocycle $\{T_{\alpha\beta}\}$, described in § 1.4.2. Indeed, if E' is another bundle with the same cocycle, then $E \cong E'$ by means of the vb-isomorphism $f \equiv \{f_\alpha\}$, where $f_\alpha(u) = \left((\tau'_{\alpha,x})^{-1} \circ \tau_{\alpha,x} \right)(u)$, for every $u \in E_{U_\alpha}$ with $\pi(u) = x \in U_\alpha$, since now $h_\alpha(x) = \mathrm{id}_\mathbb{E}$ (constantly), for every $x \in U_\alpha$.

Remark. In a more sophisticated way, the preceding relation between isomorphic vector bundles and cohomologous cocycles leads to the following cohomological classification:

Within an isomorphism, we obtain the equality

$$\mathcal{VB}_B(\mathbb{E})/_\sim = H^1(B, \mathfrak{GL}(\mathbb{E})),$$

where $\mathcal{VB}_B(\mathbb{E})/_\sim$ is the quotient of $\mathcal{VB}_B(\mathbb{E})$ with respect to the equivalence relation induced by vb-isomorphisms, and $\mathfrak{GL}(\mathbb{E})$ is the sheaf of germs of smooth $\mathrm{GL}(\mathbb{E})$-valued maps on B.

We recall that $\mathcal{VB}_B(\mathbb{E})$ is the set of (Banach) vector bundles over B, of fibre type $\mathbb{E}$. The right-hand side of the identification is the *1st cohomology group of B with coefficients in $\mathfrak{GL}(\mathbb{E})$*. Briefly, $H^1(B, \mathfrak{GL}(\mathbb{E}))$ is the union of $H^1(\mathfrak{U}, \mathfrak{GL}(\mathbb{E}))$, where $\mathfrak{U}$ is running through the set of all proper open covers of B. Each set $H^1(\mathfrak{U}, \mathfrak{GL}(\mathbb{E}))$ consists of all the cohomologous 1-cocycles $g_{\alpha\beta}\colon U_{\alpha\beta} \to \mathrm{GL}(\mathbb{E})$ $(U_\alpha, U_\beta \in \mathfrak{U})$ identified now with the sections of $\mathfrak{GL}(\mathbb{E})$ over $U_{\alpha\beta}$.

For relevant details on the cohomological classification of fibred spaces we refer to [Gro58] and [Hir66]. For the general theory of sheaves and sheaf cohomology we refer also to [Dow62], [DP97], [God73] and [War83].

1.4.4 Some useful constructions and examples

a) *Fibre product and direct sum*

Let $\ell_k = (E_k, B, \pi_k) \in \mathcal{VB}_B$ be vector bundles of fibre type $\mathbb{E}_k$ $(k = 1, 2)$. Their **fibre product** is the vector bundle $(E_1 \times_B E_2, B, \pi)$, where

$$E_1 \times_B E_2 := \{(u_1, u_2) \in E_1 \times E_2 \ : \ \pi_1(u_1) = \pi_2(u_2)\},$$
$$\pi(u_1, u_2) := \pi_1(u_1) = \pi_2(u_2).$$

Clearly, $(E_1 \times_B E_2)_x = E_{1,x} \times E_{2,x}$, for every $x \in B$. Moreover, by intersecting trivializing covers of ℓ_1 and ℓ_2, we may take the corresponding trivializing covers $\{(U_\alpha, \tau_i^1)\}_{i \in I}$ and $\{(U_\alpha, \tau_i^2)\}_{i \in I}$ inducing the trivializations (U_α, τ_α) of $E_1 \times_B E_2$, where the maps

$$\tau_\alpha \colon \pi^{-1}(U_\alpha) \longrightarrow U_\alpha \times \mathbb{E}_1 \times \mathbb{E}_2$$

are given by

$$\tau(u_1, u_2) := \left(\pi(u_1, u_2) = x, \tau_{\alpha,x}^1(u_1), \tau_{\alpha,x}^2(u_2)\right).$$

On the other hand, setting

$$E_1 \oplus E_2 := \bigcup_{x \in B} E_{1,x} \oplus E_{2,x},$$

we obtain the **direct** or **Whitney sum** $(E_1 \oplus E_2, B, \pi)$, whose projection and trivializations are defined as in the case of the fibre product.

Finite direct sums of vector bundles can be identified with their finite fibre products, as a result of the analogous identification of vector spaces.

b) *The pull-back of a vector bundle*

If $Y \to B$ is a smooth map, the **pull-back** of $\ell = (E, B, \pi)$ by f is the vector bundle $f^*(\ell) = (f^*(E), Y, f^*(\pi) \equiv \pi')$, with

$$f^*(E) \equiv Y \times_B E := \{(y, u) \in Y \times E \; : \; f(y) = \pi(u)\},$$
$$f^*(\pi) \equiv \pi' := \mathrm{pr}_1 |_{f^*(E)} \colon f^*(E) \longrightarrow Y.$$

The fibres of $f^*(\ell)$ are identified with the fibre type $\mathbb{E}$ of ℓ, since

$$f^*(\ell)_y = \{y\} \times E_{f(y)} \cong E_{f(y)}, \qquad y \in Y.$$

Moreover, if

$$f' \equiv \pi^*(f) := \mathrm{pr}_2 |_{f^*(E)} \colon f^*(E) \to E,$$

then the pair (f', f) is a vb-morphism of $f^*(\ell)$ into ℓ.

For later reference, we note that each trivialization (U, τ) of E,

$$\tau \colon E_U = \pi^{-1}(U) \longrightarrow U \times \mathbb{E},$$

induces the trivialization $(f^{-1}(U), \tau^*)$ of $f^*(E)$, with

$$\tau^* \colon (\pi')^{-1}(U) = f^{-1}(U) \times_U E_U \longrightarrow f^{-1}(U) \times \mathbb{E}$$

defined by

$$\tau^*(y, u) := (y, \tau_{f(y)}(u)) = (y, \tau(f(y), u)).$$

Accordingly, if $\{T_{\alpha\beta}\}$ is the cocycle of E, with respect to the trivializing cover $\{(U_\alpha, \tau_\alpha)\}$ $(\alpha \in I)$ of E, the corresponding cocycle of $f^*(E)$ is $\{T_{\alpha\beta} \circ f\}$, with respect to $\{(f^{-1}(U_\alpha), \tau_\alpha^*)\}_{\alpha \in I}$.

The pull-back has the following **universal property**: For each vector bundle $\bar{\ell} = (\bar{E}, \bar{\pi}, Y)$ and each vb-morphism $(\bar{f}, f) \colon \bar{\ell} \to \ell$, there is a unique smooth map $\tilde{\pi} \colon \bar{E} \to f^*(E)$ such that $(\tilde{\pi}, \mathrm{id}_Y)$ is a vb-morphism of $\bar{\ell}$ into $f^*(\ell)$ and $f' \circ \tilde{\pi} = \bar{f}$. In fact, it suffices to take $\tilde{\pi} = (\bar{\pi}, \bar{f})$.

The universal property is depicted in the next diagram.

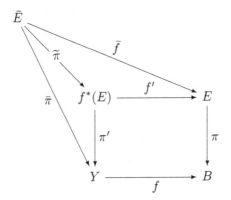

c) *Linear map bundles*

Let $\ell = (E, B, \pi)$ and $\ell' = (E', B, \pi')$ be vector bundles of fibre type $\mathbb{E}$ and $\mathbb{E}'$, respectively. We define the set of linear maps (see also the notations of §1.1.1)

$$L(E, E') := \bigcup_{x \in B} \mathcal{L}(E_x, E'_x)$$

and the projection

$$L \equiv L_{\pi, \pi'} : L(E, E') \longrightarrow B : f \mapsto L(f) := x, \ \text{if} \ f \in \mathcal{L}(E_x, E'_x).$$

Then the triple $(L(E, E'), B, L)$ is a vector bundle, a particular case of a **linear map bundle** (more generally, we can take bundles over different bases). For details we also refer to [AR67]. Here we only mention the vb-charts of $L(E, E')$: Choosing vb-charts $(U_\alpha, \phi_\alpha, \Phi_\alpha)$ and $(U_\alpha, \phi'_\alpha, \Phi'_\alpha)$ of ℓ and ℓ', respectively (over the same open cover $\{U_\alpha\}_{\alpha \in I}$ of B), with

$$\Phi_\alpha : \pi^{-1}(U_\alpha) \longrightarrow \phi_\alpha(U_\alpha) \times \mathbb{E}, \quad \Phi'_\alpha : (\pi')^{-1}(U_\alpha) \longrightarrow \phi'_\alpha(U_\alpha) \times \mathbb{E}',$$

we obtain the vb-chart $\left(L_\alpha^{-1}(U_\alpha), \phi_\alpha, L_\alpha\right)$, where the map

$$L_\alpha : L^{-1}(U_\alpha) \longrightarrow \phi_\alpha(U_\alpha) \times \mathcal{L}(\mathbb{E}, \mathbb{E}')$$

is given by

$$L_\alpha(f) := (\phi_\alpha(x), \lambda_\alpha(x)); \qquad f \in \mathcal{L}(E_x, E'_x),$$

with $\lambda_\alpha(x) \in \mathcal{L}(\mathbb{E}, \mathbb{E}')$ defined in turn by

$$\lambda_\alpha(x)(v) := \left(\mathrm{pr}_2 \circ \Phi'_\alpha \circ f \circ \Phi_\alpha^{-1}\right)(\phi(x), v), \qquad v \in \mathbb{E}.$$

d) *Multilinear map bundles*

The previous construction extends to k-linear maps. More precisely: Let $\ell_i = (E_i, B, \pi_i)$ $(i = 1, \ldots, k)$ be vector bundles of fibre type $\mathbb{E}_i$, and let $\ell' = (E', B, \pi')$ be a vector bundle of fibre type $\mathbb{E}'$. The k-**linear map bundle** consists of the triple $\big(L_k(E_1 \times \cdots \times E_k, E'), B, L^k\big)$, where

$$L_k(E_1 \times \cdots \times E_k, E') := \bigcup_{x \in B} \mathcal{L}_k(E_{1,x} \times \cdots \times E_{k,x}, E'_x)$$

$[\mathcal{L}_k(E_{1,x} \times \cdots \times E_{k,x}, E'_x)$ is the space of continuous k-linear maps between the indicated Banach spaces], and

$$L^k \colon L_k(E_1 \times \cdots \times E_k, E') \longrightarrow B \colon f \mapsto L(f) := x,$$

if $f \in \mathcal{L}_k(E_{1,x} \times \cdots \times E_{k,x}, E'_x)$.

Choosing vb-charts $(U_\alpha, \phi_\alpha, \Phi_\alpha^i,)$ and $(U_\alpha, \phi'_\alpha, \Phi'_\alpha)$ of ℓ_i $(i = 1, \ldots, k)$ and ℓ', respectively, we define the vb-chart $\big((L^k)_\alpha^{-1}(U_\alpha), \phi_\alpha, L_\alpha\big)$, with

$$L_\alpha^k \colon (L^k)^{-1}(U_\alpha) \longrightarrow \phi_\alpha(U_\alpha) \times \mathcal{L}_k(\mathbb{E}_1 \times \cdots \times \mathbb{E}_k, \mathbb{E}'),$$

given by

$$L_\alpha^k(f) := \big(\phi_\alpha(x), \lambda_\alpha^k(x)\big),$$

while $\lambda_\alpha^k(x) \in \mathcal{L}_k(\mathbb{E}_1 \times \cdots \times \mathbb{E}_k, \mathbb{E}')$ is defined by

$$\lambda_\alpha^k(x)(v_1, \ldots, v_k) :=$$
$$\Big(\mathrm{pr}_2 \circ \Phi'_\alpha \circ f \circ \big(\Phi_{1,\alpha}^{-1} \times \cdots \times \Phi_{k,\alpha}^{-1}\big)\Big)\big((\phi(x), v_1), \ldots, (\phi(x), v_k)\big).$$

e) *Alternating map bundles*

Analogously to the preceding bundle, for $E_1 = \cdots = E_k = E$, we construct the k-**alternating (antisymmetric) map bundle** whose total space is

$$A_k(E, E') := \bigcup_{x \in B} \mathcal{A}_k(E_x, E'_x),$$

with $\mathcal{A}_k(E_x, E'_x)$ denoting the space of continuous k-alternating maps of $E_x \times \cdots \times E_x$ (k factors) into E'_x.

A particular case, which will be frequently encountered, occurs when $E = TB$ and $E' = B \times \mathfrak{g}$, the latter being the total space of the trivial bundle over B with fibre the Lie algebra of a Banach Lie group G. Then

$$A_k(TB, B \times \mathfrak{g}) \equiv A_k(TB, B \times \mathfrak{g}) := \bigcup_{x \in B} \mathcal{A}_k(T_x B, \mathfrak{g}),$$

after the identification $\{x\} \times \mathfrak{g} \equiv \mathfrak{g}$. The smooth sections of $A_k(B, \mathfrak{g})$ are the $\mathfrak{g}$-**valued differential k-forms** on B, already discussed in § 1.2.5.

f) *Jets of sections*

Let (E, B, π) be a Banach vector bundle of fibre type $\mathbb{E}$, over the Banach manifold B of respective model $\mathbb{B}$. By a (global) **section** of E we mean a *smooth* map $\xi: B \to E$ such that $\pi \circ \xi = \mathrm{id}_B$. We denote by $\Gamma(E) \equiv \Gamma(B, E)$ the $\mathcal{C}^\infty(B, \mathbb{R})$-module of smooth sections of E. Analogously, if U is an open subset of B, the module of smooth sections of E over U is denoted by $\Gamma(U, E)$.

Given a $\xi \in \Gamma(E)$ and a vb-chart (U, ϕ, Φ) of ℓ, we define its local representation to be $\Phi \circ \xi \circ \phi^{-1}: \phi(U) \to \phi(U) \times \mathbb{E}$ and the corresponding **local principal part** $\xi_\phi: \phi(U) \to \mathbb{E}$ with

$$(1.4.11) \qquad \left(\Phi \circ \xi \circ \phi^{-1}\right)(x) = (x, \xi_\phi(x)), \qquad x \in \phi(U).$$

If $(U_\alpha, \phi_\alpha, \Phi_\alpha)$ is an indexed vb-chart, then we set $\xi_\alpha := \xi_{\phi_\alpha}$; hence,

$$(1.4.11') \qquad \left(\Phi_\alpha \circ \xi \circ \phi_\alpha^{-1}\right)(x) = (x, \xi_\alpha(x)), \qquad x \in \phi_\alpha(U_\alpha).$$

Our next goal, roughly speaking, is to partition the sections of a vector bundle into a kind of equivalence classes and provide the quotient space with a vector bundle structure.

For our purpose we first introduce the following notations: If $\mathcal{L}_s^k(\mathbb{B}, \mathbb{E})$ is the space of continuous symmetric k-linear maps of $\mathbb{B}^k$ into $\mathbb{E}$, then

$$P^k(\mathbb{B}, \mathbb{E}) := \mathbb{E} \times \mathcal{L}_s(\mathbb{B}, \mathbb{E}) \times \mathcal{L}_s^2(\mathbb{B}, \mathbb{E}) \times \cdots \times \mathcal{L}_s^k(\mathbb{B}, \mathbb{E})$$

is the Banach space of $\mathbb{E}$-valued **polynomials of degree k** on $\mathbb{B}$. For an open $A \subseteq \mathbb{B}$, an $a \in A$, and a smooth map $f: A \to \mathbb{E}$, we denote by $p^k f(a) \in P^k(\mathbb{B}, \mathbb{E})$ the polynomial

$$p^k f(a) := (f(a), Df(a), \ldots, D^k f(a)).$$

Fix a vector bundle $\ell = (E, B, \pi)$ as above. Let $x_1, x_2 \in B$ and ξ_1, ξ_2 local sections of E whose domains contain the points x_1 and x_2, respectively. We define the following equivalence relation:

$$(\xi_1, x_1) \sim_k (\xi_2, x_2) \quad \Leftrightarrow$$

$$\exists \text{ vb-chart } (U, \phi, \Phi): \begin{cases} x_1 = x_2 \in U, \text{ and} \\ p^k \xi_{1,\phi}(\phi(x_1)) = p^k \xi_{2,\phi}(\phi(x_2)) \end{cases}$$

By appropriate restrictions, we may assume that the domains of the sections coincide with the domain of the chart.

It is easily shown that $\sim_k$ is an equivalence relation, independent of the choice of the vb-chart satisfying the above conditions. We denote by

$$j_x^k \xi \quad \text{the equivalence class of } (\xi, x),$$
$$J^k(E) \text{ the derived quotient space,}$$
$$\pi^k \quad \text{the projection } J^k(E) \longrightarrow B \colon j_x^k \xi \mapsto x.$$

Then $J^k(\ell) := (J^k E, B, \pi^k)$ is a Banach vector bundle of fibre type $P^k(\mathbb{B}, \mathbb{E})$, called the *k-jet bundle of sections* of ℓ. Its elements are the *k-jets of (local) sections* of ℓ.

If (U, τ) is a trivialization of ℓ, with corresponding vb-chart (U, ϕ, Φ), we obtain the trivialization (U, τ^k) of $J^k(\ell)$, where

$$\tau^k \colon \left(\pi^k\right)^{-1}(U) \longrightarrow U \times P^k(\mathbb{B}, \mathbb{E}) \colon j_x^k \xi \mapsto \left(x, p^k \xi_\phi(\phi(x))\right).$$

The corresponding vb-chart is (U, ϕ^k, Φ), with

$$\Phi^k \colon \left(\pi^k\right)^{-1}(U) \longrightarrow \phi(U) \times P^k(\mathbb{B}, \mathbb{E}) \colon j_x^k \xi \mapsto \left(\phi(x), p^k \xi_\phi(\phi(x))\right).$$

Details and additional material can be found in [AR67] and [Bou71]. The latter source treats also the general case of jets of smooth maps.

1.4.5 Exact sequences

Let $\ell_k = (E_k, B, \pi_k)$ $(k = 1, 2)$ be Banach vector bundles of respective fibre type $\mathbb{E}_1$ and $\mathbb{E}_2$. If $f \colon E_1 \to E_2$ is a vb-morphism, then the sequence

$$(1.4.12) \qquad\qquad 0 \longrightarrow E_1 \xrightarrow{\ f\ } E_2$$

is called **exact** if, for every $x \in B$, the map $f_x := f|_{E_{1,x}} \colon E_{1,x} \longrightarrow E_{2,x}$ is injective and its image $f_x(E_{1,x})$ has a closed complement in $E_{2,x}$.

Equivalently, there is an open cover $\mathcal{C}$ of B and, over each $U \in \mathcal{C}$, there are corresponding trivializations

$$\tau_1 \colon \pi_1^{-1}(U) \longrightarrow U \times \mathbb{E}_1, \quad \tau_2 \colon \pi_2^{-1}(U) \longrightarrow U \times \mathbb{E}_2$$

such that $\mathbb{E}_2 = \mathbb{E}_1 \times \mathbb{F}$ ($:\mathbb{F}$ Banach space) and the diagram

$$
\begin{array}{ccc}
\pi_1^{-1}(U) & \xrightarrow{\ f\ } & \pi_2^{-1}(U) \\
\Big\downarrow{\scriptstyle \tau_1} & & \Big\downarrow{\scriptstyle \tau_2} \\
U \times \mathbb{E}_1 & \xrightarrow[inc]{} & U \times \mathbb{E}_1 \times \mathbb{F}
\end{array}
$$

is commutative, with $inc(x, u) := (x, u, 0)$ the inclusion map.

Analogously, if $\ell_k = (E_k, B, \pi_k)$ $(k = 2, 3)$ are Banach vector bundles of respective fibre type $\mathbb{E}_2$ and $\mathbb{E}_3$, and $g\colon E_2 \to E_3$ is a vb-morphism, then the sequence

$$(1.4.13) \qquad\qquad E_2 \xrightarrow{\ g\ } E_3 \longrightarrow 0$$

is called **exact** if, for every $x \in B$, the map $g_x\colon E_{2,x} \longrightarrow E_{3,x}$ is surjective and its kernel $\ker(g_x)$ has a closed complement in $E_{2,x}$.

Equivalently, there is an open cover $\mathcal{C}$ of B and, over each $U \in \mathcal{C}$, there are corresponding trivializations

$$\tau_2\colon \pi_2^{-1}(U) \longrightarrow U \times \mathbb{E}_2, \ \ \tau_3\colon \pi_3^{-1}(U) \longrightarrow U \times \mathbb{E}_3,$$

such that $\mathbb{E}_2 = \mathbb{E}_3 \times \mathbb{F}$ and the diagram

$$
\begin{array}{ccc}
\pi_2^{-1}(U) & \xrightarrow{\ \ g\ \ } & \pi_3^{-1}(U) \\
\Big\downarrow{\scriptstyle \tau_2} & & \Big\downarrow{\scriptstyle \tau_3} \\
U \times \mathbb{E}_3 \times \mathbb{F} & \xrightarrow[\ \ p\ \ ]{} & U \times \mathbb{E}_3
\end{array}
$$

is commutative, where $p(x, u, v) := (x, u)$.

In particular, the sequence

$$(1.4.14) \qquad\qquad 0 \longrightarrow E_1 \xrightarrow{\ f\ } E_2 \xrightarrow{\ g\ } E_3 \longrightarrow 0$$

is **exact** if the sequences (1.4.12) and (1.4.13) are exact and

$$\mathrm{Im}(f) = \bigcup_{x \in B} \mathrm{Im}(f_x) = \bigcup_{x \in B} \ker(g_x) = \ker(g).$$

Both $\mathrm{Im}(f)$ and $\ker(g)$ are Banach subbundles of E_2. Recall that a set $S \subset E$ is a **subbundle** of (E, B, π) if there exists an exact sequence $0 \to E' \to E$ such that $S = f(E')$.

1.4.6 The exact sequence associated to a vector bundle

Let $\ell = (E, B, \pi)$ be a Banach vector bundle. In the formalism of (linear) connections, we shall encounter the following exact sequence of vector bundles associated to ℓ:

$$(1.4.15) \qquad 0 \longrightarrow VE \xrightarrow{\ j\ } TE \xrightarrow{\ T\pi!\ } \pi^*(TB) \longrightarrow 0$$

where

• VE is the **vertical subbundle** of the tangent bundle (TE, E, τ_E), whose fibres are given by

$$V_u E := (VE)_u = \ker\left(T\pi!|_{T_u E}\right) = \ker(T_u \pi) = T_u\left(\pi^{-1}(x)\right),$$

for every $u \in E$ with $\pi(u) = x$. The last equality is easily proved by considering $T_u\left(\pi^{-1}(x)\right)$ as a subspace of $T_u E$ and using local trivializations (as a matter of fact, this is particular case of an analogous result for the tangent spaces of the fibres of a submersion).

• The morphism j is the natural inclusion.

• $\pi^*(TB) = E \times_B TB$ (: the pull-back of TB by π; see § 1.4.4).

• $T\pi!$ is the vb-morphism defined by the universal property of the pull-back, as pictured in the next diagram.

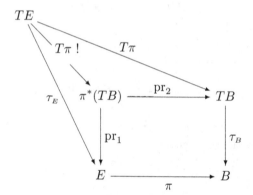

The following vb-isomorphism is standard:

(1.4.16) $$VE \cong E \times_B E,$$

(see. for instance, [Die72, problem 11, p. 136]). Therefore, there is a **canonical map** $r\colon VE \to E$ as in the diagram

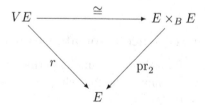

so that (r, π) is a vb-morphism between $(VE, E|_{VE}, \tau_E)$ and (E, B, π).

1.5 Connections on vector bundles

Details of the material included here can be found mainly in [Eli67], [FK72] and [Vil67].

1.5.1 General definitions

Let $\ell = (E, B, \pi)$ be a Banach vector bundle of fibre type $\mathbb{E}$. A (**vb**) **connection** on ℓ is a splitting of the exact sequence (1.4.15); that is, an exact sequence of vector bundles

(1.5.1) $$0 \longrightarrow \pi^*(TB) \xrightarrow{\ C\ } TE \xrightarrow{\ V\ } VE \longrightarrow 0$$

such that

$$T\pi! \circ C = \mathrm{id}_{\pi^*(TB)} \quad \text{and} \quad V \circ j = \mathrm{id}_{VE}.$$

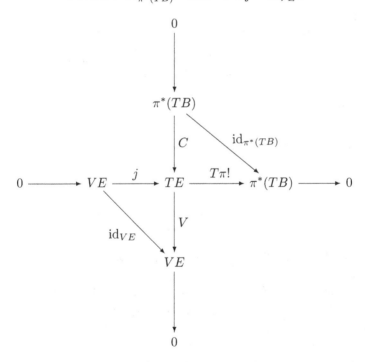

As is well known, it suffices to know either C or V. The splitting implies the decomposition

(1.5.2) $$TE = VE \oplus HE,$$

where $HE := \mathrm{Im}(C)$ is the **horizontal subbundle** of TE.

Given a connection as above, its **connection map** is defined to be

$$(1.5.3) \qquad\qquad K := r \circ V \colon TE \longrightarrow E,$$

where $r \colon VE \to E$ is the canonical morphism defined in the end of §1.4.6. Obviously, the pair (K, π) is a vb-morphism of (TE, E, τ_E) into (E, B, π). Clearly, a tangent vector $u \in TE$ is **horizontal**, i.e. $u \in HE$, if and only if $K(u) = 0$, thus

$$(1.5.4) \qquad\qquad HE = \ker(K).$$

Let $(U_\alpha, \phi_\alpha, \Phi_\alpha)$ be a vb-chart of E [see (1.4.3$'$) end the ensuing definition]. Following (1.1.7) with the appropriate modifications, the corresponding vb-chart of TE is $\bigl(\tau_E^{-1}(\pi^{-1}(U_\alpha)), \Phi_\alpha, \widetilde{\Phi}_\alpha\bigr)$, where

$$(1.5.5) \qquad \begin{aligned} \widetilde{\Phi}_\alpha &\colon \tau_E^{-1}\bigl(\pi^{-1}(U_\alpha)\bigr) \longrightarrow \phi_\alpha(U_\alpha) \times \mathbb{E} \times \mathbb{B} \times \mathbb{E} \colon \\ X &\equiv [(\gamma, u)] \longmapsto \bigl(\Phi_\alpha(u), \overline{\Phi}_\alpha(X)\bigr) = \bigl(\Phi_\alpha(u), (\Phi_\alpha \circ \gamma)'(0)\bigr), \end{aligned}$$

if $X \equiv [(\gamma, u)] \in T_u E$, $u \in \pi^{-1}(U_\alpha)$, and γ is a smooth curve in E with $\gamma(0) = u$. Then the local representation of K is the smooth map

$$(1.5.6) \quad K_\alpha \equiv K_{U_\alpha} \colon \Phi_\alpha \circ K \circ \widetilde{\Phi}_\alpha^{-1} \colon \phi_\alpha(U_\alpha) \times \mathbb{E} \times \mathbb{B} \times \mathbb{E} \longrightarrow \phi_\alpha(U_\alpha) \times \mathbb{E}$$

given by

$$(1.5.7) \qquad K_\alpha(x, \lambda, y, \mu) = (x, \mu + \kappa_\alpha(x, \lambda).y)$$

(recall that line dots as above replace parentheses), where

$$\kappa_\alpha \colon \phi_\alpha(U_\alpha) \times \mathbb{E} \to \mathcal{L}(\mathbb{B}, \mathbb{E})$$

is a smooth map, called the **local component** of K (relative to the chosen charts).

A map $K \colon TE \to E$ is the connection map of a connection on E if and only if K is locally given by (1.5.7) (see [Vil67, Lemma 1]).

1.5.2 Linear connections

Since (K, π) is a vb-morphism between (TE, E, τ_E) and (E, B, π), the restriction of K to the fibre $T_u E = \tau_E^{-1}(u)$, for every $u \in E$, is a continuous linear map. On the other hand, TE is also equipped with the vector bundle structure $T(\ell) = (TE, TB, T\pi)$, obtained by applying the tangent functor to (E, B, π). In this case, we construct the diagram

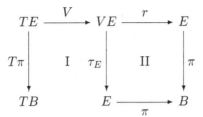

where the sub-diagram I does not close in a natural way. Therefore, $K = r \circ V$ is not necessarily a vb-morphism with respect to the vector bundle structure of $T(\ell)$, and the restrictions of K to the fibres of $T(\ell)$ are not necessarily continuous linear maps.

A connection on $\ell = (E, B, \pi)$ will be called **linear** if the connection map K is linear on the fibres of $T(E)$. Of course, even in this case, the above sub-diagram I does not necessarily close.

If K is a linear connection, then each local component κ_α is continuous linear with respect to the second variable, i.e.

$$\kappa_\alpha(x, \cdot) \in \mathcal{L}(\mathbb{E}, \mathcal{L}(\mathbb{B}, \mathbb{E})), \qquad x \in \phi_\alpha(U_\alpha).$$

1.5.3 The Christoffel symbols of a linear connection

Let K be a *linear* connection on E with local components κ_α, $\alpha \in I$, defined in the preceding subsection. For each index α, K determines a smooth map,

(1.5.8) $$\Gamma_\alpha \colon \phi_\alpha(U_\alpha) \longrightarrow \mathcal{L}(\mathbb{E}, \mathcal{L}(\mathbb{B}, \mathbb{E})),$$

given by

(1.5.8′) $$\Gamma_\alpha(x).\lambda := \kappa_\alpha(x, \lambda), \qquad (x, \lambda) \in \phi_\alpha(U_\alpha) \times \mathbb{E}.$$

The maps $\{\Gamma_\alpha\}_{\alpha \in I}$ are called the (local) **Christoffel symbols** (or *Christoffel maps*) of K.

Because of the (Banach space) identification of $\mathcal{L}(\mathbb{E}, \mathcal{L}(\mathbb{B}, \mathbb{E}))$ with the space of continuous bilinear $\mathbb{E}$-valued maps on $\mathbb{B} \times \mathbb{E}$, $\mathcal{L}_2(\mathbb{E}, \mathbb{B}; \mathbb{E}))$, (see, e.g., [Lan99, Proposition 2.4]), Γ_α identifies with a smooth map of the form $\phi_\alpha(U_\alpha) \to \mathcal{L}_2(\mathbb{E}, \mathbb{B}; \mathbb{E}))$. Also, applying the *symmetry* $s \colon \mathbb{E} \times \mathbb{B} \to \mathbb{B} \times \mathbb{E}$, we obtain the smooth map

(1.5.9) $$\widetilde{\Gamma}_\alpha \colon \phi_\alpha(U_\alpha) \longrightarrow \mathcal{L}_2(\mathbb{B}, \mathbb{E}; \mathbb{E}),$$

which is the local Christoffel symbol of K in the sense of [FK72], with

$$(1.5.9') \qquad \widetilde{\Gamma}_\alpha(x)(y,\lambda) = (\Gamma_\alpha(x).\lambda)(y),$$

for every $(x,\lambda,y) \in \phi_\alpha(U_\alpha) \times \mathbb{E} \times \mathbb{B}$.

Still, after the toplinear identification $\mathcal{L}_2(\mathbb{B},\mathbb{E};\mathbb{E}) \equiv \mathcal{L}(\mathbb{B},\mathcal{L}(\mathbb{E}))$, we may consider the map

$$(1.5.10) \qquad \overline{\Gamma}_\alpha : \phi_\alpha(U_\alpha) \longrightarrow \mathcal{L}(\mathbb{B},\mathcal{L}(\mathbb{E})),$$

such that

$$(1.5.10') \qquad (\overline{\Gamma}_\alpha(x).y)(\lambda) = \widetilde{\Gamma}_\alpha(x)(y,\lambda).$$

Therefore, a linear connection satisfies the equalities

$$
\begin{aligned}
K_\alpha(x,\lambda,y,\mu) &= \big(x, \mu + (\Gamma_\alpha(x).\lambda)(y)\big) \\
&= \big(x, \mu + \widetilde{\Gamma}_\alpha(x)(y,\lambda)\big) \\
&= \big(x, \mu + (\overline{\Gamma}_\alpha(x).y)\lambda\big) \\
&= \big(x, \mu + \boldsymbol{\kappa}_\alpha(x,\lambda).y\big),
\end{aligned}
$$

for every $(x,\lambda,y,\mu) \in \phi_\alpha(U) \times \mathbb{E} \times \mathbb{B} \times \mathbb{E}$ and every $\alpha \in I$.

Since the preceding maps generalize the ordinary Christoffel symbols of a linear connection (viz. covariant derivation) on a finite-dimensional smooth manifold (see, for instancce, [KN68]), we use for all of them the term Christoffel symbols, instead of the more appropriate Christoffel maps.

| We shall use the same symbol Γ_α to refer to any one of (1.5.8), (1.5.9) and (1.5.10), clarifying each time the range of the symbol involved.

The use of a specific type of Christoffel symbol (map) will be dictated by concrete needs.

Given two vb-charts $(U_\alpha, \phi_\alpha, \Phi_\alpha)$ and $(U_\beta, \phi_\beta, \Phi_\beta)$ with $U_{\alpha\beta} \neq \emptyset$, we obtain the following compatibility condition of the local components of an arbitrary connection K [see also equalities (1.4.8)–(1.4.10)]

$$(1.5.11) \qquad
\begin{aligned}
\boldsymbol{\kappa}_\beta(x,\lambda).y = {}& G_{\beta\alpha}((\phi_{\alpha\beta})(x))\big[DG_{\alpha\beta}(x)(y,\lambda) + \\
& + \boldsymbol{\kappa}_\alpha\big(\phi_{\alpha\beta}(x), G_{\alpha\beta}(x).\lambda\big)(D\phi_{\alpha\beta}(x).y)\big],
\end{aligned}
$$

for every $(x,\lambda,y) \in \phi_\beta(U_{\alpha\beta}) \times \mathbb{E} \times \mathbb{B}$. Here, for the sake of convenience, we have set

$$\phi_{\alpha\beta} := \phi_\alpha \circ \phi_\beta^{-1}.$$

In particular, if K is linear, then the compatibility condition of the Christoffel symbols e.g. $\{\Gamma_\alpha \colon \phi_\alpha(U_\alpha) \to \mathcal{L}_2(\mathbb{B}, \mathbb{E}; \mathbb{E})\}_{\alpha \in I}$ [see (1.5.9)] is

$$(1.5.12) \qquad \begin{aligned} \Gamma_\beta(x) = G_{\beta\alpha}\big(\phi_{\alpha\beta}(x)\big) \circ \big[DG_{\alpha\beta}(x) + \\ + \Gamma_\alpha\big(\phi_{\alpha\beta}(x)\big) \circ \big(D\phi_{\alpha\beta}(x) \times G_{\alpha\beta}(x)\big)\big] \end{aligned}$$

for every $x \in \phi_\beta(U_{\alpha\beta})$. Analogous relations hold for the other types of Christoffel symbols.

Anticipating a later application in § 1.7.4 below, we transcribe (1.5.12) in terms of the transition functions $\{T_{\alpha\beta} \colon U_{\alpha\beta} \to \mathrm{GL}(\mathbb{E})\}_{\alpha,\beta \in I}$ of E [see (1.4.2)] in the following way:

$$(1.5.13) \qquad \begin{aligned} \Gamma_\beta(\phi_\beta(x))\big(\overline{\phi}_\beta\big) = \big(T_{\alpha\beta}^{-1} dT_{\alpha\beta}\big)_x(v) + \\ + \mathrm{Ad}\big(T_{\alpha\beta}^{-1}(x)\big)\big(\Gamma_\alpha(\phi_\alpha(x)).\overline{\phi}_\alpha(v)\big), \end{aligned}$$

now for every $x \in U_{\alpha\beta}(!)$, $v \in T_x B$, and after the natural identification $\overline{\mathrm{id}}_{\mathrm{GL}(\mathbb{E})} = \overline{\mathrm{id}}_{\mathrm{GL}(\mathbb{E}),1_\mathbb{E}} \colon T_{1_\mathbb{E}}(\mathrm{GL}(\mathbb{E})) \xrightarrow{\simeq} \mathcal{L}(\mathbb{E})$, in virtue of (1.1.4). Here we view the Christoffel symbols as maps $\{\Gamma_\alpha \colon \phi_\alpha(U_\alpha) \to \mathcal{L}(\mathbb{B}, \mathcal{L}(\mathbb{E}))\}$, $\alpha \in I$. Recall that the first summand in the right-hand side of (1.5.13) is the left Maurer-Cartan differential of $T_{\alpha\beta}$ (see § 1.2.6), while Ad is the adjoint representation of $\mathrm{GL}(\mathbb{E})$.

The proof of both (1.5.11) and (1.5.12) [or (1.5.13)] is based on elementary computations.

Because of the bijective correspondence between linear connection K and families of compatible smooth maps $\{\Gamma_\alpha\}_{\alpha \in I}$, we write

$$K \equiv \{\Gamma_\alpha\}_{\alpha \in I},$$

for whatever form of the Christoffel symbols.

1.5.4 Linear connections and covariant derivations

Before proceeding, we recall that $\mathcal{X}(B)$ is the set of smooth vector fields on B (see § 1.1.9), and $\Gamma(E)$ is the set of smooth sections of a vector bundle $\ell = (E, B, \pi)$ (§ 1.4.4(f)).

A linear connection induces a **covariant derivation**

$$\nabla \colon \mathcal{X}(B) \times \Gamma(E) \longrightarrow \Gamma(E) \colon (X, \xi) \mapsto \nabla_X \xi := K \circ T\xi \circ X.$$

As in the case of an ordinary covariant derivation on a finite-dimensional manifold, ∇ is $C^\infty(B, \mathbb{R})$-linear (hence, also $\mathbb{R}$-linear) with respect to the first variable, whereas, with respect to the second variable, it is $\mathbb{R}$-linear and satisfies the **Leibniz condition**

$$(1.5.14) \qquad \nabla_X(f\xi) = f\nabla_X\xi + X(f)\cdot\xi,$$

for every $f \in C^\infty(B,\mathbb{R})$ and every $\xi \in \Gamma(E)$. Moreover, the local principal part of ∇, with respect to a vb-chart $(U_\alpha, \phi_\alpha, \Phi_\alpha)$, is given by

$$(1.5.15) \qquad \left(\nabla_X\xi\right)_\alpha(x) = D\xi_\alpha(x)\left(X_\alpha(x)\right) + \Gamma_\alpha(x)\left(X_\alpha(x), \xi_\alpha(x)\right).$$

Here X_α is the local principal part of X, with respect to the chart (U_α, ϕ_α) of B [see (1.1.14) and (1.1.14′)], and ξ_α is the local principal part of ξ, with respect to $(U_\alpha, \phi_\alpha, \Phi_\alpha)$ [see (1.4.11) and (1.4.11′)]. The maps $\{\Gamma_\alpha\}$ are also the **Christoffel symbols** of ∇.

• *Warning.* It should be noted that, unlike the finite-dimensional case, an operator ∇, satisfying only the aforementioned linearity condition and (1.5.14), does not determine a linear connection in the sense of §1.5.2; hence, linear connections and covariant derivations are not equivalent notions in the infinite-dimensional framework. However, if an operator ∇ satisfies also condition (1.5.15) over every local trivialization, with compatible $\{\Gamma_\alpha\}_{\alpha\in I}$ [in the sense of (1.5.12)], then a covariant derivation identifies with a linear connection.

1.5.5 Parallel displacement and holonomy groups

Let (E, B, π) be a Banach vector bundle endowed with a *linear* connection K. If $\gamma\colon [0,1] \to B$ is a smooth curve, then a **section of E along** γ is a smooth curve $\xi\colon [0,1] \to E$ such that $\pi \circ \xi = \gamma$. The set of such sections is denoted by $\Gamma_\gamma(E)$. The choice of the interval $[0,1]$ is only for the sake of convenience and does not restrict of the generality.

A section $\xi \in \Gamma_\gamma(E)$ is called **parallel** with respect to K if

$$(1.5.16) \qquad \nabla_\gamma \xi := K \circ T\xi \circ \partial = 0,$$

with $\partial \equiv \dfrac{d}{dt}$ denoting the basic vector field of $\mathbb{R}$. In virtue of (1.5.4) and (1.5.15) [see also (1.1.11)], $\dot{\xi}(t) := T_t\xi(\partial_t) \in H_{\xi(t)}E$; hence, ξ is a *horizontal* curve in E. Locally, over a vb-chart $(U_\alpha, \phi_\alpha, \Phi_\alpha)$ of E, (1.5.16) leads to

$$(1.5.17) \qquad \xi'_\alpha(t) = -\Gamma_\alpha(\phi_\alpha(\gamma(t)))\left((\phi_\alpha \circ \gamma)'(t), \xi_\alpha(t)\right),$$

where $\xi_\alpha\colon [0,1] \to \mathbb{E}$ now denotes the principal part of the local representation $\Phi_\alpha \circ \xi\colon [0,1] \to \phi_\alpha(U_\alpha) \times \mathbb{E}$ of ξ [compare with the general case of (1.4.11) and (1.4.11′)]. Differential equation (1.5.17) is linear of type $x' = A(t) \cdot x$ with $A(t) = -\Gamma_\alpha(\phi_\alpha(\gamma(t)))\big((\phi_\alpha \circ \gamma)'(t), \cdot\big) \in \mathcal{L}(\mathbb{E}, \mathbb{E})$. By the general theory of such equations (in Banach spaces) we prove that, for any $u \in \mathbb{E}$, there exists a unique parallel section ξ_u along γ, such that $\xi_u(0) = u$.

Under the previous notations, the **parallel displacement** or **translation** along the curve γ is defined to be the map

$$(1.5.18) \qquad \tau_\gamma \colon E_{\gamma(0)} \longrightarrow E_{\gamma(1)} \colon u \mapsto \xi_u(1).$$

In virtue of the properties of the *resolvent* of (1.5.17), we find that

- τ_γ is an isomorphism of Banach spaces whose inverse is $\tau_\gamma^{-1} = \tau_{\gamma^{-1}}$;
- $\tau_{\gamma_2 * \gamma_1} = \tau_{\gamma_2} \circ \tau_{\gamma_1}$.

As usual, γ^{-1} is the *inverse* (or *reverse*) of γ and $\gamma_2 * \gamma_1$ the *product* (or *juxtaposition, composition*) of γ_1 followed by γ_2.

Accordingly, the **holonomy group** of the linear connection K, **with reference point** $b \in B$, denoted $^K\Phi_b$, is defined by

$$^K\Phi_b := \{\tau_\gamma \colon E_b \to E_b\},$$

for all smooth curves $\gamma \colon [0,1] \to B$ with $\gamma(0) = \gamma(1) = b$. Similarly, the **restricted holonomy group** of the linear connection K, **with reference point** $b \in B$, is

$$^K\Phi_b^0 := \{\tau_\gamma \colon E_b \to E_b\},$$

for all *homotopic to zero* smooth curves $\gamma \colon [0,1] \to B$ such that $\gamma(0) = \gamma(1) = b$. The left superscript K is set in order to distinguish the present holonomy groups from those induced by connections on principal bundles, as discussed in later sections.

By appropriate identifications, the groups $^K\Phi_b$ and $^K\Phi_b^0$ can be realized as Banach-Lie subgroups of $\mathrm{GL}(\mathbb{E})$. For details we refer to [Max72] and the brief exposition of § 1.9 below.

1.5.6 Related linear connections

Let (f, h) be a vb-morphism between the vector bundles $\ell = (E, B, \pi)$ and $\ell' = (E', B', \pi')$. Two connections K and K' on E and E', respectively, are said to be (f, h)-**related** if

$$(1.5.19) \qquad K' \circ Tf = f \circ K;$$

in other words, the next diagram commutes.

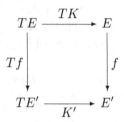

For various applications, it is useful to express (1.5.19) locally. To this end assume that the structures of ℓ and ℓ' are determined by the families of vb-charts $\{(U_\alpha, \phi_\alpha, \Phi_\alpha)\}_{\alpha \in I}$ and $\{(V_\beta, \psi_\beta, \Psi_\beta)\}_{\beta \in J}$, respectively, where $\{(U_\alpha, \phi_\alpha) \equiv (U_\alpha, \phi_\alpha, \mathbb{B})\}_{\alpha \in I}$ and $\{(V_\beta, \psi_\beta) \equiv (V_\beta, \psi_\beta, \mathbb{B}')\}_{\beta \in J}$ are charts of B and B', respectively. For a pair of vb-charts $(U_\alpha, \phi_\alpha, \Phi_\alpha)$ and $(V_\beta, \psi_\beta, \Psi_\beta)$ with $h(U_\alpha) \subseteq V_\beta$ (ensured by the definition of a vb-morphism), we obtain the commutative diagram:

$$
\begin{array}{ccc}
E_{U_\alpha} = \pi^{-1}(U_\alpha) & \xrightarrow{\ \ f\ \ } & (\pi')^{-1}(V_\beta) = E'_{V_\beta} \\
\Big\downarrow{\Phi_\alpha} & & \Big\downarrow{\Psi_\beta} \\
\phi_\alpha(U_\alpha) \times \mathbb{E} & \xrightarrow[\Psi_\beta \circ f \circ \Phi_\alpha^{-1}]{} & \psi_\beta(V_\beta) \times \mathbb{E}'
\end{array}
$$

As a result, the local representation of f takes the form

(1.5.20) $\qquad \left(\Psi_\beta \circ f \circ \Phi_\alpha^{-1} \right)(x, \lambda) = \left(h_{\beta\alpha}(x), f_{\beta\alpha}^{\#}(x).\lambda \right),$

for every $(x, \lambda) \in \phi_\alpha(U_\alpha) \times \mathbb{E}$, where

(1.5.21) $\qquad\qquad h_{\beta\alpha} := \psi_\beta \circ h \circ \phi_\alpha^{-1},$

(1.5.22) $\qquad\qquad f_{\beta\alpha}^{\#} \colon \phi_\alpha(U_\alpha) \longrightarrow \mathcal{L}(\mathbb{E}, \mathbb{E}').$

The map $f_{\beta\alpha}^{\#}$, denoting the **local principal part** of f (with respect to the previous local representation), is given by [see also (1.4.1) and (1.4.5)]

(1.5.23) $\qquad f_{\beta\alpha}^{\#}(x) = \Psi_{\beta, h(b)} \circ f_b \circ \Phi_{\alpha, b}^{-1} = \tau'_{\beta, h(b)} \circ f_b \circ \tau_{\alpha, b}^{-1},$

if $\pi(b) = x$. Moreover, differentiation of (1.5.20) yields

(1.5.24)
$$
D\left(\Psi_\beta \circ f \circ \Phi_\alpha^{-1} \right)(x, \lambda).(y, \mu) = \\
= \left(Dh_{\beta\alpha}(x).y,\ f_{\beta\alpha}^{\#}(x).\mu + (Df_{\beta\alpha}^{\#}(x).y).\lambda \right).
$$

Similarly, using the vb-charts $(U_\alpha, \Phi_\alpha, \widetilde{\Phi}_\alpha)$, and $(V_\beta, \Psi_\beta, \widetilde{\Psi}_\beta)$ of the tangent bundles (TE, E, τ_E) and $(TE', E', \tau_{E'})$, respectively [see (1.5.5)], we check that the corresponding local representation of $Tf \colon TE \to TE'$, namely

$$(1.5.25) \quad \widetilde{\Psi}_\beta \circ Tf \circ \widetilde{\Phi}_\alpha^{-1} \colon \phi_\alpha(U_\alpha) \times \mathbb{E} \times \mathbb{B} \times \mathbb{E} \longrightarrow \psi_\beta(V_\beta) \times \mathbb{E}' \times \mathbb{B}' \times \mathbb{E}',$$

has, in virtue of (1.5.20) and (1.5.24), the expression

$$(1.5.26) \quad \begin{aligned} (\widetilde{\Psi}_\beta \circ Tf \circ \widetilde{\Phi}_\alpha^{-1})(x, \lambda, y, \mu) = \\ = \Big(h_{\beta\alpha}(x), f_{\beta\alpha}^{\#}(x).\lambda, Dh_{\beta\alpha}(x).y, f_{\beta\alpha}^{\#}(x).\mu + \\ + (Df_{\beta\alpha}^{\#}(x).y).\lambda \Big), \end{aligned}$$

for every $(x, \lambda, y, \mu) \in \phi_\alpha(U_\alpha) \times \mathbb{E} \times \mathbb{B} \times \mathbb{E}$.

Now the equivalent form of (1.5.19), in terms of the local components of the connections, is essentially found from the commutative diagram on the next page. Indeed, evaluating the equality

$$K'_\beta \circ \big(\widetilde{\Psi}_\beta \circ Tf \circ \widetilde{\Phi}_\alpha^{-1} \big) = \big(\Psi_\beta \circ f \circ \Phi_\alpha^{-1} \big) \circ K_\alpha$$

at any $(x, \lambda, y, \mu) \in \phi_\alpha(U_\alpha) \times \mathbb{E} \times \mathbb{B} \times \mathbb{E}$, and applying equalities (1.5.20), (1.5.26), together with (1.5.7), we obtain:

$$(1.5.27) \quad \begin{aligned} \kappa'_\beta \big(h_{\beta\alpha}(x), f_{\beta\alpha}^{\#}(x).\lambda \big)(Dh_{\beta\alpha}(x).y) = \\ = f_{\beta\alpha}^{\#} \big(\kappa_\alpha(x, \lambda).y \big) - \big(Df_{\beta\alpha}^{\#}(x).y \big).\lambda, \end{aligned}$$

for every $(x, \lambda, y, \mu) \in \phi_\alpha(U_\alpha) \times \mathbb{E} \times \mathbb{B} \times \mathbb{E}$

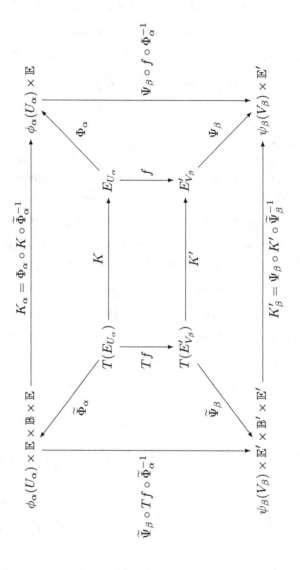

In particular, if K and K' are *linear* connections, then (1.5.27) can be expressed via the respective Christoffel symbols. For the sake of completeness, we write down the relatedness condition in terms of the Christoffel symbols discussed in 1.5.3:

- If $\Gamma_\alpha \colon \phi_\alpha(U_\alpha) \to \mathcal{L}(\mathbb{E}, \mathcal{L}(\mathbb{B}, \mathbb{E}))$, $\Gamma'_\beta \colon \psi_\beta(V_\beta) \to \mathcal{L}(\mathbb{E}', \mathcal{L}(\mathbb{B}', \mathbb{E}'))$, then

$$
\begin{aligned}
(1.5.28) \qquad & \big[\Gamma'_\beta(h_{\beta\alpha}(x))\big(f^{\#}_{\beta\alpha}(x).\lambda\big)\big]\big(Dh_{\beta\alpha}(x)\big) = \\
& = f^{\#}_{\beta\alpha}\big((\Gamma_\alpha(x).\lambda).y\big) - \big(Df^{\#}_{\beta\alpha}(x).y\big).\lambda.
\end{aligned}
$$

- If $\Gamma_\alpha \colon \phi_\alpha(U_\alpha) \to \mathcal{L}(\mathbb{B}, \mathcal{L}(\mathbb{E}))$, $\Gamma'_\beta \colon \psi_\beta(V_\beta) \to \mathcal{L}(\mathbb{B}', \mathcal{L}(\mathbb{E}'))$, then

$$
\begin{aligned}
(1.5.29) \qquad & \big(\Gamma'_\beta(h_{\beta\alpha}(x))\big(Dh_{\beta\alpha}(x).y\big)\big) \circ f^{\#}_{\beta\alpha}(x) = \\
& = f^{\#}_{\beta\alpha} \circ \big(\Gamma_\alpha(x).y\big) - Df^{\#}_{\beta\alpha}(x).y.
\end{aligned}
$$

- If $\Gamma_\alpha \colon \phi_\alpha(U_\alpha) \to \mathcal{L}_2(\mathbb{B}, \mathbb{E}; \mathbb{E})$, $\Gamma'_\beta \colon \psi_\beta(V_\beta) \to \mathcal{L}_2(\mathbb{B}', \mathbb{E}'; \mathbb{E}')$, then

$$
(1.5.30) \qquad \Gamma'_\beta(h_{\beta\alpha}(x)) \circ \big(Dh_{\beta\alpha}(x) \times f^{\#}_{\beta\alpha}\big) = f^{\#}_{\beta\alpha} \circ \Gamma_\alpha(x) - Df^{\#}_{\beta\alpha}(x).
$$

Equalities (1.5.28)–(1.5.30) hold for every $(x, \lambda, y) \in \phi_\alpha(U_\alpha) \times \mathbb{E} \times \mathbb{B}$.

When we deal with bundles over the same base B and vb-morphisms (f, id_B), all the preceding local expressions have considerably simplified variants. Note that, in this case, we can always find vb-charts $(U_\alpha, \phi_\alpha, \Phi_\alpha)$ and $(U_\alpha, \phi_\alpha, \Phi'_\alpha)$ over the same open cover $\{U_\alpha\}_{\alpha \in I}$ of B. Thus $h_{\beta\alpha} = \mathrm{id}_{\phi_\alpha(U_\alpha)}$ and, for simplicity, we write

$$
(1.5.31) \qquad f^{\#}_\alpha := f^{\#}_{\alpha\alpha} \colon \phi_\alpha(U_\alpha) \longrightarrow \mathcal{L}(\mathbb{E}, \mathbb{E}').
$$

1.6 Banach principal bundles

Basic material can be found in [Bou67], [KM97]. For finite-dimensional principal bundles, which are very similar to the Banach case (and are treated in many books), we refer e.g. to [KN68], [Nab00].

1.6.1 The structure of a principal bundle

A **principal bundle** is a quadruple $\ell = (P, G, B, \pi)$, where P and B are smooth Banach manifolds, $\pi \colon P \to B$ a smooth map, and G a Banach-Lie group acting on P (from the right), such that: For every $x \in B$, there is an open $U \subset B$, with $x \in U$, and a diffeomorphism $\Psi \colon U \times G \to \pi^{-1}(U)$ satisfying the following properties:

(**PB. 1**) $\pi \circ \Psi = \mathrm{pr}_1$, in other words, the next diagram is commutative

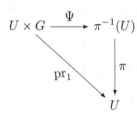

(**PB. 2**) For every $y \in U$ and $g, g' \in G$,

$$\Psi\big((y,g) \cdot g'\big) = \Psi(y, g \cdot g') = \Psi(y,g) \cdot g',$$

that is, Ψ is an **equivariant map** with respect to (the action of) G. Briefly, Ψ is a G-equivariant map.

In the preceding equality, $(y,g) \cdot g'$ denotes the obvious action of G on the right of $U \times G$. The same equality means that, $\Psi(y,g) \cdot g' = \Psi(y, g \cdot g') \in \pi^{-1}(U)$, for every $g' \in G$. Therefore, for every $p \in \pi^{-1}(U)$ and $g \in G$, it follows that $p \cdot g \in \pi^{-1}(U)$. If we set

$$(1.6.1) \qquad\qquad \Phi := \Psi^{-1},$$

then Φ is also G-equivariant; namely,

$$(1.6.2) \qquad \Phi(p \cdot g) = \Phi(p) \cdot g; \qquad p \in \pi^{-1}(U), g \in G.$$

A pair of the form (U, Φ) or (U, Ψ) determines a (local) **trivialization** of P. A family $\mathcal{C} = \{(U_\alpha, \Phi_\alpha)\}_{\alpha \in I}$, of local trivializations, where $\{(U_\alpha\}_{\alpha \in I}$ is an open cover of B, will be called a **trivializing cover** of P.

In a standard terminology, P is the **total space**, B the **base**, π the **projection** and G the **structure group** of the bundle.

▶ If there is no danger of confusion, we refer to a principal bundle $\ell = (P, G, B, \pi)$ either by ℓ or P.

Immediate consequences of the definitions are the following fundamental properties of a principal bundle:

- The projection π is a submersion.
- The fibres $\pi^{-1}(x)$, $x \in B$, are non empty regular submanifolds of P. In particular, if B is a Hausdorff space, then every $\pi^{-1}(x)$ is a closed submanifold of P.
- For every $x \in B$, $\pi^{-1}(x) = p \cdot G$, if p is any element of P with $\pi(p) = x$.

- If $p \in \pi^{-1}(x)$ and $g \in G$, then $p \cdot g \in \pi^{-1}(x)$, thus G acts on the right of each fibre.
- The action of G on P is free.
- The action of G on the fibres is freely transitive.

In particular, if (U, Φ) is a local trivialization of P, then

- $\Phi\big(\pi^{-1}(x)\big) = \{x\} \times G$, for every $x \in U$.
- The map

$$(1.6.3) \qquad \Phi_x := \mathrm{pr}_2 \circ \Phi\big|_{\pi^{-1}(x)} : \pi^{-1}(x) \longrightarrow G$$

  is a G-equivariant diffeomorphism.
- $\Phi_x^{-1}(g) = \Psi(x, g)$ for every $g \in G$.
- By means of Φ_x, the fibre $\pi^{-1}(x)$ has the structure of a Banach-Lie group and Φ_x becomes an isomorphism of Lie groups.

1.6.2 Morphisms of principal bundles

A **morphism** between the principal bundles $\ell_i = (P_i, G_i, B_i, \pi_i)$ ($i = 1, 2$) (**pb-morphism**, for short) is a triple (f, φ, h) where $f \colon P_1 \to P_2$ and $h \colon B_1 \longrightarrow B_2$ are smooth maps and $\varphi \colon G_1 \to G_2$ a morphism of Banach-Lie groups, satisfying the following conditions:

(PBM. 1) $\qquad\qquad \pi_2 \circ f = h \circ \pi_1;$

in other words, the diagram

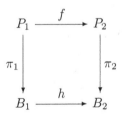

is commutative, and

(PBM. 2) $\qquad f(p \cdot g) = f(p) \cdot \varphi(g), \qquad (p, g) \in P_1 \times G_1.;$

that is, f is equivariant with respect to (the actions of) G and G'.

The map $h \colon B_1 \to B_2$ is completely determined by f and φ.

A pb-morphism (f, φ, h) is an **isomorphism** if f, h are diffeomorphisms and φ is an isomorphism of Banach-Lie groups, If $B_1 = B_2 = B$ and $h = \mathrm{id}_B$, then $(f, \varphi, \mathrm{id}_B)$ is called a B-**morphism**. If $G_1 = G_2 = G$ and $\varphi = \mathrm{id}_G$, we call (f, id_G, h) a G-**morphism**. Finally, if $B_1 = B_2 =$

B, $G_1 = G_2 = G$, $h = \mathrm{id}_B$ and $\varphi = \mathrm{id}_G$, then $(f, \mathrm{id}_G, \mathrm{id}_B)$ is said to be a G-B-**morphism**.

Every G-B-morphism is a pb-isomorphism.

1.6.3 Natural sections and transition maps

The sections of a principal bundle are defined in the usual way. We denote by $\Gamma(U, P)$ the set of smooth sections of a principal bundle $\ell = (P, G, B, \pi)$ over an open $U \subset B$.

If $\mathcal{C} = \{(U_\alpha, \Phi_\alpha) \mid \alpha \in I\}$ is a trivializing cover of $\ell = (P, G, B, \pi)$, the **natural sections** of P, with respect to $\mathcal{C}$, are the maps [see also (1.6.1)]

$$(1.6.4) \qquad s_\alpha : U_\alpha \longrightarrow P : x \mapsto \Psi_\alpha(x, e) = \Phi_\alpha^{-1}(x, e).$$

An arbitrary section $s \in \Gamma(U, P)$ induces a trivialization (U, Φ), with respect to which s is the corresponding natural. Indeed, it suffices to set

$$(1.6.5) \qquad \Psi(x, g) := s(x) \cdot g, \qquad (x, g) \in U \times G.$$

Therefore, $\Phi := \Psi^{-1}$ is given by

$$(1.6.5') \qquad \Phi(p) := (\pi(p), g); \qquad p \in \pi^{-1}(U),$$

where g is determined by the equality $p = s(\pi(p)) \cdot g$.

As a result, there is a bijection between trivializations and smooth sections of a principal bundle. In particular, a principal bundle admits global sections if and only if it is *trivial*, i.e. isomorphic to the trivial bundle $(B \times G, G, B, \mathrm{pr}_1)$.

A useful tool is the map connecting elements of the same fibre. More precisely, let us consider a principal bundle P as before. In analogy to the fibre product of vector bundles defined in § 1.4.4(a), we define the following particular **fibre product**

$$P \times_B P = \{(p, q) \in P \times P : \pi(p) = \pi(q)\},$$

and the smooth map

$$(1.6.6) \qquad k : P \times_B P \longrightarrow G : q = p \cdot k(p, q), \qquad (p, q) \in P \times_B P;$$

in other words, since p and q belong to the same fibre of P, $k(p, q)$ is the unique element of G such that $q = p \cdot k(p, q)$. The smoothness of k is checked locally by observing that, with respect to a trivialization (U, Φ) of the bundle P,

$$k(p, q) = \left(\mathrm{pr}_2(\Phi(p))\right)^{-1} \cdot \left(\mathrm{pr}_2(\Phi(q))\right),$$

for every $(p, q) \in \pi^{-1}(U) \times_U \pi^{-1}(U)$.

Given a trivializing cover $\mathcal{C} = \{(U_\alpha, \Phi_\alpha)\}_{\alpha \in I}$ of $\ell = (P, G, B, \pi)$, the **transition maps** or **functions of** ℓ (with respect to $\mathcal{C}$) are the smooth maps

$$(1.6.7) \qquad g_{\alpha\beta} \colon U_{\alpha\beta} \longrightarrow G \colon x \mapsto \left(\Phi_{\alpha,x} \circ \Phi_{\beta,x}^{-1}\right)(e),$$

for all $\alpha, \beta \in I$ and $U_{\alpha\beta} \neq \emptyset$.

The following equations provide equivalent ways to define the transition functions:

$$(1.6.8) \qquad \left(\Phi_\alpha \circ \Phi_\beta^{-1}\right)(x, g) = (x, g_{\alpha\beta}(x) \cdot g),$$

$$(1.6.9) \qquad s_\beta(x) = s_\alpha(x) \cdot g_{\alpha\beta}(x),$$

for every $(x, g) \in U_{\alpha\beta} \times G$.

The transition functions form a **cocycle** (more precisely, a 1-cocycle); that is,

$$(1.6.10) \qquad g_{\alpha\gamma} = g_{\alpha\beta} \cdot g_{\beta\gamma}$$

which means that

$$g_{\alpha\gamma}(x) = g_{\alpha\beta}(x) \cdot g_{\beta\gamma}(x), \qquad x \in U_{\alpha\beta\gamma}.$$

It follows that

$$(1.6.11) \qquad g_{\alpha\alpha} = e \quad \text{and} \quad g_{\beta\alpha} = g_{\alpha\beta}^{-1}.$$

The last equality means that $g_{\beta\alpha}(x) = g_{\alpha\beta}(x)^{-1}$, for every $x \in U_{\alpha\beta}$ [compare with the cocycle of a vector bundle (1.4.7)].

As in the case of a vector bundle (see § 1.4.2), a principal bundle is completely determined by its cocycles. We recall that, given a cocycle $\{g_{\alpha\beta} \colon U_{\alpha\beta} \to G \,|\, \alpha \in I\}$ over an open cover $\{U_\alpha\}_{\alpha \in I}$ of B, we consider the set

$$S = \bigcup_{\alpha, \beta \in I} \left(\{\alpha\} \times U_\alpha \times G\right),$$

and define the equivalence relation:

$$(\alpha, x, g) \sim (\beta, x', g') \quad \Leftrightarrow \quad x = x' \quad \text{and} \quad g' = g_{\beta\alpha}(x) \cdot g.$$

Setting $P = S/{\sim}$, we obtain the principal bundle $\ell = (P, G, B, \pi)$, where $\pi([(\alpha, x.g)]) := x$, and the local trivializations $\Phi_\alpha \colon \pi^{-1}(U_\alpha) \to U_\alpha \times G$ are given by

$$\Phi_\alpha[(\beta, x, g)] := (x, g_{\alpha\beta}(x) \cdot g).$$

The bundle thus constructed is unique up to isomorphism. This a particular case of the following general result (see the vb-analog in § 1.4.3):

> *Two principal bundles (P, G, B, π) and (P', G, B, π') are G-B-isomorphic if and only they have* **cohomologous cocycles** $\{g_{\alpha\beta}\}$ *and* $\{g'_{\alpha\beta}\}$, *respectively, over an open cover* $\{U_\alpha\}_{\alpha \in I}$ *of B.*

Two cocycles, as above, are said to be cohomologous if there is a family of smooth maps $\{h_\alpha \colon U_\alpha \to G \,|\, \alpha \in I\}$ such that

$$g'_{\alpha\beta} = h_\alpha \cdot g_{\alpha\beta} \cdot h_\beta^{-1} \quad \text{on} \quad U_{\alpha\beta}.$$

Indeed, assume first that the bundles are G-B-isomorphic. If $\{s_\alpha\}$ and $\{s'_\alpha\}$ are the natural sections of P and P', respectively (over a common open cover $\{U_\alpha\}$ of B), then we define h_α as the unique smooth map satisfying $f \circ s_\alpha = s'_\alpha \cdot h_\alpha$. The equivariance of f and equality (1.6.9) now imply that the cocycles are cohomologous.

Conversely, assume that the cocycles are cohomologous. We define $f \colon P \to P'$ by setting

$$f(p) := s'_\alpha(x) \cdot h_\alpha \cdot g_\alpha(p),$$

for every $p \in P$, with $\pi(p) = x \in U_\alpha$, where $g_\alpha(p)$ is the unique element of G determined by $p = s_\alpha(x) \cdot g_\alpha(p)$. The assumption ensures that f is a well-defined map such that $\pi' \circ f = \pi$. Its smoothness follows from equality $f|_{\pi^{-1}(U_\alpha)} = (s'_\alpha \circ \pi) \cdot (h_\alpha \circ \pi) \cdot g_\alpha$ and the smoothness of g_α resulting, in turn, from equality $g_\alpha = k \circ (\mathrm{id}_P, s_\alpha \circ \pi)$, where k is the smooth map (1.6.6).

In particular, if two bundles as above have the same cocycles, then they are G-B-isomorphic by means of the map f, given by $f(p) = s'_\alpha(x) \cdot g_\alpha(p)$, if $\pi(p) = x \in U_\alpha$.

Analogously to the classification of vector bundles, discussed towards the end of § 1.4.3, we have the identification

$$\mathfrak{P}_B(G) \equiv H^1(B, \mathfrak{G}),$$

where $\mathfrak{P}_B(G)$ is the set of equivalence classes of principal bundles over B, with structure group G, and $\mathfrak{G}$ is the sheaf of germs of G-valued smooth maps on B.

1.6.4 The pull-back of a principal bundle

Let $\ell = (P, G, B, \pi)$ be a principal bundle and $h \colon B' \to B$ a smooth map. The **pull-back of ℓ by h** is the principal bundle $h^*(\ell) = (h^*(P), G, B',$

π^*), where

$$h^*(P) = B' \times_B P := \{(x',p) \in B' \times P : h(x') = \pi(p)\},$$
$$\pi^* := \mathrm{pr}_1 |_{h^*(P)} : h^*(P) \longrightarrow B',$$
$$h^* := \mathrm{pr}_2 |_{h^*(P)} : h^*(P) \longrightarrow P.$$

The action of G on the total space $h^*(P)$ is given by

$$\delta^* : h^*(P) \times G \longrightarrow h^*(P) : ((x',p),g) \mapsto (x',p \cdot g).$$

The fibres of $h^*(\ell)$ are isomorphic to the fibres of ℓ (and both isomorphic to G), while (h^*, id_G, h) is a G-morphism of $h^*(\ell)$ into ℓ. We add that each local trivialization (U_α, Φ_α) of P determines the trivialization $(f^{-1}(U_\alpha), \Phi_\alpha^*)$ of $h^*(P)$, where

$$\Phi_\alpha^* : h^{-1}(U_\alpha) \times_{U_\alpha} \pi^{-1}(U_\alpha) \longrightarrow U_\alpha \times G : (b',p) \mapsto \Phi_{\alpha,h(b')}(p)$$

[see (1.6.3)]. As a result, the corresponding cocycle $\{g_{\alpha\beta}^*\}_{\alpha,\beta \in I}$ of $f^*(P)$ is given by $\{g_{\alpha\beta}^* = g_{\alpha\beta} \circ h\}_{\alpha,\beta \in I}$.

The **universal property** in the category of principal bundles now reads: If $\ell_1 = (P_1, G, B', \pi_1)$ is a principal bundle and $(f, id_G, h) : \ell_1 \to \ell$ a pb-morphism, then there is a unique G-B'-(iso)morphism $(\bar{\pi}, id_G, id_{B'})$ from ℓ_1 onto ℓ^* such that $f = h^* \circ \bar{\pi}$. We obtain now the following diagram:

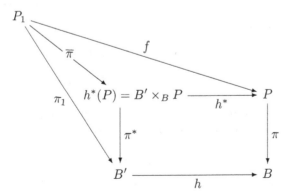

Clearly, π^* and h^* are the projections to the first and second factor, respectively.

1.6.5 The frame bundle of a vector bundle

Let $\ell = (E, B, \pi_E)$ be a vector bundle of fibre type the Banach space $\mathbb{E}$. Following [Bou67, n° 7.10.1], the set

$$P(E) := \{(x, f) : x \in B, \ f \in \mathcal{L}is(\mathbb{E}, E_x)\}$$

is an open submanifold of the linear map bundle $L(B \times \mathbb{E}, E)$ [see § 1.4.4(c)]. $\mathrm{GL}(\mathbb{E})$ acts on the right of $P(E)$ by setting

$$(x, f) \cdot g := (x, f \circ g); \qquad (x, f) \in P(E), g \in \mathrm{GL}(\mathbb{E}).$$

If $\pi_P \colon P(E) \to B$ is the map with $\pi_P(x, f) := x$, then the quadruple

$$\ell(E) := (P(E), \mathrm{GL}(\mathbb{E}), B, \pi_P)$$

is a principal bundle, called the **frame bundle of E**. The term **bundle (of linear) frames** is also in use. Equivalently, we may write

$$P(E) := \bigcup_{x \in B} \mathcal{L}is(\mathbb{E}, E_x).$$

The local structure of $P(E)$ is as follows: Let (U_α, τ_α) be a local trivialization of E with $\tau_\alpha \colon \pi_E^{-1}(U_\alpha) \to U_\alpha \times \mathbb{E}$. Then we obtain a local section $\sigma_\alpha \colon U_\alpha \to P(E)$ of $P(E)$ by setting [recall the notation of (1.4.3)]

$$(1.6.12) \qquad \sigma_\alpha(x) := (x.\tau_{\alpha,x}^{-1}), \qquad x \in U.$$

In virtue of (1.5.7) and (1.6.5′), equality (1.6.12) determines the local trivialization of $P(E)$

$$(1.6.13) \qquad \begin{aligned} \Psi_\alpha \colon U_\alpha \times \mathrm{GL}(\mathbb{E}) &\longrightarrow \pi_P^{-1}(U_\alpha): \\ \Psi_\alpha(x, g) &= \sigma_\alpha(x) \cdot g = (x, \tau_{\alpha,x}^{-1}) \cdot g = (x, \tau_x^{-1} \circ g), \end{aligned}$$

whose inverse $\Phi_\alpha \colon \pi_P^{-1}(U_\alpha) \to U_\alpha \times \mathrm{GL}(\mathbb{E})$ is given by

$$(1.6.14) \qquad \Phi_\alpha(x, f) = (x, \tau_{\alpha,x} \circ f), \quad \text{if } f \in \mathcal{L}is(\mathbb{E}, E_x).$$

Clearly, σ_α can be thought of as the natural section of $P(E)$ with respect to (U_α, Ψ_α).

The preceding local structure of $P(E)$, derived from that of E, implies that the transition functions of $P(E)$ and E coincide; that is,

$$(1.6.15) \qquad g_{\alpha\beta} = T_{\alpha\beta} \colon U_{\alpha\beta} \longrightarrow \mathrm{GL}(\mathbb{E}); \qquad \alpha, \beta \in I.$$

where $\{T_{\alpha\beta}\}$ is the cocycle of E, with respect to the trivializing cover $\{(U_\alpha, \tau_\alpha)\}_{\alpha,\beta \in I}$ (of E), and $\{g_{\alpha\beta}\}$ is the cocycle of $P(E)$, with respect to the corresponding trivializing cover $\{(U_\alpha, \Phi_\alpha)\}_{\alpha,\beta \in I}$ (we refer also to § 1.4.2 and § 1.6.3).

The initial vector bundle E is related with $P(E)$ in the following way, which is a particular case of an associated bundle discussed in the next subsections. Namely, we define on $P(E) \times \mathbb{E}$ the equivalence relation

$$(x', f', u') \sim (x, f, u) \quad \Leftrightarrow$$
$$x' = x \quad \text{and} \quad \exists\, g \in \text{GL}(\mathbb{E}) : (f', u') = (f \circ g, g^{-1}(u))$$

[g is uniquely determined by the free action of $\text{GL}(\mathbb{E})$ on $P(E)$]. Let

$$\bar{E} := P(E) \times \mathbb{E}/\sim,$$
$$\bar{\pi} : \bar{E} \longrightarrow B : \bar{\pi}([(x, f, u)]) := x.$$

Then $(\bar{E}, B, \bar{\pi})$ is a vector bundle isomorphic to (E, B, π_E). In fact, if we are given a trivializing cover $\{(U_\alpha, \Phi_\alpha)\}$ of $P(E)$, with corresponding natural sections $\{\sigma_\alpha\}$, we define the trivializations [see also (1.6.12)]

(1.6.16)
$$\bar{\tau}_\alpha^{-1} : U_\alpha \times \mathbb{E} \longrightarrow \bar{\pi}^{-1}(U_\alpha):$$
$$\bar{\tau}_\alpha^{-1}(x, u) := [(\sigma_\alpha(x), u)] = [(x, \tau_{\alpha,x}^{-1}, u)];$$

thus, by simple computations,

(1.6.17) $$\bar{\tau}_\alpha([(x, f, u)]) = (x, (\tau_{\alpha,x} \circ f)(u)).$$

The desired vb-isomorphism is provided by the map

(1.6.18) $$F : \bar{E} \longrightarrow E : [(x, f, u)] \mapsto f(u).$$

It is well-defined for if $[(x, f, u)] = [(x', f', u')]$, then $x = x'$ and $f' = f \circ g$, $u' = g^{-1}(x)$, for a (uniquely determined) $g \in GL(\mathbb{E})$. Therefore $f'(u') = f(u)$. Moreover,

1. $\pi_E \circ F = \bar{\pi}$.

2. F is injective: Let $[(x, f, u)], [(x', f', u')] \in \bar{E}$ with $f'(u') = f(u)$. Since

$$x' = \pi_P([(x', f', u')]) = \pi_E(f'(u')) = \pi_E(f(u)) = \pi_P([(x, f, u)]) = x,$$

it follows that $f, f' : \mathbb{E} \to E_x$. Setting $g := f^{-1} \circ f'$, we check that

$$[(x', f', u')] = [(x, f \circ g, g^{-1}(u))] = [(x, f, u)].$$

3. F is surjective: Let an arbitrary $h \in E_x$. If $\tau_{\alpha,x} : E_x \to \mathbb{E}$ is the Banach space isomorphism induced by a trivialization (U_α, τ_α) of E, with $x \in U_\alpha$, then $[(x, \tau_{\alpha,x}^{-1}, \tau_{\alpha,x}(h))]$ is mapped to h by F.

It remains to show that F (and analogously F^{-1}) satisfies condition (VBM. 2) of §1.4.3. Indeed, for an arbitrary $x_0 \in B$, we consider a trivialization (U_α, τ_α) of E, $x_0 \in U_\alpha$ and the trivialization $(U_\alpha, \bar{\tau}_\alpha)$ given by (1.6.17) (also ultimately determined by (U_α, τ_α), according to the local structure of $P(E)$ and $\bar{E}$). Then, $F(\bar{\pi}^{-1}(U_\alpha)) \subseteq \pi_E^{-1}(U_\alpha)$ and

$$(\tau_\alpha \circ F \circ \bar{\tau}_\alpha^{-1})(x, u) = (x, u), \qquad (x, u) \in U \times \mathbb{E}.$$

Therefore, restricted to the fibres over $x \in U$, we see that the map

$$U \ni x \longmapsto \tau_{\alpha, x} \circ F \circ \bar{\tau}_{\alpha, x}^{-1} = \mathrm{id}_\mathbb{E}$$

is smooth, thus proving the claim.

1.6.6 Associated bundles

Given a principal bundle, we intend to associate to it principal and vector bundles by means of appropriate Lie group homomorphisms.

a) *Principal bundles associated by Lie group morphisms*

Let $\ell = (P, G, B, \pi)$ be a Banach principal bundle, and let $\varphi \colon G \to H$ be a morphism of Banach Lie groups. Then G acts on the *left* of H by

$$G \times H \longrightarrow H \colon (g, h) \mapsto g \cdot h := \varphi(g) \cdot h,$$

and on the *right* of $P \times H$ by setting

$$(p, h) \cdot g := \big(p \cdot g, \varphi(g^{-1}) \cdot h\big); \qquad (p, h) \in P \times H, \, g \in G.$$

The previous action induces the following equivalence relation on $P \times H$:

$$(p', f') \sim (p, h) \quad \Leftrightarrow \quad \exists\, g \in G : (p', h') = \big(p \cdot g, \varphi(g^{-1}) \cdot h\big).$$

The resulting quotient space, denoted by $P \times^G H$, has a differential structure making the natural projection

$$\kappa \colon P \times H \longrightarrow P \times^G H \colon (p, h) \mapsto [(p, h)]$$

a submersion. Thus, a map $g \colon P \times^G H \to Y$ (Y: smooth manifold) is smooth if and only if so is $g \circ p$.

Regarding the smooth structure of the above quotient space, we refer to [Bou67, n°s 6.6.1, 6.5.1], in conjunction with [Bou67, n$^\circ$ 5.9.5], the latter containing more results on the structure of quotient manifolds induced by equivalence relations. Detailed proofs (in a more general setting) can be found in [KM97, § 37.12] and [Die72, §§ 16.10.3, 16.14.7].

Another way to define a smooth structure on $P \times^G H$, by gluing local data, will be described below. First, let H act on the right of $P \times^G H$ by setting

$$[(p, h)] \cdot h' = [(p, h \cdot h')]; \qquad p \in P, (h, h') \in H \times H,$$

and define the projection

$$\pi_H \colon P \times^G H \longrightarrow B \colon [(p,h)] \mapsto \pi_H([(p,h)]) := \pi(p).$$

Then the quadruple $\varphi(\ell) := (P \times^G H, H, B, \pi_H)$ is called the principal bundle **associated to ℓ by the Lie group morphism** $\varphi \colon G \to H$.

The local structure of $\varphi(\ell)$ is derived from that of ℓ as follows: Let $\{(U_\alpha, \Phi_\alpha)\}_{\alpha,\beta \in I}$ be a trivializing cover of P, $\Phi_\alpha \colon \pi^{-1}(U_\alpha) \to U_\alpha \times G$, and natural sections $s_\alpha \colon U_\alpha \to P$ given by $s_\alpha(x) = \Psi_\alpha(x,e) := \Phi_\alpha^{-1}(x,e)$. Then we define the trivializations

$$(1.6.19) \quad \overline{\Psi}_\alpha \colon U_\alpha \times H \longrightarrow \pi_H^{-1}(U_\alpha) \colon (x,h) \mapsto \overline{\Psi}(x,h) := [(s_\alpha(x),h)]).$$

It is easy to check that $\overline{\Psi}_\alpha$ is an equivariant (with respect to G and H) bijection, whose inverse $\overline{\Phi}_\alpha \colon \pi_H^{-1}(U_\alpha) \to U_\alpha \times H$ is given by

$$(1.6.20) \qquad\qquad \overline{\Phi}_\alpha([(p,h)]) := (x, \varphi(g) \cdot h),$$

with $x = \pi(p)$, and $g \in G$ determined by $p = s_\alpha(x) \cdot g$.

The previous local structure induces also a smooth structure on the quotient $P \times^G H$: Each $\overline{\Psi}_\alpha$ determines a smooth structure on $\pi_H^{-1}(U_\alpha)$ by transferring that of $U_\alpha \times H$. Since the local smooth structures coincide on the overlappings, we obtain a smooth structure on the quotient by gluing the local structures together.

The relations between the local trivializations of $\varphi(\ell)$ and ℓ lead to the following equalities, connecting the respective cocycles and the local sections of the aforementioned bundles:

$$(1.6.21) \qquad\qquad \bar{g}_{\alpha\beta} = \varphi \circ g_{\alpha\beta} \colon U_{\alpha\beta} \longrightarrow H,$$
$$(1.6.22) \qquad\qquad \bar{s}_\alpha(x) = [(s_\alpha(x), e)]; \qquad x \in U_\alpha,$$

for all indices $\alpha, \beta \in I$.

The bundles ℓ and $\varphi(\ell)$ are also related by the natural map

$$(1.6.23) \qquad \bar{\kappa} \colon P \longrightarrow P \times^G H \colon p \mapsto \bar{\kappa}(p) := [(p,e)],$$

which, in fact, determines the B-morphism $(\kappa, \varphi, \mathrm{id}_B)$ between them. Then (1.6.22) is rewritten as

$$(1.6.22') \qquad\qquad \bar{s}_\alpha(x) = [(s_\alpha(x), e)], \qquad x \in U_\alpha.$$

It is worth adding that $P \times^G H$ (together with κ) has the following universal property: If $(F, \varphi, \mathrm{id}_B)$ is a B-morphism of ℓ into another principal bundle $\ell' = (P', H, B, \pi')$, then there a unique H-B-isomorphism

$(\theta, \mathrm{id}_H, \mathrm{id}_B)$ of $\varphi(\ell)$ onto ℓ', such that $F = \theta \circ \bar{\kappa}$, as pictured also in the next diagram:

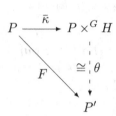

Actually, θ is given by

$$(1.6.24) \qquad \theta([(p, h)]) = F(p) \cdot h, \qquad [(p, h)] \in P \times^G H.$$

Its smoothness is a consequence of the manifold structure on the quotient space mentioned earlier. The other properties of θ are clear.

b) *Vector bundles associated by representations of the structure group*
Let again $\ell = (P, G, B, \pi)$ be a Banach principal bundle, and let φ be a representation of G into a Banach space $\mathbb{E}$, i.e. $\varphi \colon G \to \mathrm{GL}(\mathbb{E})$ is a morphism of Banach-Lie groups. We construct a vector bundle associated to ℓ by specializing the process of the foregoing case a) as follows:

We first define a left action of $\mathrm{GL}(\mathbb{E})$ on $\mathbb{E}$

$$\mathrm{GL}(\mathbb{E}) \times \mathbb{E} \colon (g, u) \mapsto g \cdot u := \varphi(g)(u),$$

and the action of $\mathrm{GL}(\mathbb{E})$ on the right of $P \times \mathbb{E}$ by

$$(p, u) \cdot g := \big(p \cdot g, \varphi(g^{-1})(u)\big); \qquad (p, u) \in P \times \mathbb{E}, \, g \in G.$$

We obtain the quotient space $E := P \times^G \mathbb{E}$ and the map $\pi_E \colon E \to B$ with $\pi_E([(p, u)]) := \pi(p)$. Then (E, B, π_E) is the **vector bundle associated to P by** $\varphi \colon G \to \mathrm{GL}(\mathbb{E})$, of fibre type $\mathbb{E}$.

The local structure of E is defined as follows: As in case a), we consider a trivializing cover $\{(U_\alpha, \Phi_\alpha)\}_{\alpha \in I}$ of P, and the corresponding natural sections $s_\alpha \colon U_\alpha \to P$. Then we define the maps

$$(1.6.25) \ \tau_\alpha^{-1} \colon U_\alpha \times \mathbb{E} \longrightarrow \pi_E^{-1}(U_\alpha) \colon (x, u) \mapsto \tau_\alpha^{-1}(x, u) := [(s_\alpha(x), u)]);$$

thus $\{(U_\alpha, \tau_\alpha)\}_{\alpha \in I}$ is trivializing cover of E. We clarify that τ_α is defined analogously to (1.6.20); namely,

$$(1.6.26) \qquad\qquad \tau_\alpha([(p, u)]) = (x, \varphi(g)(u)),$$

where $x = \pi(p)$, and $g \in G$ is determined by the equality $p = s_\alpha(x) \cdot g$.

Moreover, the corresponding cocycle $\{T_{\alpha\beta}\}$ of E is connected with the cocycle $\{g_{\alpha\beta}\}$ of P by

$$(1.6.27) \qquad T_{\alpha\beta} = \varphi \circ g_{\alpha\beta} \colon U_{\alpha\beta} \longrightarrow \mathrm{GL}(\mathbb{E}); \qquad \alpha, \beta \in I.$$

Applying the previous results to the particular case of the frame bundle $P(E)$ of a vector bundle E, and the morphism $\mathrm{id}_{\mathrm{GL}(\mathbb{E})} \colon \mathrm{GL}(\mathbb{E}) \to \mathrm{GL}(\mathbb{E})$, we see that the vector bundle $P(E) \times^{\mathrm{GL}(\mathbb{E})} \mathbb{E}$, associated to $P(E)$ by $\mathrm{id}_{\mathrm{GL}(\mathbb{E})}$, coincides with the bundle $\bar{E}$ defined in § 1.6.5. Therefore, by (1.6.18), E is isomorphic to $P(E) \times^{\mathrm{GL}(\mathbb{E})} \mathbb{E}$.

c) *Interrelations*

We consider, once again, a Banach principal bundle (P, G, B, π) and a Banach-Lie group morphism (representation) $\varphi \colon G \to \mathrm{GL}(\mathbb{E})$. Then, in virtue of case a), we obtain the principal bundle

$$\left(P_\varphi := P \times^G \mathrm{GL}(\mathbb{E}), \mathrm{GL}(\mathbb{E}), B, \pi_\varphi\right)$$

(for convenience, we set $\pi_\varphi = \pi_{\mathrm{GL}(\mathbb{E})}$), and, in virtue of case b), the vector bundle

$$\left(E_\varphi := P \times^G \mathbb{E}, B, \pi\right).$$

In turn, the latter determines the principal bundle of frames (see § 1.6.5)

$$(P(E_\varphi), \mathrm{GL}(\mathbb{E}), B, \bar{\pi}).$$

We wish to connect the three principal bundles P, P_φ and $P(E_\varphi)$. Already, we know that P, P_φ are related by the B-morphism $(\kappa, \phi, \mathrm{id}_B)$ determined by (1.6.23), now taking the form $\kappa(p) = [(p, \mathrm{id}_\mathbb{E})]$. On the other hand, if $\{g_{\alpha\beta}\}$ is the cocycle of P, then equalities (1.6.21), (1.6.25) and (1.6.15) imply that the corresponding cocycles of P_φ, E_φ and $P(E_\varphi)$ coincide with $\{\varphi \circ g_{\alpha\beta}\}$. Therefore P_φ and $P(E_\varphi)$ are $\mathrm{GL}(\mathbb{E})$-B-isomorphic. However, based on the discussion at the end of § 1.6.6(b), we can single out a concrete isomorphism, because there is also a natural B-morphism of P into $P(E_\varphi)$. With this in mind, we first define the map

$$(1.6.28) \qquad F \colon P \longrightarrow P(E_\varphi) \colon p \mapsto (x, \widetilde{p}),$$

where $x := \pi(p)$ and $\widetilde{p} \colon \mathbb{E} \to E_{\varphi, x}$ is the Banach space isomorphism given by $\widetilde{p}(u) := [(p, u)]$ (recall that $[(p, u)] \in E_\varphi = P \times^G \mathbb{E}$).

The smoothness of F is checked locally: If p_0 is an arbitrary point in P with $\pi(p_0) = x_0$, we choose a local trivialization (U, Φ) of P with $x_0 \in U$, and consider the corresponding trivializations (U, τ) of E_φ and $(U, \underline{\Phi})$ of $P(E_\varphi)$, given by the analogs of (1.6.25) and (1.6.14), respectively (for convenience we drop the index α from the latter). Since $F(\pi^{-1}(U)) \subseteq \pi_\varphi^{-1}(U)$, we obtain the local representation $\underline{\Phi} \circ F \circ \Phi^{-1}$, as in the diagram

$$
\begin{array}{ccc}
\pi^{-1}(U) & \xrightarrow{\quad F \quad} & \pi_\varphi^{-1}(U) \\[2mm]
\Big\downarrow{\scriptstyle \Phi} & & \Big\downarrow{\scriptstyle \underline{\Phi}} \\[2mm]
U \times G & \xrightarrow[\underline{\Phi} \,\circ\, F \,\circ\, \Phi^{-1}]{} & U \times \mathrm{GL}(\mathbb{E})
\end{array}
$$

Therefore, for every $(x, g) \in U \times \mathbb{E}$,

$$
\left(\underline{\Phi} \circ F \circ \Phi^{-1}\right)(x, g) = \left(\underline{\Phi} \circ F\right)(\Psi(x, g)) =
$$
$$
= (\underline{\Phi} \circ F)(s(x) \cdot g) = \left(x, \tau_x \circ \widetilde{s(x) \cdot g}\right).
$$

Because, for every $u \in \mathbb{E}$,

$$
\left(\tau_x \circ \widetilde{s(x) \cdot g}\right)(u) = \tau_x([(s(x) \cdot g, u)])
$$
$$
= (\mathrm{pr}_2 \circ \tau)([(s(x), \varphi(g)(u))])
$$
$$
= \mathrm{pr}_2(x, \varphi(g)(u)) = \varphi(g)(u),
$$

it follows that $\left(\underline{\Phi} \circ F \circ \Phi^{-1}\right)(x, g) = (x, \varphi(g))$, from which we deduce the smoothness of F at (an arbitrary) $p_0 \in P$.

On the other hand, it is immediate that $\bar{\pi} \circ F = \pi$, while

$$
F(p \cdot g) = \left(\pi(p \cdot g), \widetilde{p \cdot g}\right) = \left(\pi(p), \widetilde{p} \circ \varphi(g)\right) = F(p) \cdot \varphi(g),
$$

for every $p \in p$ and $g \in G$; that is, F is equivariant with respect to the actions of G and $\mathrm{GL}(\mathbb{E})$ on P and $P(E_\varphi)$. As a consequence of the previous arguments, $(F, \varphi, \mathrm{id}_B)$ is a B-morphism of P into $P(E_\varphi)$.

The desired $\mathrm{GL}(\mathbb{E})$-B-isomorphism between P_φ and $P(E_\varphi)$ is provided by the universal property of the quotient manifold; namely, according to (1.6.24), (1.6.28) and the action of $\mathrm{GL}(\mathbb{E})$ on $P(E_\varphi)$,

$$
(1.6.29) \qquad \theta \colon P_\varphi \xrightarrow{\;\cong\;} P(E_\varphi) \colon [(p, g)] \mapsto \theta([(p, g)]) := (\pi(p), \widetilde{p} \circ g).
$$

Summarizing, we obtain the following diagram [in analogy to the one

given in part a)]

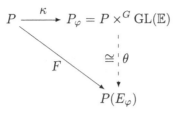

with κ, F and θ given, respectively, by (1.6.23), (1.6.28) and (1.6.29).

For later use, let us relate also the natural sections of the previously mentioned bundles, over corresponding trivializations. To this end, we denote by

- $\{s_\alpha\}$ the natural sections of P, with respect to the trivializing cover $\{(U_\alpha, \Phi_\alpha)\}_{\alpha \in I}$;

- $\{s_\alpha^\varphi\}$ the natural sections of P_φ, with respect to the trivializing cover $\{(U_\alpha, \overline{\Phi}_\alpha)\}_{\alpha \in I}$, defined as in case a).

- $\{\sigma_\alpha^\varphi\}$ the natural sections of $P(E_\varphi)$, with respect to the trivializing cover $(U_\alpha, \underline{\Phi}_\alpha)_{\alpha \in I}$, derived from the study of the smoothness of (1.6.28).

Then, adapting (1.6.22) to the present settings, we have that

$$(1.6.30) \qquad s_\alpha^\varphi(x) = [s_\alpha(x), \mathrm{id}_{\mathbb{E}}] = \kappa(s_\alpha(x)); \qquad x \in U_\alpha, \, \alpha \in I.$$

Taking into account the earlier definition of $\underline{\Phi}_\alpha$ (where the subscript α was then omitted), we check that

$$(1.6.31) \qquad \sigma_\alpha^\varphi(x) = \left(x, \widetilde{s_\alpha(x)}\right) = F(s_\alpha(x)); \qquad x \in U_\alpha, \, \alpha \in I,$$

where $\widetilde{s_\alpha} \colon \mathbb{E} \to (P \times^G \mathbb{E})_x$ is the Banach space isomorphism given by $\widetilde{s_\alpha}(u) = [(s_\alpha, u)]$. Therefore, applying θ to the preceding equality, we also have

$$(1.6.32) \qquad \sigma_\alpha^\varphi(x) = \theta\big(s_\alpha^\varphi(x)\big); \qquad x \in U_\alpha, \, \alpha \in I.$$

The latter equality is in accordance with the fact that the cocycles of P_φ and $P(E_\varphi)$ coincide, as explained before introducing the map (1.6.28).

1.6.7 The exact sequence associated to a principal bundle

Given a principal bundle $\ell = (P, G, B, \pi)$, we associate to it the following exact sequence of vector bundles (over P), used to define connections on

ℓ (see §1.7.1 below):

(1.6.33) $0 \longrightarrow P \times \mathfrak{g} \xrightarrow{\ \nu\ } TP \xrightarrow{\ T\pi!\ } \pi^*(TB) \longrightarrow 0$

where

• $P \times \mathfrak{g}$ is the trivial vector bundle, $\mathfrak{g}$ denoting the Lie algebra of G, identified with $T_e G$.

• The vb-morphism ν is defined by

$$\nu(p, X) := X_p^*, \qquad (p, X) \in P \times \mathfrak{g}.$$

We recall that X^* is the fundamental (Killing) vector field associated to $X \equiv X_e$ by the action $\delta \colon P \times G \to G$ (see §1.3.2). The morphism ν is an immersion and $\nu(P \times \mathfrak{g}) = VP$. Here VP is the **vertical subbundle** of the tangent bundle (TP, P, τ_P), whose fibres are given by

$$V_p P := (VP)_p = \ker\big(T_p \pi! |_{T_p P}\big) = \ker(T_p \pi) = T_p(\pi^{-1}(x)),$$

for every $p \in P$ with $\pi(p) = x$. Clearly, the restriction of ν to the fibre $p \times \mathfrak{g} \equiv \mathfrak{g}$ gives the linear isomorphism

$$\nu_p \colon \mathfrak{g} \xrightarrow{\ \cong\ } V_p P;$$

as a result,

(1.6.34) $\nu_p(X) = X_p^* = T_e \delta_p(X), \qquad X \in \mathfrak{g}.$

We notice that the partial map $\delta_p \colon G \to \pi^{-1}(x)$, $x = \pi(p)$ is a diffeomorphism, whose inverse is given by $\delta_p^{-1}(q) = k(p, q)$ [see (1.6.6) for the definition of k].

• $\pi^*(TB) = P \times_B TB$ (: the pull-back of TB by $\pi \colon P \to B$; see §1.4.4).

• $T\pi!$ is the vb-morphism defined by the universal property of the pull-back, as pictured below.

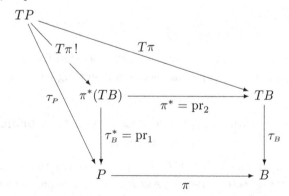

A crucial property of the vector bundles involved in (1.6.33) is that G acts naturally on them. As a matter of fact, we have for every $g \in G$:

(1.6.35)
$$
\begin{aligned}
(p, X_e) \cdot g &:= \big(p \cdot g, \mathrm{Ad}(g^{-1})(X_e)\big); & (p, X_e) &\in P \times \mathfrak{g}, \\
u \cdot g &:= TR_g(u); & u &\in TP, \\
(p, v) \cdot g &:= (p \cdot g, v); & (p, v) &\in \pi^*(TB).
\end{aligned}
$$

We recall that $R_g \colon P \to P$ is the right translation of P by $g \in G$ (see § 1.3.1), i.e. $R_g(p) = p \cdot g \equiv pg$, and Ad is the adjoint representation of G defined in § 1.2.4.

Taking into account the previous actions and the fact that $T\pi! = (\tau_P, \tau_B)$, it follows that the vb-morphisms ν and $T\pi!$ are also G-**equivariant**, that is

(1.6.36)
$$
\begin{aligned}
\nu\big((p, X_e) \cdot g\big) &= T_p R_g\big(\nu(p, X_e)\big) = \nu(p, X_e) \cdot g, \\
T\pi!(u \cdot g) &= \big(p \cdot g, T\pi(u)\big) = \big(p, T\pi(u)\big) \cdot g = T\pi!(u) \cdot g,
\end{aligned}
$$

for every $(p, X_e) \in P \times \mathfrak{g}$, $(p, v) \in \pi^*(TB)$ and every $g \in G$. Note that from the first of (1.6.36), the actions (1.6.35) and the definition of X^* (§1.3.2), we see that

$$
T_g R_g(X_p^*) = \big(\mathrm{Ad}(g^{-1})(X_e)\big)_{p \cdot g}^*,
$$

for every $g \in G$ and $p \in P$.

1.7 Connections on principal bundles

Connections on principal bundles can be handled in many equivalent ways. Here we define connections as splittings of the exact sequence (1.6.33), as $\mathfrak{g}$-valued connection forms on the total space of the bundle, or as a family of $\mathfrak{g}$-valued local connection forms over a (trivializing) open cover of the base space. The previous approaches will be used interchangeably.

1.7.1 Principal bundle connections as splitting G-morphisms

A **connection** on the principal bundle $\ell = (P, G, B, \pi)$ is a G-**splitting** of the exact sequence (1.6.33). This means that there is an exact sequence of vector bundles

(1.7.1)
$$
0 \longrightarrow \pi^*(TB) \xrightarrow{\ C\ } TP \xrightarrow{\ V\ } P \times \mathfrak{g} \longrightarrow 0
$$

such that C, V are G-equivariant morphisms, and

$$T\pi! \circ C = \mathrm{id}_{\pi^*(TB)} \quad \text{and} \quad V \circ \nu = \mathrm{id}_{VE}$$

(see also [Pen69]). The definition is illustrated by the diagram on the next page.

The reader may have noticed that the morphisms of (1.7.1) are denoted by the same symbols used in the splitting sequence (1.5.1). From the context it will be understood which of the two sequences or connections we are referring to.

As in the case of connections on vector bundles (discussed in §1.5.1), it suffices to know either C or V. The splitting implies the decomposition

$$(1.7.2) \qquad\qquad TP = VP \oplus HP,$$

where $HE := \mathrm{Im}(C)$ is the **horizontal subbundle** of TP. The vertical

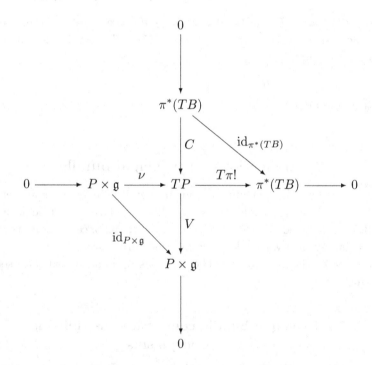

subbundle VP of TP has already been defined in §1.6.7. It follows that

$$(1.7.3) \qquad\qquad HP = C\left(\pi^*(TB)\right) = \ker V,$$
$$(1.7.4) \qquad\qquad T_p R_g(H_p P) = H_{pg} P,$$

for every $p \in P$ and $g \in G$. On the other hand, the equalities $T\pi! \circ C = \mathrm{id}_{\pi^*(TB)}$ and $T\pi! = (\tau_P, T\pi)$ imply that

$$(1.7.5) \qquad u^h = C(\tau_P(u), T\pi(u)); \qquad u \in T_pP),$$

where u^h is the **horizontal component** of u after the decomposition (1.7.2).

1.7.2 Connection forms

Global connection forms provide a useful tool to handle connections on a principal bundle. There is no essential difference between the finite and infinite-dimensional case.

Let $\ell = (P, G, B, \pi)$ be a principal bundle and let $L(TP, P \times \mathfrak{g})$ be the linear map bundle, whose fibre over a $p \in P$ is the space of continuous linear maps $\mathcal{L}(T_pP, \mathfrak{g})$ [see § 1.4.4(c)]. The smooth sections of the previous bundle are called $\mathfrak{g}$-**valued 1-forms** on P and their set is denoted by $\Lambda^1(P, \mathfrak{g})$.

A **connection form** of ℓ is a form $\omega \in \Lambda^1(P, \mathfrak{g})$ satisfying the following conditions:

$(\boldsymbol{\omega}.\ \mathbf{1}) \qquad\qquad \omega(X^*) = X; \qquad X \in \mathfrak{g},$

$(\boldsymbol{\omega}.\ \mathbf{2}) \qquad\qquad R_g^*\omega = \mathrm{Ad}\big(g\big)\omega; \qquad g \in G.$

The preceding equalities, evaluated at any $p \in P$ give, respectively:

$$\omega_p(X_p^*) = X \equiv X_e,$$
$$\omega_{pg}(T_pR_g(u)) = \mathrm{Ad}\big(g^{-1}\big).\omega_p(u),$$

for every $X \in \mathfrak{g}$, $g \in G$ and $u \in T_pP$. The line dot inserted above replaces obvious parentheses and should not be confused with center dots indicating multiplication or action of a group.

A connection form ω is related with the G-equivariant morphism V of the splitting sequence (1.7.1) by

$$\omega_p(u) = (\mathrm{pr}_2 \circ V)(u); \qquad p \in P,\ u \in T_pP.$$

Therefore, (1.7.3) implies that

$$(1.7.6) \qquad\qquad H_pP = \ker \omega_p, \qquad p \in P.$$

Let $u \in T_pP$ with $u = u^v + u^h$ where u^v denotes the **vertical component of** u. If we set $\omega_p(u) = A$, then (1.7.6) implies that $\omega_p(u^v) = A$.

Since, $\omega_p(A_p^*) = A$, it follows that $\omega_p(A_p^* - u^v) = 0$, or $V_pP \ni A_p^* - u^v \in H_pP$, thus $A_p^* = u^v$. In other words,

> $\omega_p(u)$ *identifies with the element of the Lie algebra* $\mathfrak{g}$ *of* G *whose corresponding fundamental vector field coincides at* p *with the vertical component of* u.

This is another way to define ω (see [KN68]). Moreover, using (1.7.5), we find that

$$(1.7.7) \qquad C(p, T_p\pi(u)) = u - \nu_p(\omega_p(u)); \qquad p \in P, u \in T_pP.$$

Equivalently,

$$(1.7.8) \qquad C(p, v) = u - \nu_p(\omega_p(u)); \qquad (p, v) \in P \times_B TB,$$

where $u \in T_pP$ is any vector such that $T_p\pi(u) = v$.

1.7.3 Local connection forms

Local connection forms are particularly useful because they involve only the base space and the structure group of the bundle.

Let $\ell = (P, G, B, \pi)$ be a principal bundle with trivializing cover $\mathcal{C} = \{(U_\alpha, \Phi_\alpha) \,|\, \alpha \in I\}$ and the corresponding natural sections $\{s_\alpha\}$ defined by (1.6.4). Assume that ℓ is equipped with a connection whose connection form is $\omega \in \Lambda^1(P, \mathfrak{g})$. Then the 1-forms

$$(1.7.9) \qquad \omega_\alpha := s_\alpha^*\omega \in \Lambda^1(U_\alpha, \mathfrak{g}), \qquad \alpha \in I$$

are called the **local connection forms** of the given connection ($\equiv \omega$), with respect to the trivializing cover $\mathcal{C}$. Therefore,

$$\omega_{\alpha,x}(v) = \omega_{s_\alpha(x)}(T_x s_\alpha(v)); \qquad x \in U_\alpha, v \in T_x U_\alpha \equiv T_x B.$$

The local connection forms satisfy the compatibility condition

$$(1.7.10) \qquad \omega_\beta = \mathrm{Ad}(g_{\alpha\beta}^{-1})\omega_\alpha + g_{\alpha\beta}^{-1}dg_{\alpha\beta}; \qquad \alpha, \beta \in I,$$

over $U_{\alpha\beta}$. The second summand on the right-hand side of (1.7.10) is the left Maurer-Cartan differential of $g_{\alpha\beta} \in C^\infty(U_{\alpha\beta}, G)$ defined in § 1.2.6. More explicitly, (1.7.10) evaluated at any $x \in U_{\alpha\beta}$ and $v \in T_x B$, yields:

$$\omega_{\beta,x}(v) = \mathrm{Ad}(g_{\alpha\beta}(x)^{-1}).\omega_{\alpha,x}(v) + T_x(\lambda_{g_{\alpha\beta}(x)^{-1}} \circ g_{\alpha\beta})(v).$$

Recall that λ_g is the left translation of G and the line dot replaces parentheses.

Conversely, let $\mathcal{C} = \{(U_\alpha, \Phi_\alpha) \,|\, \alpha \in I\}$ be a trivializing cover of a principal bundle ℓ. A family of 1-forms $\{\omega_\alpha \in \Lambda^1(U_\alpha, \mathfrak{g}) \mid \alpha \in I\}$, satisfying the compatibility condition (1.7.10), determines a unique connection form $\omega \in \Lambda^1(P, \mathfrak{g})$, whose local connection forms coincide with the given $\{\omega_\alpha\}$. Indeed, for each $\alpha \in I$, we define the map $g_\alpha \colon \pi^{-1}(U_\alpha) \to G$ given by $g_\alpha(p) = (\mathrm{pr}_2 \circ \Phi_\alpha)(p)$. It follows that g_α is a smooth map such that

$$p = s_\alpha(\pi(p)) \cdot g_\alpha(p), \qquad p \in \pi^{-1}(U_\alpha)$$

(the preceding equality can be also used to define g_α). Then, for every $p \in P$ with $\pi(p) \in U_\alpha$, and every $u \in T_p P$, we set

$$(1.7.11) \qquad \omega_p(u) := \mathrm{Ad}\left(g_\alpha(p)^{-1}\right) . (\pi^* \omega_\alpha)_p(u) + \left(g_\alpha^{-1} dg_\alpha\right)_p (u).$$

Condition (1.7.10) ensures that ω is a well-defined $\mathfrak{g}$-valued differential 1-form on P. It turns out that ω is a connection form.

For the analog of (1.7.11) in the case of a Lie group G acting on the left of P see [SW72, p. 129]. We refer also to [Ble81, pp. 32–33], [KN68, p. 66] and [Pha69, pp. 227–228] for other ways to define ω from $\{\omega_\alpha\}$.

Since ω is completely known by its local connection forms, we may write

$$\omega \equiv \{\omega_\alpha\}_{\alpha \in I}.$$

The proof of (1.7.10) and the fact that (1.7.11) is well-defined are based on certain arguments and computations which will be also used later. More precisely, assume that σ and s are two sections of P over the same open subset U of B. Then there is a unique smooth map $g \colon U \to G$ such that $\sigma = s \cdot g = \delta \circ (s, g)$, where δ is the action of G on P. Then, for every $x \in U$ and $v \in T_x B$,

$$
\begin{aligned}
T_x\sigma(v) &= T_{s(x)}\delta_{g(x)}(T_x s(v)) + T_{g(x)}\delta_{s(x)}(T_x g(v)) = \\
(1.7.12) \quad &= T_{s(x)}R_{g(x)}(T_x s(v)) + \left(T_e \delta_{s(x) \cdot g(x)} \circ T_{g(x)}\lambda_{g(x)^{-1}}\right)(T_x g(v)) \\
&= T_{s(x)}R_{g(x)}(T_x s(v)) + T_e \delta_{\sigma(x)}\left((g^{-1}dg)_x(v)\right).
\end{aligned}
$$

Setting $(g^{-1}dg)_x(v) = A \in \mathfrak{g}$, we have that

$$T_e \delta_{\sigma(x)}\left((g^{-1}dg)_x(v)\right) = \nu_{\sigma(x)}\left((g^{-1}dg)_x(v)\right) = A^*_{\sigma(x)};$$

hence,

$$(1.7.13) \qquad T_x\sigma(v) = T_{s(x)}R_{g(x)}(T_x s(v)) + A^*_{\sigma(x)}.$$

Applying ω to the latter, we obtain

$$\omega_{\sigma(x)}(T_x\sigma(v)) = \big(R^*_{g(x)}\omega\big)_{s(x)}(T_x\sigma(v)) + \omega_{\sigma(x)}\big(A^*_{\sigma(x)}\big)$$

(1.7.14)
$$= \mathrm{Ad}\big(g(x)^{-1}\big).\omega_{s(x)}(T_x s(v)) + A$$

$$= \mathrm{Ad}\big(g(x)^{-1}\big).\omega_{s(x)}(T_x s(v)) + \big(g^{-1}dg\big)_x(v).$$

Equivalently, for every $x \in U$ and $v \in T_x B$,

$$(\sigma^*\omega)_x(v) = \mathrm{Ad}\big(g(x)^{-1}\big).(s^*\omega)_x(v) + \big(g^{-1}dg\big)_x(v).$$

In summary [see also the comments following (1.7.10)],

(1.7.15) $$\sigma^*\omega = \mathrm{Ad}\big(g^{-1}\big)(s^*\omega) + g^{-1}dg$$

for every $\sigma, s \in \Gamma(U, P)$ with $\sigma = s \cdot g$.

1.7.4 Connections on the frame bundle

Let $\ell = (E, B, \pi_E)$ be a vector bundle of fibre type $\mathbb{E}$ and the corresponding principal bundle of frames $\ell(E) = (P(E), \mathrm{GL}(\mathbb{E}), B, \pi_P)$, defined in §1.6.5. Let also $\{(U_\alpha, \phi_\alpha)\}_{\alpha \in I}$ be an atlas of B over which we define the local trivializations of E and $P(E)$.

Assume first that E admits a linear connection $K \equiv \{\Gamma_\alpha\}_{\alpha \in I}$ with Christoffel symbols viewed as smooth maps of the form $\Gamma_\alpha \colon \phi_\alpha(U_\alpha) \to \mathcal{L}(\mathbb{B}, \mathcal{L}(\mathbb{E}))$ [see (1.5.10′)]. For every $x \in U_\alpha$ and every $v \in T_x B$, we set

(1.7.16) $$\omega_{\alpha, x}(v) := \Gamma_\alpha(\phi_\alpha(x))\big(\overline{\phi}_\alpha(v)\big),$$

where $\overline{\phi}_\alpha \equiv \overline{\phi}_{\alpha, x} \colon T_x B \to \mathbb{B}$ is the isomorphism (1.1.4). More explicitly, for every x, v as before, and every $u \in \mathbb{E}$,

(1.7.17) $$(\omega_{\alpha, x}(v))(u) = \big(\Gamma_\alpha(\phi_\alpha(x)).\overline{\phi}_\alpha(v)\big)(u).$$

It turns out that $\omega_\alpha \in \Lambda^1(\omega_\alpha, \mathcal{L}(\mathbb{E}))$, for every $\alpha \in I$. Since the transition functions of E and $P(E)$ coincide [see (1.6.5)], the compatibility condition of the Christoffel symbols (1.5.13) implies the compatibility condition (1.7.10), thus $\{\omega_\alpha\}_{\alpha \in I}$ determine a connection (form) ω on P (with local connection forms $\{\omega_\alpha\}$).

Conversely, assume that $P(E)$ admits a connection $\omega \equiv \{\omega_\alpha\}$. We set

(1.7.18) $$\Gamma_\alpha(z).y := [(\psi^*_\alpha\omega_\alpha)_z](y) \equiv [(\psi^*_\alpha\omega_\alpha)_z]\big(\overline{\mathrm{id}}_{\mathbb{B}, z}(y)\big)$$

for every $z \in \phi_\alpha(U_\alpha)$, $y \in \mathbb{B}$, with $\psi_\alpha = \phi_\alpha^{-1}$ and $\overline{\mathrm{id}}_{\mathbb{B}, z} \colon T_z\mathbb{B} \xrightarrow{\simeq} \mathbb{B}$. Now (1.7.10) implies (1.5.13), thus $\{\Gamma_\alpha\}_{\alpha \in I}$ determine a linear connection K, with Christoffel symbols the previous family.

It is obvious that (1.7.8) is the inverse of (1.7.7). Moreover, the association $K \leftrightarrow \omega$, by the described procedure, establishes a bijective correspondence between linear connections on E and connections on the principal bundle of frames $P(E)$.

It is also possible to relate the linear connections on E with the connections on $P(E)$ using the splittings C of the exact sequences (1.5.1) and (1.7.1) (see, for instance, [Pen69] and [Vas82]). However, this approach will not be pursued here.

1.7.5 Related connections on principal bundles

Let $\ell = (P, G, B, \pi)$ and $\ell' = (P', G', B', \pi')$ be principal bundles endowed with the connections ω and ω', respectively. If (f, φ, h) is a pb-morphism of ℓ into ℓ', then the connections ω and ω' are said to be (f, φ, h)-**related** if one of the following equivalent conditions hold (see also [Vas78(a)]):

$$f^* \omega' = \overline{\varphi} \cdot \omega,$$

$$Tf(u^v) = (Tf(u))^{v'},$$

$$(1.7.19) \qquad Tf(u^h) = (Tf(u))^{h'},$$

$$V' \circ Tf = C \circ (f \times \overline{\varphi}),$$

$$C' \circ (f \times Th) = Tf \circ C.$$

More explicitly: The first condition means that

$$(f^* \omega')_p(u) = \omega'_{f(p)}(T_p f(u)) = \overline{\varphi}(\omega_p(u)); \qquad p \in P, \, u \in T_p P.$$

Here $\overline{\varphi}$ is the Lie algebra morphism induced by φ, identified with $T_e \varphi$ [in virtue of (1.2.3)]. Recall that C, V are the splittings of the exact sequence (1.7.1) corresponding to ω. The superscripts v and h indicate, respectively, the vertical and horizontal components of $u \in TP$, after the decomposition (1.7.2). The dashed quantities refer to the bundle (P', G', B', π').

Another equivalent condition in terms of *parallel displacements* is given by (1.9.5).

Assume now that $\ell = (P, G, B, \pi)$ and $\ell' = (P', G', B, \pi')$ are principal bundles over the same base, equipped with the respective connections ω and ω'. Taking local trivializations over the same open cover $\mathcal{C} = \{U_\alpha \,|\, \alpha \in I\}$ of B, we consider the natural local sections $\{s_\alpha\}_{\alpha \in I}$ and $\{s'_\alpha\}_{\alpha \in I}$ of P and P', respectively, as well as the local connection forms $\{\omega_\alpha\}_{\alpha \in I}$ and $\{\omega'_\alpha\}_{\alpha \in I}$ corresponding to ω and ω'.

We prove two propositions, referring also to [Vas13] for further results on related connections and their applications.

Proposition 1.7.1 *Let $(f, \varphi, \mathrm{id}_B)$ be a pb-morphism of $\ell = (P, G, B, \pi)$ into $\ell' = (P', G', B, \pi')$. Two connections ω and ω' on ℓ and ℓ', respectively, are $(f, \varphi, \mathrm{id}_B)$-related if and only if*

$$(1.7.20) \qquad \overline{\varphi}\omega_\alpha = \mathrm{Ad}(h_\alpha^{-1})\omega_\alpha' + h_\alpha^{-1}dh_\alpha; \qquad \alpha, \beta \in I,$$

where $\{h_\alpha \colon U_\alpha \to G' \,|\, \alpha \in I\}$ are smooth maps defined by

$$f(s_\alpha(x)) = s_\alpha'(x) \cdot h_\alpha(x), \qquad x \in U_\alpha.$$

Proof Assume first that ω and ω' are $(f, \varphi, \mathrm{id}_B)$-related. Then, for every $x \in U_\alpha$ and $v \in T_x B$,

$$[s_\alpha^*(f^*\omega')]_x(v) = [(f \circ s_\alpha)^*\omega']_x(v)$$

or, applying (1.7.15),

$$(1.7.21) \quad [s_\alpha^*(f^*\omega')]_x(v) = \mathrm{Ad}\left(h_\alpha(x)^{-1}\right).\omega_{\alpha,x}(v) + \left(h_\alpha^{-1}dh_\alpha\right)_x(v).$$

Similarly,

$$(1.7.22) \qquad [s_\alpha^*(\overline{\varphi}\omega)]_x(v) = \overline{\varphi}\left((s_\alpha^*\omega)_x(v)\right) = \overline{\varphi}\left(\omega_{\alpha,x}(v)\right).$$

In virtue of the assumption, equalities (1.7.21) and (1.7.22) lead now to the desired condition (1.7.20).

Conversely, assume that (1.7.20) holds. To proceed, we shall need the following two equalities whose verification is immediate:

$$(1.7.23) \qquad \overline{\varphi} \circ \mathrm{Ad}\left(g_\alpha(p)^{-1}\right) = \mathrm{Ad}\left((\varphi \circ g_\alpha)(p)^{-1}\right) \circ \overline{\varphi},$$

$$(1.7.24) \qquad \overline{\varphi}\left(g_\alpha^{-1}dg_\alpha\right) = (\varphi \circ g_\alpha)^{-1}d(\varphi \circ g_\alpha),$$

for any smooth map $g_\alpha \colon \pi^{-1}(U_\alpha) \to G$.

Now, on $\pi^{-1}(U_\alpha)$ we determine a smooth map $g_\alpha \colon \pi^{-1}(U_\alpha) \to G$ such that $p = s_\alpha(\pi(p)) \cdot g_\alpha(p)$, for all $p \in \pi^{-1}(U_\alpha)$. Then, for every $p \in \pi^{-1}(U_\alpha)$ and $u \in T_p P$, (1.7.23) and (1.7.24) applied to (1.7.11) imply that

$$\begin{aligned}
\overline{\varphi}(\omega_p(u)) &= \overline{\varphi} \circ \mathrm{Ad}\left(g_\alpha(p)^{-1}\right).(\pi^*\omega_\alpha)_p(u) + \overline{\varphi}\left(g_\alpha^{-1}dg_\alpha\right)_p(u) \\
&= \mathrm{Ad}\left(\varphi(g_\alpha(p))^{-1}\right) \circ \overline{\varphi}.(\pi^*\omega_\alpha)_p(u) \\
&\qquad + \left((\varphi \circ g_\alpha)^{-1}d(\varphi \circ g_\alpha)\right)_p(u) \\
&= \mathrm{Ad}\left(\varphi(g_\alpha(p))^{-1}\right) \circ \overline{\varphi}.\omega_{\alpha,x}(T_p\pi(u)) \\
&\qquad + \left((\varphi \circ g_\alpha)^{-1}d(\varphi \circ g_\alpha)\right)_p(u),
\end{aligned}$$

or, by (1.7.20) and setting $\pi(p) = x$,

$$(1.7.25) \quad \overline{\varphi}(\omega_p(u)) = \mathrm{Ad}\big(\varphi(g_\alpha(p))^{-1}\big)\big[\mathrm{Ad}\big(h_\alpha(x)^{-1}\big).\omega'_{\alpha,x}(T_p\pi(u)) \\ + (h_\alpha^{-1}dh_\alpha)_p(u]\big] + \big((\varphi \circ g_\alpha)^{-1}d(\varphi \circ g_\alpha)\big)_p(u).$$

On the other hand, $(f^*\omega')_p(u) = \omega'_{f(p)}(T_p(u))$. To express ω' by local connection forms [analogously to (1.7.11)], we define $g'_\alpha : \pi'^{-1}(U_\alpha) \to G'$ such that $p' = s'_\alpha(\pi'(p')) \cdot g'_\alpha(p')$, for every $p' \in \pi'^{-1}(U_\alpha)$. Then, for $p' = f(p)$, the latter equality yields

$$f(p) = s'_\alpha(\pi'(f(p))) \cdot g'_\alpha(f(p)) = s'_\alpha(x) \cdot g'_\alpha(f(p)).$$

Since also

$$f(p) = f(s_\alpha(x) \cdot g_\alpha(p)) = f(s_\alpha(x)) \cdot \varphi(g_\alpha(p))) = s'_\alpha(x) \cdot h_\alpha(x) \cdot \varphi(g_\alpha(p)),$$

it follows that $g'_\alpha(f(p)) = h_\alpha(x) \cdot \varphi(g_\alpha(p))$, or

$$(1.7.26) \qquad g'_\alpha \circ f|_{\pi^{-1}(U_\alpha)} = (h_\alpha \circ \pi) \cdot (\varphi \circ g_\alpha)$$

over $\pi^{-1}(U)$. Therefore,

$$(f^*\omega')_p(u) = \omega'_{f(p)}(T_pf(u)) = \\ = \mathrm{Ad}\big(g'_\alpha(f(p))^{-1}\big).(\pi'^*\omega'_\alpha)_{f(p)}(T_pf(u)) + \big(g'^{-1}_\alpha dg'_\alpha\big)_p(T_pf(u)) \\ = \mathrm{Ad}\big(g'_\alpha(f(p))^{-1}\big).(\pi'^*\omega'_\alpha)_{f(p)}(T_pf(u)) + \big((g'_\alpha \circ f)^{-1}d(g'_\alpha \circ f)\big)_x(u).$$

Applying (1.7.26), together with the Maurer-Cartan differential for the product of maps, we transform the last series of equalities into

$$(f^*\omega')_p(u) = \mathrm{Ad}\big(\varphi(g_\alpha(p))^{-1}\big) \circ \mathrm{Ad}\big(h_\alpha(x)^{-1}\big).\omega'_{\alpha,x}(T_p\pi(u)) \\ + \big[\big((h_\alpha \circ \pi) \cdot (\varphi \circ g_\alpha)\big)^{-1}d\big((h_\alpha \circ \pi) \cdot (\varphi \circ g_\alpha)\big)\big]_p(u) \\ = \mathrm{Ad}\big(\varphi(g_\alpha(p))^{-1}\big) \circ \mathrm{Ad}\big(h_\alpha(x)^{-1}\big).\omega'_{\alpha,x}(T_p\pi(u)) \\ + \big((\varphi \circ g_\alpha)^{-1}d((\varphi \circ g_\alpha))\big)_p(u) \\ + \mathrm{Ad}\big(\varphi(g_\alpha(f(p))^{-1}\big) \cdot \big((h_\alpha \circ \pi)^{-1}d(h_\alpha \circ \pi)\big)_p(u) \\ = \mathrm{Ad}\big(\varphi(g_\alpha(p))^{-1}\big)\big[\mathrm{Ad}\big(h_\alpha(x)^{-1}\big).\omega'_{\alpha,x}(T_p\pi(u)) \\ + (h_\alpha^{-1}dh_\alpha)_x(u)\big] + \big((\varphi \circ g_\alpha)^{-1}d((\varphi \circ g_\alpha))\big)_p(u).$$

Comparing the preceding with (1.7.25), we finally obtain the first of (1.7.19), which proves the statement. $\square$

Remark. Proposition 1.7.1 can be easily extended to principal bundles ℓ and ℓ' over diffeomorphic bases B and B', respectively. In this case we may consider morphisms of the form (f, φ, h), where $h \colon B \to B'$ is a diffeomorphism.

The following result will be systematically used in the next subsection.

Proposition 1.7.2   *Let* $(f, \varphi, \mathrm{id}_B)$ *be a pb-morphism of* $\ell = (P, G, B, \pi)$ *into* $\ell' = (P', G', B, \pi')$. *If* ω *is a connection on* ℓ, *then there exists a unique connection* ω' *on* ℓ', $(f, \varphi, \mathrm{id}_B)$-*related with* ω.

Proof   As before, we consider trivializations over the same open cover $\mathcal{C} = \{U_\alpha\}_{\alpha \in I}$ of B. The connection (form) ω is completely known from its local connection forms $\{\omega_\alpha\}_{\in I}$. If $\{h_\alpha\}_{\alpha \in I}$ are the smooth maps of Proposition 1.7.1, we define the local forms

$$(1.7.27) \qquad \omega'_\alpha := \mathrm{Ad}(h_\alpha).(\overline{\varphi}\omega_\alpha) - dh_\alpha h_\alpha^{-1}, \qquad \alpha \in I.$$

By quite lengthy computations, in the spirit of the proof of Proposition 1.7.1, based also on the interplay between left and right Maurer-Cartan differentials, we prove that

$$\omega'_\beta = \mathrm{Ad}(g'_{\alpha\beta}{}^{-1})\omega'_\alpha + g'_{\alpha\beta}{}^{-1}dg'_{\alpha\beta}; \qquad \alpha, \beta \in I.$$

This is precisely the analog of (1.7.10), ensuring the existence of a connection (form) ω' on ℓ', as we have described in the second part of §1.7.3.
    Since (1.7.27) transforms into

$$\overline{\varphi}\omega_\alpha = \mathrm{Ad}(h_\alpha^{-1})\omega'_\alpha + h_\alpha^{-1}dh_\alpha; \qquad \alpha \in I,$$

Proposition 1.7.1 implies that ω and ω' are $(f, \varphi, \mathrm{id}_B)$-related.
    It remains to show that ω' is unique. The easiest way to see this is to use the corresponding splittings C and C' of the connections. More explicitly, since ω and ω' are $(f, \varphi, \mathrm{id}_B)$-related, by the equivalent conditions (1.7.19) we obtain that

$$C' \circ (f \times \mathrm{id}_B) = Tf \circ C.$$

Analogously, if there is a connection $\bar{\omega} \equiv \bar{C}$, also $(f, \varphi, \mathrm{id}_B)$-related with ω,

$$\bar{C} \circ (f \times \mathrm{id}_B) = Tf \circ C,$$

thus $\bar{C} \circ (f \times \mathrm{id}_B) = C' \circ (f \times \mathrm{id}_B)$; that is,

$$(1.7.28) \qquad \bar{C}(f(p), v) = C'(f(p), v); \qquad (p, v) \in P \times_B TB.$$

Let now any $(p', u) \in P' \times_B TB$. If $x := \pi'(p') = \tau_B(u)$, we choose an arbitrary $p \in \pi^{-1}(x)$, thus there is a $g' \in G'$ such that $p' = f(p) \cdot g'$. As a result, in virtue of the action of G' on $(\pi')^*(TB)$ and the G'-equivariance of C' and $\bar{C}$ [see also equalities (1.6.35)], as well as in conjunction with (1.7.28), we obtain the following equalities concluding the proof:

$$\bar{C}(p', u) = \bar{C}(f(p) \cdot g', u) = \bar{C}(f(p), u) \cdot g'$$
$$:= T_{f(p) \cdot g'} R'_{g'} \big(\bar{C}(f(p), u) \big) = T_{f(p) \cdot g'} R'_{g'} \big(C'(f(p), u) \big)$$
$$= C'(f(p) \cdot g', u) = C'(p', u). \qquad \square$$

For further details on related connections and their applications we refer to [Vas13].

1.7.6 Connections on associated bundles

We fix a principal bundle $\ell = (P, G, B, \pi)$ and a connection ω on it. If $\varphi \colon G \to H$ is a morphism of (Banach-) Lie groups, then as in § 1.6.6(a), we obtain the associated principal bundle $\varphi(\ell) = (P \times^G H, H, B, \pi_H)$ and the canonical morphism $(\kappa, \varphi, \mathrm{id}_B)$ defined by (1.6.23). Therefore, by Proposition 1.7.2, $\varphi(\ell)$ admits a uniquely defined connection, say, ω^φ, which is $(\kappa, \varphi, \mathrm{id}_B)$-related with ω; that is $\kappa^* \omega^\varphi = \overline{\varphi}\omega$. The latter condition, in terms of local connection forms, is equivalent with $\overline{\varphi}\omega_\alpha = \omega_\alpha^\varphi$, $\alpha \in I$. This is the case, because the local connection forms ω_α^φ of ω^φ are induced by the natural local sections of $\varphi(\ell)$ given by (1.6.22$'$), thus the maps h_α of Proposition 1.7.1 are now identified with the unit of H.

Specializing to the particular case of a representation of G into a Banach space $\mathbb{E}$, $\varphi \colon G \to \mathrm{GL}(\mathbb{E})$, we obtain the analogous associated bundle $\varphi(\ell) = \big(P_\varphi := P \times^G \mathrm{GL}(\mathbb{E}), \mathrm{GL}(\mathbb{E}), B, \pi_\varphi \big)$ [see § 1.6.6(c)]; hence, ω induces a connection ω^φ as in the previous general case. The corresponding equality of local connection forms is obtained by applying (1.6.30).

The same representation determines the principal bundle (of frames) $\bar{\ell} = (P(E_\varphi), \mathrm{GL}(\mathbb{E}), B, \bar{\pi})$ along with the canonical morphism $(F, \varphi, \mathrm{id}_B)$, where F is given by (1.6.28). We obtain a unique connection, say, $\bar{\omega}$, such that $F^* \bar{\omega} = \overline{\phi}\omega$. Equivalently $\overline{\varphi}\omega_\alpha = \bar{\omega}_\alpha$, because again $h_\alpha \equiv \mathrm{id}_\mathbb{E} \in \mathrm{GL}(\mathbb{E})$, as a consequence of equality (1.6.31).

Finally we see that ω^φ and $\bar{\omega}$ are $(\theta, \mathrm{id}_{\mathrm{GL}(\mathbb{E})}, \mathrm{id}_B)$-related [see (1.6.29) for the definition of θ] since $\omega_\alpha^\varphi = \bar{\omega}_\alpha$.

1.8 The curvature of a principal connection

We review a few facts about the curvature of a connection on a principal fibre bundle needed in our treatment.

1.8.1 Curvature forms

Let ω be a connection on $\ell = (P, G, B, \pi)$. The **curvature form** of ω is the $\mathfrak{g}$-valued 2-form Ω on P, i.e. $\Omega \in \Lambda^2(P, \mathfrak{g})$, defined by

$$(1.8.1) \qquad \Omega = D\omega := d\omega \circ (h \times h),$$

where $h \colon TP \to HP$ is the vb-morphism assigning to each tangent vector of P its horizontal component. Ω satisfies Cartan's *(second) structure equation*

$$(1.8.2) \qquad \Omega = d\omega + \frac{1}{2}[\omega, \omega].$$

Equalities (1.8.1) and (1.8.2) are equivalent.

Clearly, Ω is *horizontal*, i.e. $\Omega(X, Y) = 0$, if one of the vector fields X, Y of P is vertical, and

$$(1.8.3) \qquad R_g^*\Omega = \mathrm{Ad}(g^{-1})\Omega$$

(*G-equivariance* of Ω). The last two properties characterize Ω as a **tensorial form of adjoint type**.

On the other hand, if X and Y are horizontal vector fields of P, then the annihilation of ω on the horizontal subbundle and (1.8.2) imply that

$$\Omega(X, Y) = d\omega(X, Y) + \frac{1}{2}[\omega(X), \omega(Y)] =$$
$$= X(\omega(Y)) - Y(\omega(X)) - \omega([X, Y]) = -\omega([X, Y]),$$

which shows that Ω is the obstruction to the integrability of the horizontal subbundle of TP.

Moreover, Ω satisfies the **Bianchi identity**

$$(1.8.4) \qquad d\Omega = [\Omega, \omega],$$

equivalently written in the form

$$(1.8.4') \qquad D\Omega = 0,$$

with D being defined as in (1.8.1).

Given a trivializing cover $\mathcal{C} = \{(U_\alpha, \Phi_\alpha) \,|\, \alpha \in I\}$ of P and the corresponding natural sections $\{s_\alpha\}$ defined by (1.6.4), the **local curvature forms** of Ω (with respect to $\mathcal{C}$) are

$$(1.8.5) \qquad \Omega_\alpha := s_\alpha^*\Omega \in \Lambda^1(U_\alpha, \mathfrak{g}), \qquad \alpha \in I.$$

An immediate consequence of the structure equation (1.8.2) and the properties of the pull-back are the **local structure equations**

$$(1.8.6) \qquad \Omega_\alpha = d\omega_\alpha + \frac{1}{2}[\omega_\alpha, \omega_\alpha], \qquad \alpha \in I.$$

On the other hand, the compatibility condition

$$(1.8.7) \qquad \Omega_\beta = \mathrm{Ad}(g_{\alpha\beta}^{-1})\Omega_\alpha; \qquad \alpha, \beta \in I,$$

holds over $U_{\alpha\beta}$. Indeed, for every $x \in U_{\alpha\beta}$ and every $u, v \in T_x B$,

$$\Omega_{\beta,x}(u,v) = \Omega_{s_\beta(x)}\big(T_x s_\beta(u), T_x s_\beta(v)\big).$$

Since $s_\beta = s_\alpha \cdot g_{\alpha\beta}$, by the formulas leading to (1.7.15), we obtain

$$T_x s_\beta(u) = T_{s_\alpha(x)} R_{g_{\alpha\beta}(x)}(T_x s_\alpha(u)) + A^*_{s_\beta(x)},$$

and similarly for $T_x s_\beta(v)$. Therefore, the G-equivariance and horizontality of Ω imply that

$$\Omega_{\beta,x}(u,v) =$$
$$= \Omega_{s_\alpha(x) \cdot g_{\alpha\beta}(x)}\big(T_{s_\alpha(x)} R_{g_{\alpha\beta}(x)}(T_x s_\alpha(u)), T_{s_\alpha(x)} R_{g_{\alpha\beta}(x)}(T_x s_\alpha(v))\big)$$
$$= \big(R^*_{g_{\alpha\beta}(x)}\Omega\big)_{s_\alpha(x)}(T_x s_\alpha(u), T_x s_\alpha(v))$$
$$= \mathrm{Ad}\big(g_{\alpha\beta}(x)^{-1}\big).\Omega_{s_\alpha(x)}(T_x s_\alpha(u), T_x s_\alpha(v))$$
$$= \mathrm{Ad}\big(g_{\alpha\beta}(x)^{-1}\big).\Omega_{\alpha,x}(u,v),$$

which proves (1.8.7).

As is expected, Ω is completely known by the local forms $\{\Omega_\alpha\}$. More precisely, let $\omega \equiv \{\omega_\alpha\}_{\alpha \in I}$ be a connection on P, and let $\{\theta_\alpha \in \Lambda^1(U_\alpha), \mathfrak{g}\}_{\alpha \in I}$ be a family of 2-forms satisfying

$$(1.8.8) \qquad \theta_\beta = \mathrm{Ad}(g_{\alpha\beta}^{-1})\theta_\alpha,$$

$$(1.8.9) \qquad \theta_\alpha = d\omega_\alpha + \frac{1}{2}[\omega_\alpha, \omega_\alpha].$$

Then $\{\theta_\alpha\}$ determine a 2-form $\Theta \in \Lambda^1(P, \mathfrak{g}\}$ which coincides with the curvature Ω of ω.

To see this we need the following (general) formulas satisfied by the Maurer-Cartan differentials. Namely, with the notations of §1.2.6,

$$(1.8.10) \qquad d(\mathrm{Ad}(g)\theta) = [D^r g, \mathrm{Ad}(g)\theta] + \mathrm{Ad}(g).d\theta,$$

or, equivalently,

$$(1.8.11) \qquad d\big((g^{-1})\theta\big) = -[D^l g, \mathrm{Ad}(g^{-1})\theta] + \mathrm{Ad}(g^{-1}).d\theta,$$

for every smooth map $g \colon U \to G$ ($U \subseteq B$ open) and every $\theta \in \Lambda^1(U, \mathfrak{g})$.

The proof of both formulas is a bit technical and involves the differentiation of the exterior product of the $\mathrm{Aut}(\mathfrak{g})$-valued 0-form $\mathrm{Ad}(g)$ and the $\mathfrak{g}$-valued 1-form θ. A very detailed proof, regardless the dimension of the manifolds at hand, can be found in the monograph [KJ80], whereas [Nic95] contains a proof valid for finite-dimensional $\mathfrak{g}$-valued forms.

The second formula needed is

(1.8.12) $$d(D^l g) = -\frac{1}{2}[D^l g, D^l g].$$

This is a direct consequence of the expression of the (left) Maurer-Cartan differential as the pull-back of the (left) canonical Maurer-Cartan form on G by g and the corresponding Maurer-Cartan equation.

Going back to the proof of our claim, we define the form $\Theta|_{\pi^{-1}(U_\alpha)}$ by setting $\Theta(p) := \mathrm{Ad}(g_\alpha^{-1}(p)).(\pi^*\theta_\alpha)_p$, where $g_\alpha \colon \pi^{-1}(U_\alpha) \to G$ is the smooth map determined by $p = s_\alpha(x) \cdot g_\alpha(p)$, with $x = \pi(p)$ [see the analogous construction of a connection ω by the local connection forms $\{\omega_\alpha\}$ and compare with (1.7.11)].

Equalities (1.8.8) and $s_\beta = s_\alpha \cdot g_{\alpha\beta}$ imply that Θ is well-defined. The next step is to show that $\Theta = d\omega + \frac{1}{2}[\omega, \omega]$. Indeed, writing, for convenience, (1.7.11) in the form

(1.8.13) $$\omega|_{\pi^{-1}(U_\alpha)} = \mathrm{Ad}(g_\alpha^{-1}).(\pi^*\omega_\alpha) + D^l g_\alpha,$$

and omitting restrictions, we have

(1.8.14)
$$\begin{aligned}
d\omega &= d(\mathrm{Ad}(g_\alpha^{-1}).\pi^*\omega_\alpha + D^l g_\alpha) \\
&= d(\mathrm{Ad}(g_\alpha^{-1}).\pi^*\omega_\alpha) + d(D^l g_\alpha).
\end{aligned}$$

In virtue of (1.8.11) and (1.8.13), the first summand in the right-hand side of (1.8.14) transforms into

$$\begin{aligned}
d(\mathrm{Ad}(g_\alpha^{-1}).\pi^*\omega_\alpha) &= \\
&= -[D^l g_\alpha, \mathrm{Ad}(g_\alpha^{-1}).\pi^*\omega_\alpha] + \mathrm{Ad}(g_\alpha^{-1}).d(\pi^*\omega_\alpha) \\
&= -[D^l g_\alpha, \mathrm{Ad}(g_\alpha^{-1}).\pi^*\omega_\alpha] + \mathrm{Ad}(g_\alpha^{-1}).\pi^*\left(\theta_\alpha - \frac{1}{2}[\omega_\alpha, \omega_\alpha]\right) \\
&= -[D^l g_\alpha, \omega - D^l g_\alpha] + \mathrm{Ad}(g_\alpha^{-1}).\pi^*\theta_\alpha \\
&\qquad\qquad - \frac{1}{2}[\mathrm{Ad}(g_\alpha^{-1}).\pi^*\omega_\alpha, \mathrm{Ad}(g_\alpha^{-1}).\pi^*\omega_\alpha] \\
&= -[D^l g_\alpha, \omega] + [D^l g_\alpha, D^l g_\alpha] + \Theta - \frac{1}{2}[\omega - D^l g_\alpha, \omega - D^l g_\alpha] \\
&= \Theta - \frac{1}{2}[\omega, \omega] + \frac{1}{2}[D^l g_\alpha, D^l g_\alpha].
\end{aligned}$$

Substituting (1.8.12) and the preceding in (1.8.14), we obtain

$$\Theta = d\omega + \frac{1}{2}[\omega, \omega] = \Omega,$$

as claimed.

1.8.2 Flat connections

As usual, a connection ω on $\ell = (P, G, B, \pi)$ is called **flat** if $\Omega = 0$. Some useful equivalent conditions are the following:

i) The horizontal subbundle HP of TP is (completely) integrable.

ii) For every $x \in B$, there is an (open) neighbourhood U of x and a trivialization $\Phi \colon \pi^{-1}(U) \to U \times G$ of P such that $\omega|_{\pi^{-1}(U)} = \Phi^*\omega^\circ|_{U \times G}$, where ω° is the **canonical flat connection** on $B \times G$. We recall that

$$\omega^\circ_{(x,g)}(u, v) := T_g\lambda_{g^{-1}}(v) = (\mathrm{pr}_2^*\,\alpha)_{(x,g)}(u, v),$$

for every $(x, g) \in B \times G$ and $(u, v) \in T_xB \times T_gG$, if α denotes the left Maurer-Cartan form of G.

iii) For every $x \in B$, there is an open neighbourhood U of x and a (smooth) section $s \colon U \to P$ of P such that $s^*\omega = 0$.

iv) P reduces to a bundle with structure group G_d (the group G considered with the discrete structure).

The following properties are also well known:

a) If B is *1-dimensional*, then every connection on P is flat.

b) If B is *simply connected* and P admits a flat connection ω, then P is trivial. Moreover, there is a (global) section $s \colon B \to P$ such that $s^*\omega = 0$; in other words, s is a *horizontal section*.

For relevant proofs, valid for finite-dimensional and Banach bundles, we refer to [Dup78], [KN68], [Kos60] and [Pha69].

1.9 Holonomy groups

Let $\ell = (P, G, B, \pi)$ be a principal bundle endowed with a connection ω. If $\alpha \colon I = [0, 1] \to B$ is a smooth curve, a **horizontal lifting** of α is a smooth curve $\widehat{\alpha} \colon I \to P$ projecting to α, with horizontal tangent (velocity) vectors; that is, according to the notation (1.1.11),

$$\pi \circ \widehat{\alpha} = \alpha,$$

(1.9.1)

$$\dot{\widehat{\alpha}}(t) \in H_{\widehat{\alpha}(t)}P, \quad \text{equivalently,} \quad \omega_{\widehat{\alpha}(t)}\big(\dot{\widehat{\alpha}}(t)\big) = 0.$$

For convenience we also write $\omega(\widehat{\alpha}(t))(\dot{\widehat{\alpha}}(t))$, even simpler $\omega(\dot{\widehat{\alpha}}(t))$, in place of $\omega_{\widehat{\alpha}(t)}(\dot{\widehat{\alpha}}(t))$.

We shall be mainly concerned with the horizontal lifts of piece-wise smooth curves.

Given a piece-wise smooth curve $\alpha\colon I \to B$ and any $p \in P$, there exists a unique (piece-wise smooth) horizontal lift $\widehat{\alpha}_p\colon I \to P$ with $\widehat{\alpha}_p(0) = p$.

A way to prove this is the following (see, e.g., [Kos60], [Pha69]): Denoting also by $\alpha\colon J \to B$ the smooth curve extending the initial α, where J is an open interval containing I, we consider the pull-back of P by the former α

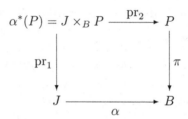

and the pull-back connection $\bar{\omega} := \mathrm{pr}_2^* \omega$ on $\alpha^*(P)$. By the properties a) and b) listed in § 1.8.2, $\bar{\omega}$ is flat and $\alpha^*(P)$ trivial, thus there is a (global) horizontal section $\sigma\colon J \to \alpha^*(P)$. The curve $\widehat{\alpha} := \mathrm{pr}_2 \circ \sigma|_I$ is a horizontal lift of α. If $\widehat{\alpha}(0) = p$, then $\widehat{\alpha}_p = \widehat{\alpha}$. If $\widehat{\alpha}(0) \neq p$, there is a unique $g \in G$ such that $p = \widehat{\alpha}(0) \cdot g$; hence, $\widehat{\alpha}_p = \widehat{\alpha} \cdot g$.

Another way, using equations with total differentials, goes as follows: If α is contained in an open set $U \subseteq B$ over which P is trivial, then there is a section $\sigma\colon U \to P$ such that $\beta = \sigma \circ \alpha$ is a lift of α with $\beta(0) = p$. Since β is not necessarily horizontal, we look for a smooth curve $g\colon I \to G$ such that $g(0) = e$ and $\gamma(t) := \beta(t) \cdot g(t)$ is a horizontal curve of P. Therefore, differentiating the latter equation and applying the horizontality condition $\omega(\dot{\gamma}(t)) = 0$ [see the analogous computations preceding (1.7.15)], we are lead to

$$g(0) = e,$$
(1.9.2)
$$-T_{g(t)}\rho_{g(t)^{-1}}(\dot{g}(t)) = \omega(\dot{\beta}(t)),$$

or, in the notations of § 1.2.6,

$$g(0) = e,$$
(1.9.3)
$$D^r g = -\beta^* \omega.$$

Equation (1.9.3) has a unique solution on I, provided that the form $-\beta^*\omega$ is integrable; that is, $d(\beta^*\omega) + \frac{1}{2}[\beta^*\omega, \beta^*\omega]$ (see the comments following (1.2.7) and the detailed proof of [Bou72, Ch. III, §6, Proposition 15]). This is the case, for if we take the pull-back bundle $\alpha^*(P)$ and the flat connection $\bar{\omega} = \mathrm{pr}_2^*\,\omega$ considered earlier, we check that the curve $\delta(t) := (t, \beta(t)) \in \alpha^*(P)$, with $t \in J$, satisfies the equality $\delta^*\bar{\omega} = \beta^*\omega$. Therefore,

$$0 = \delta^*\bar{\Omega} = \delta^*\left(d\bar{\omega} + \frac{1}{2}[\bar{\omega}, \bar{\omega}]\right) = d(\beta^*\omega) + \frac{1}{2}[\beta^*\omega, \beta^*\omega],$$

as desired.

For the solution of (1.9.2) in the finite-dimensional case we refer also to [KN68]. We notice that (1.9.2) is a particular case of the following result (see [Pen67, Proposition 1.5]), stated here for the sake of completeness (after adapting the original statement to the present setting and the final notation of § 1.1.7):

Let G be a Banach-Lie group with algebra Lie $\mathfrak{g}$. If X is a smooth manifold and $f \colon X \to \mathcal{C}^\infty(I, \mathfrak{g})$ a smooth map, then there exists a unique smooth map $g \colon X \to \mathcal{C}^\infty(I, G)$ such that

$$g(x)(0) = e,$$
$$T_{g(x)(t)}\rho_{(g(x)(t))^{-1}}\big(g(x)^{\cdot}(t)\big) = f(x)(t),$$

for every $(x, t) \in X \times I$.

To conclude our discussion on the existence of horizontal lifts, we add that if the curve $\alpha \colon I \to B$ is not entirely contained in a single open set of B defining a local trivialization of the bundle, then we cover I by a finite family $\{U_i\}$ of such sets, and we solve the corresponding equations, taking as initial condition of the i-th equation the final point of the $(i-1)$-th solution.

Having defined the horizontal lifts of a curve α, we obtain the diffeomorphism

$$\tau_\alpha \colon \pi^{-1}(\alpha(0)) \longrightarrow \pi^{-1}(\alpha(1)) \colon \tau_\alpha(p) := \widehat{\alpha}_p(1),$$

called the **parallel displacement** or **translation** along α. It is shown that $\tau_{\beta*\alpha} = \tau_\beta \circ \tau_\alpha$ and $\tau_\alpha^{-1} = \tau_{\alpha^{-1}}$, for appropriate curves (see the vb-analog in § 1.5.6).

Denoting by C_x the set of all *piecewise smooth* closed curves starting and ending at x (**loop group at** x), and by $C_x^0 \subset C_x$ the group of 0-homotopic loops, the **holonomy group with reference point** $x \in B$

is $\mathbf{\Phi}_x := \{\tau_\alpha \mid \alpha \in C_x\}$, while the **restricted holonomy group with reference point** $x \in B$ is $\mathbf{\Phi}_x^0 := \{\tau_\alpha \mid \alpha \in C_x^0\}$.

To realize these groups as subgroups of G, we choose an arbitrary point $p \in P$ with $\pi(p) = x$. Then the map $k_p \colon \{\tau_\alpha \mid \alpha \in C_x\} \to G$, defined by $\tau_\alpha(p) = p \cdot k_p(\tau_\alpha)$, determines a group homomorphism. Accordingly, the **holonomy group of** ω **with reference point** $p \in P$ is given by

$$\mathbf{\Phi}_p := \{k_p(\tau_\alpha) \mid \alpha \in C_x\}, \qquad x = \pi(p).$$

Equivalently, $\mathbf{\Phi}_p$ consists of all $g \in G$ such that p and $p \cdot g$ can be joined by a (piecewise) smooth curve.

Analogously, the **restricted holonomy group of** ω **with reference point** p is given by

$$\mathbf{\Phi}_p^0 = \left\{ k_p(\tau_\alpha) \mid \alpha \in C_x^0 \right\}.$$

The latter is a normal subgroup of $\mathbf{\Phi}_p$, and there is a natural homomorphism

(1.9.4) $h \colon \pi_1(B) \longrightarrow \mathbf{\Phi}_p / \mathbf{\Phi}_p^0,$

where $\pi_1(B) \equiv \pi_1(B, x)$, if $x = \pi(p)$ (B is assumed to be connected). For relevant proofs and other properties of the holonomy groups in the finite-dimensional case, we refer to [KN68].

Regarding now the holonomy groups in the Banach framework, the following result has been proved by L. Maxim ([Max72]):

Theorem 1.9.1   *i)* $\mathbf{\Phi}_p$ *and* $\mathbf{\Phi}_p^0$ *are Banach-Lie subgroups of* G. *In particular,* $\mathbf{\Phi}_p^0$ *is the identity component of* $\mathbf{\Phi}_p$.

ii) If the base B *is connected and paracompact, then* $\mathbf{\Phi}_p / \mathbf{\Phi}_p^0$ *is countable.*

iii) The structure group G *reduces to* $\mathbf{\Phi}_p$.

iv) $\mathbf{\Phi}_p = \{e\}$ *if and only if* P *is trivial.*

v) $\mathbf{\Phi}_p^0 = \{e\}$ *if and only if the connection is flat.*

A few comments are necessary here: In i), saying that $\mathbf{\Phi}_p$ is a *Banach-Lie subgroup* of G we mean that $\mathbf{\Phi}_p$ is a Banach-Lie group such that the natural injection $i \colon \mathbf{\Phi}_p \hookrightarrow G$ is a smooth morphism and $T_e i \colon \mathcal{L}(\mathbf{\Phi}_p) \to \mathcal{L}(G) \equiv \mathfrak{g}$ is 1–1 (see [Laz65] and [Mai62]). Therefore, $\mathbf{\Phi}_p$ here is neither a regular nor an embedded submanifold of G (compare [Bou71] and [Lan99]). Similar remarks apply to $\mathbf{\Phi}_p^0$. Also, iii) means that (P, G, B, π) reduces to the principal bundle $(P[p], \mathbf{\Phi}_p, B, \pi')$, where $P[p]$ denotes the **holonomy bundle** at p, consisting of all the points of P joined with p by a horizontal curve. $\mathbf{\Phi}_p$ acts on $P[p]$ in a natural way.

Another useful result, concerning the parallel displacements of related connections is the following:

Two connections ω and ω' on (P, G, B, π) and (P', G', B', π'), respectively, are (f, φ, h)-related if and only if

$$(1.9.5) \qquad f \circ \tau_\gamma = \tau'_{h \circ \gamma}\big|_{\pi^{-1}(\gamma(0))}.$$

Therefore, (1.9.5) is equivalent to the conditions of (1.7.5).

1.10 Classification of flat bundles

In this section we are dealing with principal bundles with a fixed *connected* base B and a fixed structure group G.

A **flat bundle** is a pair (P, ω), where (for simplicity) $P \equiv (P, G, B, \pi)$ and ω is a flat connection on P. Two flat bundles (P, ω) and (P', ω') are called **equivalent** if there is a G-B-isomorphism $(f, \mathrm{id}_G, \mathrm{id}_B)$ of P onto P' such that $\omega = f^* \omega'$. The set of equivalence classes thus obtained is denoted by $H(B, G)$.

On the other hand, in virtue of (1.9.4) and property v) of the holonomy groups (Theorem 1.9.1), every flat bundle (P, ω) determines the group homomorphism

$$(1.10.1) \qquad h_\omega : \pi_1(B) \longrightarrow G \colon [\alpha] \mapsto k_p(\tau_\alpha),$$

called the **holonomy homomorphism** of (P, ω), thus $h_\omega(\pi_1(B)) = \Phi_p$, where $\pi_1(B) \equiv \pi_1(B, x_0)$, for a fixed $x_0 \in B$, and $p \in \pi^{-1}(x_0)$ arbitrarily chosen. Recall that $\tau_\alpha(p) = p \cdot k_p(\tau_\alpha)$. The homomorphism h_ω is thought of as a Lie group homomorphism under the usual discrete smooth structure on $\pi_1(B)$.

Two (arbitrary) homomorphisms $h, h' \colon \pi_1(B) \to G$ are called **similar** (or *conjugate*) if they differ by an inner automorphism of g, i.e. there is a $g \in G$ such that $h'([\alpha]) = g \cdot h([\alpha]) \cdot g^{-1}$, for every $[\alpha] \in \pi_1(B)$. Briefly, $h' = I(g) \circ h$ (the inner)automorphism $I(g)$ is also denoted by $\mathrm{ad}(g)$). The terminology comes from the similarity of matrices in the case of $G = \mathrm{GL}(n, \mathbb{R})$. The set of equivalence classes of similar homomorphisms as above is denoted by $\mathcal{S}(B, G)$.

With the previous notations,

$$H(B, G) = \mathcal{S}(B, G), \text{ within a bijection.}$$

For a coherent proof, based on elementary properties of connections and J. Milnor's association of flat bundles with covering spaces (see [Mil58]), we refer to [Vas83]. We outline the main ideas of the proof:

We saw before that a flat bundle determines the corresponding holonomy homomorphism h_ω. Conversely, let $h\colon \pi_1(B) \to G$ be an arbitrary homomorphism. If $(\tilde{B}, \pi_1(B), B, \tilde{p})$ is the principal bundle determined by the universal covering $\tilde{B}$ of B, then h associates to $\tilde{B}$ the principal bundle

$$\tilde{\ell} := \big((\tilde{B} \times G)/\pi_1(B), G, B, \tilde{\pi}\big),$$

with $\tilde{\pi}$ given by $\tilde{\pi}([(\tilde{b}, s)]) = \tilde{p}(\tilde{b})$. The bundle $\tilde{\ell}$ admits a flat connection $\tilde{\omega}$ such that $\omega^\circ = \kappa^* \tilde{\omega}$, where ω° is the canonical flat connection of the trivial bundle $\ell^\circ = (\tilde{B} \times G, \tilde{B}, \mathrm{pr}_1)$, and $\kappa\colon \tilde{B} \times G \to (\tilde{B} \times G)/\pi_1(B)$ is the canonical map [not be confused with that given by (1.6.23)]. Obviously, $(\kappa, \mathrm{id}_G, \tilde{p})$ is a principal bundle morphism between ℓ° and $\tilde{\ell}$, while ω° and $\tilde{\omega}$ are $(\kappa, \mathrm{id}_G, \tilde{p})$-related.

The previous arguments raise now the natural question: If we start with a flat bundle (P, ω) and its holonomy homomorphism $h_\omega\colon \pi_1(B) \to G$, and we consider the flat bundle $\big((\tilde{B} \times G)/\pi_1(B), \tilde{\omega}\big)$ induced by the same h, how are the two bundles related? It turns out that they are equivalent by means of an appropriate G-B-isomorphism, determined as follows: If $\pi_1(B) \equiv \pi_1(B, x_0)$, we fix two arbitrary points $\tilde{x}_0 \in \tilde{p}^{-1}(x_0)$ and $q_0 \in \pi^{-1}(x_0)$. Then we define the map $\varrho_0\colon \tilde{B} \times G \to P$ associating to a pair $(\tilde{x}, g)$ the element $\tau_\alpha(q_0) \cdot g$, where τ_α is the parallel displacement (with respect to ω) along the curve $\alpha = \tilde{p} \circ \tilde{\alpha}$, if $\tilde{\alpha}$ is any piece-wise smooth curve joining $\tilde{x}$ with $\tilde{x}_0$. This is a well-defined smooth map satisfying the equality

$$\varrho_0\big((\tilde{x}, g) \cdot [\gamma]\big) = \varrho_0(\tilde{x}, g); \qquad (\tilde{x}, g) \in \tilde{B} \times G, [\gamma] \in \pi_1(B).$$

The action of $\pi_1(B)$ on the right of $\tilde{B} \times G$ is defined by means of h, i.e. $(\tilde{x}, g) \cdot [\gamma] = (\tilde{x} \cdot [\gamma], h([\gamma])^{-1} \cdot g)$. As a result, the map $\varrho\colon (\tilde{B} \times G)/\pi_1(B) \to P$, with $\varrho([(\tilde{x}, g)]) = \varrho_0(\tilde{x}, g)$ is a G-B-isomorphism such that $\tilde{\omega} = \varrho^* \omega$.

The classification stated above relies on the existence of the isomorphism ϱ, for each flat bundle, along with the fact that similar homomorphisms $h_i\colon \pi_1(B) \to G$ $(i = 1, 2)$ determine equivalent flat bundles $(\tilde{\ell}_i, \tilde{\omega}_i)$ and vice versa.

2

Fréchet spaces

Starting with a brief summary of the topology of Fréchet spaces and various useful examples, we then discuss the differentiability method adopted here, which is due to J.A. Leslie. It is valid for arbitrary topological vector spaces, while it remains closer to more classical methods, without recourse to any particular topologies.

Since projective systems of geometrical structures are our central theme, the main part of the present chapter is devoted to the representation of a Fréchet space as the projective limit of (a countable projective system) of Banach spaces, and questions related with the differentiability of projective systems of maps between such spaces. Of particular interest is the construction of certain spaces of continuous linear maps between Fréchet spaces so as to remain in the same category of spaces, a fact not in general true. A functional space of this kind eventually replaces (in subsequent chapters) the pathological general linear group $GL(\mathbb{F})$ of a Fréchet space $\mathbb{F}$.

Linear differential equations in the same framework is the first application of the projective limit approach expounded from this chapter onwards. It should be noted that there is not a general solvability theory for differential equations, even linear ones, in non-Banach spaces.

2.1 The topology of Fréchet spaces

In a number of cases that have significance in global analysis and physical field theory, Banach space representations break down and we need Fréchet spaces, which have weaker requirements for their topology. To see how this happens we shall look at some examples, but first we need some definitions and our main references for this are R.S. Hamilton [Ham82]

73

and K-H. Neeb [Nee06], cf. also L.A. Steen and J.A. Seebach Jnr. [SS70]. In this section we give a brief description of the topology of Fréchet spaces and point out the main differences from that of Banach spaces. In what follows X denotes a real vector space.

Definition 2.1.1   A *seminorm* on X is a real valued map $p\colon X \to \mathbb{R}$ such that

(i) $p(x) \geq 0,$

(ii) $p(x + y) \leq p(x) + p(y),$

(iii) $p(\lambda x) = |\lambda|\, p(x),$

for every $x, y \in X$ and $\lambda \in \mathbb{R}$.

The notion of a seminorm seems to be a modest generalization of a norm by being just one step away: the seminorm of a non trivial vector may be zero. However, this difference has a serious topological consequence, namely the topology induced on X by p is not necessarily Hausdorff, as we explain below.

Definition 2.1.2   A family of seminorms $\Gamma = \{p_\alpha\}_{\alpha \in I}$ on X defines a unique *topology* $\mathcal{T}_\Gamma$ compatible with the vector space structure of X. The neighbourhood base of $\mathcal{T}_\Gamma$ is determined by the family

$$\mathcal{B}_\Gamma = \{S(\Delta, \varepsilon) : \varepsilon > 0 \text{ and } \Delta \text{ a finite subset of } \Gamma\},$$

where

$$S(\Delta, \varepsilon) = \{x \in X : p(x) < \varepsilon,\ \forall\, p \in \Delta\}.$$

The basic properties of the previous topology are summarized in the next statement. For details see, e.g., to J. Dugundji [Dug75], H.H. Schaeffer [Sch80]), R. Meise-D. Vogt [MV97].

Proposition 2.1.3   *i) $(X, \mathcal{T}_\Gamma)$ is a topological vector space. In particular, $\mathcal{T}_\Gamma$ is the finest topology on X making all the seminorms of Γ continuous.*

ii) The topological vector space $(X, \mathcal{T}_\Gamma)$ is locally convex. Conversely, a topology on X is locally convex only if it is defined by a family of seminorms.

iii) $(X, \mathcal{T}_\Gamma)$ is generally not a Hausdorff space. The Hausdorff property is ensured under the following additional condition:

$$x = 0 \iff p(x) = 0,\ \text{for every } p \in \Gamma.$$

iv) If a topological vector space $(X, \mathcal{T}_\Gamma)$ is Hausdorff, then it is also metrizable if and only if the family of seminorms Γ is countable.

v) The convergence of a sequence $(x_n)_{n \in \mathbb{N}}$ in X is controlled by all the seminorms of Γ; that is,

$$x_n \longrightarrow x \iff p(x_n - x) \longrightarrow 0, \ \forall \, p \in \Gamma.$$

vi) X is complete (with respect to $\mathcal{T}_\Gamma$) if and only if every sequence $(x_n)_{n \in \mathbb{N}}$ in X, such that

$$\lim_{n.m \to \infty} p(x_n - x_m) = 0; \ \forall \, p \in \Gamma,$$

converges in X. Customarily such an $(x_n)_{n \in \mathbb{N}}$ is called a **Cauchy sequence**.

We recall that a topological vector space is *locally convex* if each point has a fundamental system of convex neighbourhoods.

The spaces carrying all the above properties form the category of Fréchet spaces with morphisms the continuous maps. Precisely:

Definition 2.1.4 A **Fréchet space** is a topological vector space $\mathbb{F}$ that is locally convex, Hausdorff, metrizable and complete.

Some typical examples of Fréchet spaces are listed below.

Examples 2.1.5

1. Every Banach space $\mathbb{E}$ is a Fréchet space where the family of seminorms contains only one element, the norm defining the topology of $\mathbb{E}$.

2. The space $\mathbb{R}^\infty = \prod_{n \in \mathbb{N}} \mathbb{R}^n$, endowed with the cartesian topology, is a Fréchet space with corresponding family of seminorms

$$\left\{ p_n(x_1, x_2, ...) = |x_1| + |x_2| + ... + |x_n| \right\}_{n \in \mathbb{N}}.$$

Metrizability can be established by setting

$$(2.1.1) \qquad d(x, y) = \sum_i \frac{|x_i - y_i|}{2^i (1 + |x_i - y_i|)}.$$

In $\mathbb{R}^\infty$ the completeness is inherited from that of each copy of the real line. For if $x = (x_i)$ is a Cauchy sequence in $\mathbb{R}^\infty$, then, for each i, (x_i^m), with $m \in \mathbb{N}$, is a Cauchy sequence in $\mathbb{R}$; hence, it converges, say, to X_i and $(X_i) = X \in \mathbb{R}^\infty$, with $d(x, X_i) \to 0$ as $i \to \infty$. We note that $\mathbb{R}^\infty$ is separable in consequence of the countable dense subset of elements having finitely many rational components and the remainder zero; second countability comes from metrizability.

3. More generally, every *countable* cartesian product of Banach spaces $\mathbb{F} = \prod_{n \in \mathbb{N}} \mathbb{E}^n$ is a Fréchet space with topology defined by the seminorms $(q_n)_{n \in \mathbb{N}}$, given by

$$q_n(x_1, x_2, \ldots) = \sum_{i=1}^{n} \|x_i\|_i \,,$$

where $\| \cdot \|_i$ denotes the norm of the i-factor $\mathbb{E}^i$.

4. The space of continuous functions $C^0(\mathbb{R}, \mathbb{R})$ is a Fréchet space with seminorms $(p_n)_{n \in \mathbb{N}}$ defined by

$$p_n(f) = \sup \big\{ \, | \, f(x) \, |, \ x \in [-n, n] \big\}.$$

5. The space of smooth functions $C^\infty(I, \mathbb{R})$, where I is a compact interval of $\mathbb{R}$, is a Fréchet space with seminorms defined by

$$p_n(f) = \sum_{i=0}^{n} \sup \big\{ \, | \, D^i f(x) \, |, \ x \in I \big\}.$$

In this respect see also [MV97].

6. The space $C^\infty(M, V)$, of smooth sections of a vector bundle V over a compact smooth Riemannian manifold M with covariant derivative ∇, is a Fréchet space with

$$(2.1.2) \qquad \|f\|_n = \sum_{i=0}^{n} \sup{}_x \, | \, \nabla^i f(x) \, |, \quad \text{for} \quad n \in \mathbb{N}.$$

7. Fréchet spaces of sections arise naturally as configurations of a physical field. Then the moduli space, consisting of inequivalent configurations of the physical field, is the quotient of the infinite-dimensional configuration space $\mathcal{X}$ by the appropriate symmetry gauge group. Typically, $\mathcal{X}$ is modelled on a Fréchet space of smooth sections of a vector bundle over a closed manifold. For example, see H. Omori [Omo70], [Omo97]. Lie groups and their algebras are important in geometry and physics for symmetry groups of diffeomorphisms, but for these difficulties quickly arise in the infinite dimensional setting. H.G. Dales [Dal00] gives a comprehensive account of the Banach case and for recent algebraic results in the Fréchet case see, for example, [CEO09, Nee09, Pir09, Tka10].

8. For a compact Riemannian manifold M with $\mathfrak{g} = \mathcal{X}(M)$ the Lie algebra of smooth vector fields on M, $\mathcal{X}(M)$ is a Fréchet space. It is (topologically) regular since disjoint points and closed sets can be separated by disjoint open sets, and has an exponential function but in general it induces no local diffeomorphism of a 0-neighbourhood in $\mathcal{X}(M)$

onto a 1-neighbourhood in $G = \text{Diff}(M)$. R.S. Hamilton [Ham82] and K-H. Neeb [Nee06] provide details of the construction to circumvent this difficulty, including explicit study of the crucial case $\text{Diff}(\mathbb{S}^1)$, where rotations illustrate the difference from the Banach case.

9. As we might anticipate from the prototype Fréchet model $\mathbb{R}^\infty$, which is the projective limit of a countable collection of copies of $\mathbb{R}$, a Fréchet space can always be represented as a projective limit of Banach spaces, which we discuss further in § 2.3.1; for more details cf. [DZ84], [FW96], [Nee06], [Pal68], [Val89], as well as [BDH86], [BMM89] and [Wen03]. S. Agethen et al. [ABB09] consider locally convex spaces which are intersections of a sequence of unions of sequences of Banach spaces of continuous functions with weighted sup-norms, by clarifying when such spaces satisfy the principle of uniform boundedness.

Some important properties of Fréchet spaces are summarized in the following statement (see also [BP75], [Dug75], [Jar81], [MV97], [Sch80]):

Proposition 2.1.6 *i) Every closed subspace of a Fréchet space is also a Fréchet space under the same family of seminorms.*

ii) The open mapping theorem as well as the Hahn-Banach theorem hold true in Fréchet spaces.

iii) The continuity of linear and bilinear maps between Fréchet spaces is checked via seminorms. More precisely, let $\mathbb{F}_1$, $\mathbb{F}_2$ and $\mathbb{F}_3$ be Fréchet spaces with topologies defined by the families of seminorms $\{p_n\}_{n\in\mathbb{N}}$, $\{q_n\}_{n\in\mathbb{N}}$ and $\{r_n\}_{n\in\mathbb{N}}$, respectively. Then a linear map $f\colon \mathbb{F}_1 \to \mathbb{F}_2$ is continuous if and only if, for every seminorm q_n of $\mathbb{F}_2$, there exists a seminorm p_m of $\mathbb{F}_1$ and a positive real M such that

$$q_n(f(x)) \leq M p_m(x),$$

for every $x \in \mathbb{F}_1$. Analogously, a bilinear map $g\colon \mathbb{F}_1 \times \mathbb{F}_2 \to \mathbb{F}_3$ is continuous if and only if, for every seminorm r_n of $\mathbb{F}_3$, there exist seminorms p_m of $\mathbb{F}_1$, q_k of $\mathbb{F}_2$ and a positive real L such that

$$r_n(g(x,y)) \leq L\, p_m(x)\, q_k(y),$$

for every $x \in \mathbb{F}_1$ and $y \in \mathbb{F}_2$.

Although a Fréchet space may be thought of as a short step away from a Banach space because of the Proposition 2.1.6, nevertheless important differences between the two categories of spaces occur:

- The space of continuous linear maps $\mathcal{L}(\mathbb{F}_1, \mathbb{F}_2)$ between two Fréchet spaces $\mathbb{F}_1$ and $\mathbb{F}_2$ is not necessarily a Fréchet space. In particular, if

$\{p_n\}_{n\in\mathbb{N}}$ and $\{q_n\}_{n\in\mathbb{N}}$ are the seminorms of $\mathbb{F}_1$ and $\mathbb{F}_2$, respectively, then $\mathcal{L}(\mathbb{F}_1, \mathbb{F}_2)$ is a Hausdorff locally convex topological vector space whose topology is derived from the family of seminorms $|\cdot|_{n,B}$ given by

$$(2.1.3) \qquad |f|_{n,B} = \sup\{q_n(f(x))\ x \in B\},$$

where $n \in \mathbb{N}$ and B is any bounded subset of $\mathbb{F}_1$ containing the zero element. This topology is complete but not metrizable since the family (2.1.3) is not countable (see also [Hye45]).

• Whereas the dual of a Banach space is a Banach space, the dual of a Fréchet space that is not Banach is never a Fréchet space. The dual of $\mathbb{R}^\infty$ consists of sequences having only finitely many nonzero elements, the dual of $C^\infty(M, \mathbb{R})$ for a compact manifold M is the space of distributions on M, and none of these duals is a Fréchet space, though they are complete locally convex topological vector spaces. This is a major source of difficulty, as we shall see in the sequel.

• The inverse function theorem is not valid in general. However, relevant modifications of it have been proposed to deal with special cases of Fréchet spaces and maps (see, e.g., [Ham82], [Nee06], [Omo70]). For example, a Fréchet space is **graded** if its topology can be defined by a collection of increasing seminorms; this can be ensured by adding to each seminorm all those below it. A **tame** linear map between graded spaces satisfies a uniform linear growth constraint through the grading [Ham82, p. 135]. A graded space is **tame** if it is a tame direct summand of a space of exponentially decreasing sequences in some Banach space. These constructions turn out to cover the cases for all nonlinear partial differential operators and most of their inverses, including those for elliptic, parabolic, hyperbolic and subelliptic operators.

Nash-Moser inverse function theorem for the category of tame Fréchet spaces (cf. [Ham82, pp. 67, 171]) states roughly that: *If the derivatives $DP(f)h = k$ of an operator P in the category have solutions in the category, then the operator P has a local inverse in the category.*

The elaboration of several examples is given in [Ham82] and further structural results are given in [AO09], [BB03], [Dub79], [KLT09], [KM90], [KM97], [KS09], [LT09], [MV85], [PV95], [Vog10], [Wol09]. In addition, the interested reader may consult the following (not exhaustive) list of more specialized aspects of the structure theory of Fréchet spaces: [Vog77], [Vog79], [Vog83], [Vog87], [Vog10] and [VW80], [VW81].

• The **general linear group** $GL(\mathbb{F})$ of a Fréchet space $\mathbb{F}$, i.e. the group of all linear isomorphisms of $\mathbb{F}$, does not admit a non trivial topology compatible with its group structure (see, e.g., [Ham82], [Les67], [Les68], [Nee06], [Omo70]).

- There is no general solvability theory for differential equations in Fréchet spaces analogous to that developed for finite-dimensional or Banach spaces. Since differential equations are very important in the framework of differential geometry, § 2.4 is devoted to a detailed study of them within the category under consideration.

2.2 Differentiability

The topological deviations between Fréchet and Banach spaces, discussed in the previous section, have direct repercussions on the issue of differentiability. In the last 40 years or so, the ordinary differentiation of finite-dimensional or Banach spaces ([Car67(a)], [Lan99]) has been extended to more general topological vector spaces, in a variety of ways, according to the particular problems and applications each author has in mind (see, e.g., [Ham82], [Lem86], [Les67], [Les68], [KM97], [Nee06], [Omo70], [Pap80], [VerE83] and [VerE85]. The latter two sources contain a remarkably extensive bibliography).

In the present work, the differentiation method proposed by J.A. Leslie is adopted since it fits well to the requirements of our differential geometric framework, without using any particular topology. In what follows we outline the basic definitions and properties of this approach. For full details the reader is referred to [Les67] and [Les68].

Let $\mathbb{F}_1$ and $\mathbb{F}_2$ be two *Hausdorff locally convex topological vector spaces*, and let U be an open subset of $\mathbb{F}_1$.

Definition 2.2.1 A continuous map $f \colon U \to \mathbb{F}_2$ is said to be **differentiable at** $x \in U$ if there exists a continuous linear map $Df(x) \colon \mathbb{F}_1 \to \mathbb{F}_2$ such that

$$R(t,v) := \begin{cases} \dfrac{f(x+tv) - f(x) - Df(x)(tv)}{t}, & t \neq 0 \\ 0, & t = 0 \end{cases}$$

is continuous at every $(0,v) \in \mathbb{R} \times \mathbb{F}_1$. The map f is said to be **differentiable** if it is differentiable at every $x \in U$. We call $Df(x)$ the **differential** (or **derivative**) **of** f **at** x. Clearly, this is a special case of the Gâteaux derivative.

As in the classical (Fréchet) differentiation, $Df(x)$ is uniquely determined.

Definition 2.2.2   A map $f: U \to \mathbb{F}_2$, as before, is called C^1-**differentiable** if it is differentiable at every point $x \in U$, and the *(total)* **differential** or *(total)* **derivative**

$$Df: U \times \mathbb{F}_1 \longrightarrow \mathbb{F}_2: (x, v) \mapsto Df(x)(v)$$

is continuous.

It is worth noticing here a non trivial difference between the above definition of differentiability and the classical one for Banach spaces: The total differential Df does not involve the space of continuous linear maps $\mathcal{L}(\mathbb{F}_1, \mathbb{F}_2)$, thus avoiding the possibility of dropping out of the working framework in case where $\mathbb{F}_1$ and $\mathbb{F}_2$ are Fréchet spaces (see the comments following Proposition 2.1.6).

The notion of C^n-differentiability ($n \geq 2$) can be defined by induction:

Definition 2.2.3   A map $f: U \to \mathbb{F}_2$ is C^n-**differentiable** on U if the following conditions hold true:
1) f is C^{n-1}-differentiable.
2) For every $x \in U$, there exists a symmetric and continuous n-linear map $D^n f(x): \mathbb{F}_1^n \to \mathbb{F}_2$ such that

$$R(t, v) :=$$
$$= \begin{cases} \frac{f(x+tv)-f(x)-Df(x)(tv)-\frac{1}{2!}D^2f(x)(tv,tv)-\cdots-\frac{1}{n!}D^nf(x)(tv,\dots,tv)}{t^n}, & t \neq 0 \\ 0, & t = 0 \end{cases}$$

is continuous at every $(0, v) \in \mathbb{R} \times \mathbb{F}_1$.
3) The differential (of order n)

$$D^n f: U \times \mathbb{F}_1^n \longrightarrow \mathbb{F}_2:$$
$$(x; v_1, v_2\dots, v_n) \longmapsto D^n f(x)(v_1, v_2\dots, v_n)$$

is continuous.

The definition of C^∞-differentiability is now obvious. Occasionally, a C^n-map in the sense of Definition 2.2.3 will be called **Leslie** C^n.

Remarks 2.2.4   1) Assuming that $\mathbb{F}_1$ and $\mathbb{F}_2$ are *Banach* spaces, then (see also [Gal96]) the ordinary (Fréchet) C^n-differentiability implies Leslie's C^n-differentiability, but the converse is not always true.

The previous incompatibility disappears in the case of smooth functions, namely:

f is Leslie C^∞ if and only if it is C^∞ in the ordinary sense

Of course, Leslie's differentiation coincides with the ordinary differentiation within the framework of finite dimensional topological vector spaces.

2) Linear and bilinear maps behave as in the classical case:

- Every *continuous linear* map $f \colon \mathbb{F}_1 \to \mathbb{F}_2$ is Leslie C^∞ with

$$Df = f \quad \text{and} \quad D^n f = 0 \quad (n \geq 2).$$

- Analogously, every *continuous bilinear* map $g \colon \mathbb{F}_1 \times \mathbb{F}_2 \to \mathbb{F}_3$ is Leslie C^∞ with

$$\begin{aligned} Dg(x,y)(a,b) &= g(x,b) + g(a,y), \\ D^2 g(x,y)(a,b,c,d) &= g(a,d) + g(c,b), \\ D^n g &= 0 \quad (n \geq 3), \end{aligned}$$

for every $x, a, c \in \mathbb{F}_1$ and $y, b, d \in \mathbb{F}_2$.

3) From Definition 2.2.1, it follows that the differential at x satisfies the standard relation

$$Df(x)(h) = \lim_{t \to 0} \frac{f(x + th) - f(x)}{t}.$$

4) If the domain of a C^r-map f is a complete Hausdorff locally convex topological space, then $D^s f$ is a uniquely determined C^{r-s}-map, for every positive integer $s \leq r$.

5) The chain rule holds for all differentiable maps $f \colon U \subset \mathbb{F}_1 \to \mathbb{F}_2$ and $g \colon V \subset \mathbb{F}_2 \to \mathbb{F}_3$, with $f(U) \subset V$ and $\mathbb{F}_1$, $\mathbb{F}_2$ as in case 2); that is,

$$D(g \circ f)(x) = Dg(f(x)) \circ Df(x),$$

for every $x \in U$.

7) Clearly, the above considerations hold à fortiori for *Fréchet* spaces.

For the sake of simplicity, throughout this work differentiability will be assumed to be of class C^∞, unless something different is explicitly stated.

2.3 Fréchet spaces as projective limits

As already mentioned in the previous sections, Fréchet spaces have fundamental differences from Banach spaces. However, many problems arising in the Fréchet category of spaces can be reduced to their Banach counterparts, because both categories are intimately related. As a matter of fact, every Fréchet space can be represented as a projective limit of Banach spaces. We give here a brief account of this approach, which will be systematically applied throughout this book.

Definition 2.3.1   Let I be a directed set, and let $\{E^i\}_{i \in I}$ be a family of topological vector spaces mutually connected by the continuous linear maps $\rho^{ji} \colon E^j \longrightarrow E^i$ $(j \geq i)$ satisfying the conditions

$$\rho^{ik} \circ \rho^{ji} = \rho^{jk}; \qquad j \geq i \geq,$$

in other words, the following diagram commutes.

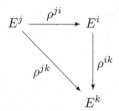

The family $\{E^i, \rho^{ji}\}_{i \in I}$ is called a **projective system** with **factors** the spaces $\{E^i\}_{i \in I}$ and **connecting morphisms** the maps $\{\rho^{ji}\}$. The **(projective) limit** of this system is the subspace of the cartesian product $\prod_{i \in I} E^i$

$$\varprojlim E^i := \left\{ (x^i)_{i \in I} : \rho^{ji}(x^j) = x^i, \ \forall \, j \geq i \right\}.$$

For simplicity, we write (x^i) instead of $(x^i)_{i \in I}$, if there no ambiguity about the index set I.

The construction of a projective limit implies the existence of the **canonical projections**

$$\rho^i \colon \varprojlim E^i \longrightarrow E^i \colon (x^i)_{i \in I} \mapsto x^i.$$

They are related with the connecting morphisms by

$$\rho^{ji} \circ \rho^j = \rho^i; \qquad j \geq i,$$

pictured also in the following commutative diagram:

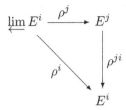

The above definitions extend to all categories of topological spaces provided that the connecting maps are morphisms of the given category; in the category sense, projective limits are called **left limits, inverse limits** or just **limits of diagrams**, see for example [Dod88].

Projective limits inherit the topological and linear structure of the cartesian product with the following additional properties (see, for instance, [Dug75]):

Proposition 2.3.2 *Let $\varprojlim E^i$ be the projective limit of a projective system of topological vector spaces. Then:*

i) The topology of the limit is the weakest (coarsest) topology making the canonical projections ρ^i continuous.

ii) If each factor E^i is a Hausdorff space, then so is the limit.

iii) If all the connecting morphisms ρ^{ji} are injective (resp. bijective), then so are all the canonical projections ρ^i.

Projective limits can be used to describe several different spaces or structures. Some basic but characteristic examples are listed below.

Examples 2.3.3

1. Let E^i ($i \in \mathbb{N}$) be a *countable* family of topological vector spaces that form a descending sequence of subspaces:

$$E^1 \supseteq E^2 \supseteq \cdots \supseteq E^n \supseteq E^{n+1} \supseteq \cdots$$

It determines a projective system with connecting morphisms the natural embeddings

$$\rho^{ji} \colon E^j \hookrightarrow E^i \qquad (j \geq i).$$

Then

$$\varprojlim E^i = \bigcap_{i \in \mathbb{N}} E^i.$$

The topology of the limit now coincides with the relative topology of the intersection as a subspace of each factor E^i.

2. The cartesian product $\prod_{i\in\mathbb{N}} E^i$ of a *countable* family of topological vector spaces coincides, both from the topological and the algebraic point of view, with the limit of the projective system formed by the partial products $\prod_{k=1}^{i} E^k$, with connected morphisms

$$\rho^{ji}: \prod_{k=1}^{j} E^k \longrightarrow \prod_{k=1}^{i} E^k : (x^1, x^2, ..., x^j) \mapsto (x^1, x^2, ..., x^i), \qquad j \geq i.$$

In this case

$$\prod_{i\in\mathbb{N}} E^i \equiv \varprojlim_{i\in\mathbb{N}} \left(\prod_{j=1}^{i} E^i \right) ;$$

therefore,

$$\left(x^i\right) = \left(x^i\right)_{i\in\mathbb{N}} \equiv \left(x^1, \left(x^1, x^2\right), \ldots, \left(x^1, x^2, \ldots, x^i\right), \ldots\right).$$

Projective limits of topological vector spaces naturally form a category whose morphisms are maps compatible with the connecting morphisms and the canonical projections of the limit.

Definition 2.3.4   Let $\{E^i; \rho^{ji}\}_{i,j\in I}$ and $\{F^i; \varphi^{ji}\}_{i,j\in I}$ be two projective systems with limits $E = \varprojlim E^i$ and $F = \varprojlim F^i$, respectively. A family $\{f^i : E^i \to F^i\}_{i\in I}$ forms a **projective system of maps** if

$$\varphi^{ji} \circ f^j = f^i \circ \rho^{ji} \qquad j \geq i.$$

Schematically, this amounts to the commutativity of the following diagram.

In this case, the **projective limit of** (f^i) is defined by

$$f := \varprojlim f^i : E \longrightarrow F : (x^i)_{i\in I} \mapsto \left(f^i(x^i)\right)_{i\in I}.$$

One easily checks that $\varprojlim f^i$ is the unique map related with the canonical projections of the limits $\varprojlim E^i$ and $\varprojlim F^i$ as in the following commutative diagram:

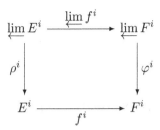

in other words

$$\varphi^i \circ f = f^i \circ \rho^i.$$

The uniqueness of $\varprojlim f^i$ essentially proves the following very frequently used conclusion.

Proposition 2.3.5 *With the assumptions of the preceding definition, consider a map* $g \colon \varprojlim E^i \to \varprojlim F^i$. *Then*

$$g = \varprojlim f^i \quad \text{if and only if} \quad \varphi^i \circ g = f^i \circ \rho^i, \ \forall \, i \in \mathbb{N}.$$

The next result is an immediate consequence of the definitions.

Proposition 2.3.6 *With the same assumptions as before, we have:*
i) If every f^i *is linear, then* f *is linear.*
ii) If every f^i *is continuous, then* f *is continuous.*
iii) If every f^i *is bijective, then* f *is bijective.*

The categories of projective limits and Fréchet spaces are closely related as it will be shown below. A first important result towards this direction is the following.

Proposition 2.3.7 *Let* $\left\{ \mathbb{E}^i, \|\cdot\|_{\mathbb{E}^i} \right\}_{i \in \mathbb{N}}$ *be a countable family of Banach spaces forming a projective system with corresponding connecting morphisms* $\rho^{ji} \colon \mathbb{E}^j \to \mathbb{E}^i$ $(j \geq i)$. *Then* $\varprojlim \mathbb{E}^i$ *is a Fréchet space.*

Proof Setting

$$(2.3.1) \qquad p_n\big((x^i)_{i \in \mathbb{N}}\big) = \sum_{i=1}^{n} \|x^i\|_{\mathbb{E}^i} \, ; \qquad n \in \mathbb{N},$$

we obtain a (countable) family of seminorms on $\varprojlim \mathbb{E}^i$. The topology induced by this family coincides with the projective limit topology. Indeed, each p_n is continuous, with respect to the projective limit topology, as the composite of the canonical projections and the norms of the factors, i.e.

$$(2.3.2) \qquad\qquad p_n = \sum_{i=1}^{n} \left(\|\cdot\|_{\mathbb{E}^i} \circ \rho^i \right).$$

On the other hand, the inequality

$$\left\| \rho^n \big((x^i)_{i \in \mathbb{N}} \big) \right\|_{\mathbb{E}^n} \le p_n \big((x^i)_{i \in \mathbb{N}} \big)$$

ensures that all the canonical projections are continuous with respect to the seminorm topology. Therefore, the two topological structures coincide.

Next we show that the space thus obtained is Hausdorff. Indeed, if

$$p_n \big((x^i)_{i \in \mathbb{N}} \big) = 0$$

for all $n \in \mathbb{N}$, then

$$\sum_{i=1}^{n} \|x^i\|_{\mathbb{E}^i} = 0 \quad \Rightarrow \quad \|x^i\|_{\mathbb{E}^i} = 0, \quad \forall\, i \in \mathbb{N}.$$

Taking into account that each factor $\mathbb{E}^i$ is a Hausdorff space, it follows directly that $\big((x^i)_{i \in \mathbb{N}} \big) = 0$.

Moreover, $\varprojlim \mathbb{E}^i$ is metrizable as derived from a countable family of seminorms (see Proposition 2.1.2) and complete. To check completeness we consider a Cauchy sequence $(x_n)_{n \in \mathbb{N}}$ in $\varprojlim \mathbb{E}^i$, where $x_n = (x_n^i)_{i \in \mathbb{N}}$. Then

$$\sum_{i=1}^{n} \|x_n^i - x_m^i\|_{\mathbb{E}^i} \xrightarrow[n,m]{} 0;$$

therefore,

$$\|x_n^i - x_m^i\|_{\mathbb{E}^i} \xrightarrow[n,m]{} 0, \quad \forall\, i \in \mathbb{N}.$$

In other words, every sequence $\left(x_n^i \right)_{n \in \mathbb{N}}$ is Cauchy in the Banach space $\mathbb{E}^i$, thus it convergence to an element $x^i \in \mathbb{E}^i$. The continuity of the connecting morphisms ρ^{ji}, and the fact that every $\mathbb{E}^i$ is a Hausdorff space imply that $\rho^{ji}(x^j) = x^i$, for every $j \ge i$, or $x = (x^i)_{i \in \mathbb{N}} \in \varprojlim \mathbb{E}^i$. Now the initial sequence (x_n) converges to x since $\rho^i(x_n) = x_n^i$ converges to x^i, for every $i \in \mathbb{N}$. $\qquad\qquad\qquad\square$

What makes the theory of projective limits really efficient in the study of Fréchet spaces is that the converse of Proposition 2.3.7 is also true.

Theorem 2.3.8 *Every Fréchet space can be identified with a projective limit of Banach spaces.*

Proof Let $\mathbb{F}$ be a Fréchet space with topology generated by a family of seminorms $(p_i)_{i \in \mathbb{N}}$. Without loss of generality, we may assume that the sequence of seminorms inducing the topology of $\mathbb{F}$ is increasing, i.e.

$$p_1 \leq p_2 \leq \cdots \leq p_i \leq p_{i+1} \leq \cdots$$

If this is not the case, we can consider the family

$$q_i := p_1 + p_2 + \cdots + p_i \; ; \qquad i \in \mathbb{N},$$

which defines the same topology on $\mathbb{F}$. We denote by $\mathbb{E}^i$ the completion of the quotient space $\mathbb{F}/\ker p_i$ $(i \in \mathbb{N})$, and by ρ^{ji} the connecting morphisms

$$\rho^{ji} \colon \mathbb{E}^j \longrightarrow \mathbb{E}^i \colon [x + \ker p_j]_j \mapsto [x + \ker p_i]_i \; ; \qquad j \geq i,$$

where the bracket $[\;]_i$ stands for the corresponding equivalence class. Then $\mathbb{F}$ coincides with the projective limit of $\{\mathbb{E}^i; \rho^{ji}\}_{i,j \in I}$ by means of the isomorphism

$$\Phi \colon \mathbb{F} \longrightarrow \varprojlim \mathbb{E}^i \colon x \mapsto \big([x + \ker p_i]_i \big)_{i \in \mathbb{N}}. \qquad \square$$

Remarks 2.3.9 1) As a byproduct of the preceding identification, the canonical projections

$$\rho^i \colon \mathbb{F} \equiv \varprojlim \mathbb{E}^i \longrightarrow \mathbb{E}^i \colon x \mapsto [x + \ker p_i]_i \; ; \qquad i \in \mathbb{N},$$

are isometries in the sense that

$$p_i(x) = \big\| [x + \ker p_i]_i \big\|_{\mathbb{E}^i} = \big\| \rho^i(x) \big\|_{\mathbb{E}^i}.$$

2) The construction of the above particular projective limit yielding $\mathbb{F}$ allows us to consider the connecting morphisms and the canonical projections of the system $\{\mathbb{E}^i\}_{i \in \mathbb{N}}$ as surjective morphisms, a fact not true in the case of arbitrary projective systems.

The representation of Fréchet spaces as projective limits is very advantageous: Questions arising within the Fréchet framework can be reduced to their counterparts in the Banach factors. In this way, obstacles set by the very structure of Fréchet spaces and obstructing the solution of many problems, can be surmounted.

A first important application of this representation is dealing with the space of continuous linear maps $\mathcal{L}(\mathbb{F}_1, \mathbb{F}_2)$ between two Fréchet spaces. As mentioned earlier, it drops out of the Fréchet category. However, if the maps under consideration can be realized also as projective limits, then $\mathcal{L}(\mathbb{F}_1, \mathbb{F}_2)$ can be replaced by a new space within the Fréchet framework.

More precisely, we assume that $\mathbb{F}_1 \equiv \varprojlim \mathbb{E}_1^i$ and $\mathbb{F}_2 \equiv \varprojlim \mathbb{E}_2^i$ are Fréchet spaces, where $\{\mathbb{E}_1^i; \rho_1^{ji}\}_{i,j \in \mathbb{N}}$ and $\{\mathbb{E}_2^i; \rho_2^{ji}\}_{i,j \in \mathbb{N}}$ are projective systems. of Banach spaces. Then we prove the following:

Theorem 2.3.10 *The space of all continuous linear maps between $\mathbb{F}_1$ and $\mathbb{F}_2$ that can be represented as projective limits,*

$$(2.3.3) \qquad \mathcal{H}(\mathbb{F}_1, \mathbb{F}_2) := \left\{ (f^i) \in \prod_{i \in \mathbb{N}} \mathcal{L}(\mathbb{E}_1^i, \mathbb{E}_2^i) : \varprojlim f^i \text{ exists} \right\},$$

is a Fréchet space. Moreover, $\mathcal{H}(\mathbb{F}_1, \mathbb{F}_2)$ is also represented as the projective limit of appropriate Banach functional spaces, and the map

$$(2.3.4) \qquad \varepsilon \colon \mathcal{H}(\mathbb{F}_1, \mathbb{F}_2) \longrightarrow \mathcal{L}(\mathbb{F}_1, \mathbb{F}_2) \colon (f^i) \mapsto \varprojlim f^i$$

is continuous linear.

Proof To prove that $\mathcal{H}(\mathbb{F}_1, \mathbb{F}_2)$ is a Fréchet space, it suffices (by Proposition 2.1.6) to show that $\mathcal{H}(\mathbb{F}_1, \mathbb{F}_2)$ is a closed subspace of $\prod_{i=1}^{\infty} \mathcal{L}(\mathbb{E}_1^i, \mathbb{E}_2^i)$, since the latter is a Fréchet space according to Example 2.1.5(3). Indeed, if $(a_n)_{n \in \mathbb{N}}$ is a sequence of elements in $\mathcal{H}(\mathbb{F}_1, \mathbb{F}_2)$, with $a_n = (f_n^1, f_n^2, \dots)$, such that $\lim_n a_n = (f^1, f^2, \dots) \in \prod_{i=1}^{\infty} \mathcal{L}(\mathbb{E}_1^i, \mathbb{E}_2^i)$, then $\lim_n f_n^i = f^i$, for every $i \in \mathbb{N}$. Equivalently, $\lim_n \|f_n^i - f^i\|_i = 0$, where $\|\|_i$ is the norm of the Banach space $\mathcal{L}(\mathbb{F}_1, \mathbb{F}_2)$. Therefore, for every $j \geq i$,

$$\left\| \rho_2^{ji} \circ f^j - f^i \circ \rho_1^{ji} \right\| \leq \left\| \rho_2^{ji} \circ (f^j - f_n^j) \right\|_{\mathcal{L}(\mathbb{F}_1^j, \mathbb{F}_2^i)} +$$
$$+ \left\| (f_n^i - f^i) \circ \rho_1^{ji} \right\|_{\mathcal{L}(\mathbb{F}_1^j, \mathbb{F}_2^i)}$$
$$\leq \left\| \rho_2^{ji} \right\| \cdot \left\| f^j - f_n^j \right\|_j + \left\| f_n^i - f^i \right\|_i \cdot \left\| \rho_1^{ji} \right\|.$$

Since the right-hand side of the inequality tends to 0, it follows that $\rho_2^{ji} \circ f^j = f^i \circ \rho_1^{ji}$ for every $i, j \in \mathbb{N}$ with $j \geq i$. Hence, $\lim_n a_n = (f^i)_{i \in \mathbb{N}} \in \mathcal{H}(\mathbb{F}_1, \mathbb{F}_2)$, which proves the claim.

For each $i \in \mathbb{N}$, we define the set

$$(2.3.5) \qquad \mathcal{H}^i(\mathbb{F}_1, \mathbb{F}_2) := \left\{ \begin{array}{l} (f^j) \in \prod_{j=1}^{i} \mathcal{L}(\mathbb{E}_1^j, \mathbb{E}_2^j) : \\[2mm] \rho_2^{jk} \circ f^j = f^k \circ \rho_1^{jk}, \quad j \geq k. \end{array} \right\}$$

Working as before, we show that each $\mathcal{H}^i(\mathbb{F}_1, \mathbb{F}_2)$ is a Banach space as a closed subspace of $\prod_{j=1}^{i} \mathcal{L}(\mathbb{E}_1^j, \mathbb{E}_2^j)$. On the other hand, the map

(2.3.6)
$$h^{ji}: \mathcal{H}^j(\mathbb{F}_1, \mathbb{F}_2) \longrightarrow \mathcal{H}^i(\mathbb{F}_1, \mathbb{F}_2):$$
$$(f^1, \ldots, f^j) \mapsto (f^1, \ldots, f^i) \qquad (j \geq i)$$

is continuous linear. Linearity is obvious, whereas continuity is a consequence of

$$\begin{aligned}
\left\| h^{ji}(f^1, \ldots, f^j) \right\|_{\mathcal{H}^i(\mathbb{F}_1, \mathbb{F}_2)} &= \left\| (f^1, \ldots, f^i) \right\|_{\mathcal{H}^i(\mathbb{F}_1, \mathbb{F}_2)} \\
&= \left\| f^1 \right\|_1 + \cdots + \left\| f^i \right\|_i \\
&\leq \left\| f^1 \right\|_1 + \cdots + \left\| f^j \right\|_j \\
&= \left\| (f^1, \ldots, f^j) \right\|_{\mathcal{H}^j(\mathbb{F}_1, \mathbb{F}_2)}.
\end{aligned}$$

It is clear that $h^{jk} = h^{ik} \circ h^{ji}$ holds for every $i, j, k \in \mathbb{N}$ with $j \geq i \geq k$, thus $\left\{ \mathcal{H}^j(\mathbb{F}_1, \mathbb{F}_2); h^{ji} \right\}$ is a projective system of Banach spaces, and (see Proposition 2.3.7) $\varprojlim \mathcal{H}^j(\mathbb{F}_1, \mathbb{F}_2)$ is a Fréchet space.

Next, we define the maps

$$h^k: \mathcal{H}(\mathbb{F}_1, \mathbb{F}_2) \longrightarrow \mathcal{H}^i(\mathbb{F}_1, \mathbb{F}_2): \left(f^i \right)_{i \in \mathbb{N}} \mapsto \left(f^1, \ldots, f^k \right), \qquad k \in \mathbb{N}.$$

They are continuous linear and satisfy the equality $h^i = h^{ji} \circ h^j$, for every $j \geq i$. As a result, we obtain the continuous linear map

$$h := \varprojlim h^i: \mathcal{H}(\mathbb{F}_1, \mathbb{F}_2) \longrightarrow \varprojlim \mathcal{H}^i(\mathbb{F}_1, \mathbb{F}_2).$$

More explicitly,

(2.3.7)
$$h\left(\left(f^i \right)_{i \in \mathbb{N}} \right) = h(f^1, f^2, \ldots) =$$
$$= \left((f^1), (f^1, f^2), (f^1, f^2, f^3), \ldots \right) = \left((f^1, \ldots, f^i) \right)_{i \in \mathbb{N}}.$$

We check that h is injective: Indeed,

$$\begin{aligned}
& h(f^1, f^2, \ldots) = h(g^1, g^2, \ldots) \\
\Rightarrow\ & h^k(f^1, f^2, \ldots) = h^k(g^1, g^2, \ldots), \quad k \in \mathbb{N} \\
\Rightarrow\ & (f^1, f^2, \ldots, f^k) = (g^1, g^2, \ldots, g^k), \quad k \in \mathbb{N} \\
\Rightarrow\ & (f^1, f^2, \ldots) = (g^1, g^2, \ldots).
\end{aligned}$$

Also, h is surjective: Let any $a = \left(a^i \right)_{i \in \mathbb{N}} \in \varprojlim \mathcal{H}^i(\mathbb{F}_1, \mathbb{F}_2)$, where

$$a^i = \left(f_i^1, \ldots, f_i^i \right) \in \mathcal{H}^i(\mathbb{F}_1, \mathbb{F}_2) \subseteq \prod_{j=1}^{i} \mathcal{L}(\mathbb{E}_1^j, \mathbb{E}_2^j).$$

By (2.3.5) and (2.3.6),

$$f_i^k \in \mathcal{L}(\mathbb{E}_1^k, \mathbb{E}_2^k); \qquad k \le i,$$
$$\rho_2^{lk} \circ f_i^l = f_i^k \circ \rho_1^{lk}; \qquad k \le l \le i,$$
$$h^{ji}\big(f_j^1, \dots, f_j^j\big) = \big(f_i^1, \dots, f_i^i\big); \qquad i \le j.$$

The last equality implies that $\big(f_j^1, \dots, f_j^i\big) = \big(f_i^1, \dots, f_i^i\big)$, thus $f_j^i = f_i^i$, for every $j \ge i$. Consequently,

$$f_1^1 = f_2^1 = f_3^1 = \cdots =: f^1 \in \mathcal{L}(\mathbb{E}_1^1, \mathbb{E}_2^1),$$
$$f_2^2 = f_3^2 = \cdots =: f^2 \in \mathcal{L}(\mathbb{E}_1^2, \mathbb{E}_2^2),$$
$$\vdots$$
$$f_i^i =: f^i \in \mathcal{L}(\mathbb{E}_1^i, \mathbb{E}_2^i).$$

Moreover,

$$\rho_2^{lk} \circ f^l = \rho_2^{lk} \circ f_l^l = f_l^k \circ \rho_1^{lk} = f^k \circ \rho_1^{lk}; \qquad k \le l,$$

thus $\big(f^1, f^2, \dots\big) \in \mathcal{H}(\mathbb{F}_1, \mathbb{F}_2)$, and

$$h\big(f^1, f^2, \dots\big) = \big(h^k\,\big(f^1, f^2, \dots\big)\big)_{i \in \mathbb{N}} =$$
$$= \big((f^1, f^2, \dots, f^k)\big)_{k \in \mathbb{N}} = \big((f_k^1, f_k^2, \dots, f_k^k)\big)_{k \in \mathbb{N}} = a.$$

The open mapping theorem (see also Proposition 2.1.6) now ensures that h is an isomorphism of Fréchet spaces; hence,

$$(2.3.8) \qquad\qquad \mathcal{H}(\mathbb{F}_1, \mathbb{F}_2) \overset{h}{\cong} \varprojlim \mathcal{H}^i(\mathbb{F}_1, \mathbb{F}_2).$$

Finally, we immediately check that ε is linear. For its continuity we take an arbitrary seminorm $|\cdot|_{(n,B)}$ of $\mathcal{L}(\mathbb{F}_1, \mathbb{F}_2)$ given by (2.1.3); that is, $|f|_{(n,B)} = \sup \{q_n(f(x))\ x \in B\}$, where B is a bounded subset of $\mathbb{F}_1$ containing 0, and $q_n = \sum_{i=1}^n \big(\|\cdot\|_{\mathbb{F}_2^i} \circ \rho_2^i\big)$ is a seminorm of $\mathbb{F}_2 = \varprojlim \mathbb{F}_2^i$ [recall equalities (2.3.1)–(2.3.2)]. Analogously, we consider the seminorms $p_n = \sum_{i=1}^n \big(\|\cdot\|_{\mathbb{F}_1^i} \circ \rho_1^i\big)$ and $r_n = \sum_{i=1}^n \big(\|\cdot\|_{\mathcal{H}^i(\mathbb{F}_1, \mathbb{F}_2)} \circ h^i\big)$ of $\mathbb{F}_1 = \varprojlim \mathbb{F}_1^i$ and $\mathcal{H}(\mathbb{F}_1, \mathbb{F}_2) \equiv \varprojlim \mathcal{H}^i(\mathbb{F}_1, \mathbb{F}_2)$, respectively. Therefore, for every $(f^i) = \big(f^i\big)_{i \in \mathbb{N}}$, we obtain:

$$\left|\varepsilon\left((f^i)\right)\right|_{(n,B)} = \left|\varprojlim (f^i)\right|_{(n,B)}$$

$$= \sup\left\{q_n\left((f^i(x^i))_{i\in\mathbb{N}}\right),\ x = (x^i) \in B\right\}$$

$$= \sup\left\{\sum_{i=1}^{n}\|f^i(x^i)\|_{\mathbb{F}_2^i},\ x = (x^i) \in B\right\}$$

$$\leq \sup\left\{\sum_{i=1}^{n}\|f^i\|_{\mathcal{H}^i(\mathbb{F}_1,\mathbb{F}_2)}\cdot\|x^i\|_{\mathbb{F}_1^i},\ x = (x^i) \in B\right\}$$

$$\leq \sup\left\{\left(\sum_{i=1}^{n}\|f^i\|_{\mathcal{H}^i(\mathbb{F}_1,\mathbb{F}_2)}\right)\cdot p_n(x),\ x \in B\right\}$$

$$= \sup\left\{r_n((f^i))\cdot p_n(x),\ x \in B\right\}$$

$$= \sup\left\{p_n(x),\ x \in B\right\}\cdot r_n((f^i)),$$

which proves the continuity of ε and concludes the proof. $\qquad\square$

As a byproduct of equalities (2.3.7) and (2.3.8), we obtain the following useful identification:

$$(2.3.9) \qquad \left(f^1, f^2, f^3, \ldots\right) \overset{h}{\equiv} \left((f^1), (f^1, f^2), (f^1, f^2, f^3), \ldots\right).$$

or, in a condensed form,

$$(2.3.9') \qquad \varprojlim(f^i) = \left(f^i\right)_{i\in\mathbb{N}} \overset{h}{\equiv} \varprojlim\left(f^1, \ldots, f^i\right).$$

We connect now the projective limits with the differentiation method we adopted in §2.2.

Proposition 2.3.11 *Let $\mathbb{F}_1$ and $\mathbb{F}_2$ be two Fréchet spaces as in the previous statement. Let also $f^i\colon U_i \to \mathbb{E}_2^i$, where $U^i \subseteq \mathbb{E}_1^i$ $(i \in \mathbb{N})$ are open sets. We assume that $U := \varprojlim U^i$, $f := \varprojlim f^i\colon U \to \mathbb{F}_2$ exist, and U is an open subset of $\mathbb{F}_1$. Then*

i) If each f^i is differentiable on U^i, then so is f, and

$$Df(x) = \varprojlim Df^i(x^i), \qquad x = (x^i) \in U.$$

ii) If each f^i is C^k, then so is f.

Proof i) Since the ordinary differentiability in Banach spaces is equivalent to that of Leslie (see Remarks 2.2.4), the assumptions imply that, for every $i \in \mathbb{N}$, there exists a continuous linear map $Df^i(x^i)\colon \mathbb{E}_1^i \to \mathbb{E}_2^i$

such that the functions

$$R^i(t,v) = \begin{cases} \dfrac{f^i(x+tv) - f^i(x) - Df^i(x)(tv)}{t}, & t \neq 0 \\[2mm] 0, & t = 0 \end{cases}$$

are continuous at every $(0,v) \in \mathbb{R} \times \mathbb{E}_1^i$. On the other hand, the fact that f is the projective limit of (f^i), yields

$$\rho_2^{ji} \circ f^j = f^i \circ \rho_1^{ji} \quad \text{and} \quad \rho_2^i \circ f = f^i \circ \rho_1^i,$$

for every $i, j \in \mathbb{N}$ with $j \geq i$, where $\rho_k^{ji} \colon \mathbb{E}_k^j \to \mathbb{E}_k^i$ $(k = 1,2)$ are the connecting morphisms of the two projective systems producing $\mathbb{F}_1$ and $\mathbb{F}_2$, respectively. Taking into account that the connecting morphisms and the canonical projections are continuous linear maps and differentiating the last formulas at any point $x = (x^i) \in U$, we have

$$
\begin{aligned}
(2.3.10) \qquad \rho_2^{ji} \circ Df^j(x^j) &= D\big(\rho_2^{ji} \circ f^j\big)(x^j) = D\big(f^i \circ \rho_1^{ji}\big)(x^j) \\
&= Df^i(\rho_1^{ji}(x^j)) = Df^i(x^i)j \quad (j \geq i).
\end{aligned}
$$

As a result, the projective limit operator $\varprojlim Df^i(x^i)$ can be defined and is continuous linear. Moreover, if

$$R(t,v) = \begin{cases} \dfrac{f(x+tv) - f(x) - Df(x)(tv)}{t}, & t \neq 0 \\[2mm] 0, & t = 0 \end{cases}$$

we check that

$$
\begin{aligned}
\big(\rho_2^{ji} \circ R^j\big)(t,v) &= \rho_2^{ji}\left(\frac{f^j(x^j+tv) - f^j(x^j) - Df^j(x^j)(tv)}{t} \right) = \\
&= \frac{\big(\rho_2^{ji} \circ f^j\big)(x^j+tv) - \big(\rho_2^{ji} \circ f^j\big)(x^j) - \big(\rho_2^{ji} \circ Df^j(x^j)\big)(tv)}{t} \\
&= \frac{f^i\big(x^i + t\rho_1^{ji}(v)\big) - f^j(x^i) - Df^i(x^i)\big(t\rho_1^{ji}(v)\big)}{t} \\
&= \Big(R^i \circ \big(\mathrm{id}_{\mathbb{R}} \times \rho_1^{ji}\big)\Big)(t,v),
\end{aligned}
$$

for all $j \geq i$. Analogously,

$$\big(\rho_2^i \circ R\big)(t,v) = \big(R^i \circ (\mathrm{id}_{\mathbb{R}} \times \rho_1^i)\big)(t,v), \qquad i \in \mathbb{N}.$$

Therefore, $\varprojlim R^i$ can be defined and coincides with R, making the latter a continuous function. This means that f is Leslie differentiable at $x = (x^i)$, and its differential is the projective limit of the differentials of its factors, as in the formula of assertion i).

Assertion ii) is now a direct consequence of equality $Df = \varprojlim Df^i$ and the fact that projective limits respect continuity. Recall that

$$Df^i \colon U^i \times \mathbb{E}_1^i \to \mathbb{E}_2^i, \quad \text{and} \quad Df \colon U \times \mathbb{F}_1 \to \mathbb{F}_2$$

are the (total) differentials of f^i and f, respectively, as in Definitions 2.2.2 and 2.2.3. $\qquad\square$

By similar arguments, the assertions of Proposition 2.3.11 can be extended to the case of C^k-differentiable maps.

If $f \colon U \subseteq \mathbb{F}_1 \to \mathbb{F}_2$ is an arbitrary smooth map between Fréchet spaces ($U \subseteq \mathbb{F}_1$ open), the total differential $Df \colon U \times \mathbb{F}_1 \to \mathbb{F}_2$ does not necessarily imply that the map

$$U \ni x \longmapsto Df(x) \in \mathcal{L}(\mathbb{F}_1, \mathbb{F}_2)$$

is smooth, as is the case of the (Fréchet) derivative in Banach spaces. In contrast, if we are dealing with projective limits of smooth maps, we obtain the following result, which will be applied later on.

Proposition 2.3.12 *With the assumptions of Proposition 2.3.11, the map*

$$U \ni x \longmapsto Df(x) = \varprojlim Df^i(x) \in \mathcal{L}(\mathbb{F}_1, \mathbb{F}_2); \qquad x = \left(x^i\right)_{i \in \mathbb{N}},$$

is smooth.

Proof We define the maps

$$R^i \colon U^i \longrightarrow \mathcal{L}(\mathbb{E}_1^i, \mathbb{E}_2^i) \colon x^i \mapsto Df^i(x^i)$$

(not to be confused with $R^i \colon \mathbb{R} \times \mathbb{E}_1^i \to \mathbb{E}_2^i$ of the previous proof), and

$$F^i \colon U^i \longrightarrow \mathcal{H}^i(\mathbb{F}_1, \mathbb{F}_2) \colon x^i \mapsto \left(R^1(\rho_1^{i1}(x^i)), R^2(\rho_1^{i2}(x^i)), \ldots, R^i(x^i)\right).$$

Each R^i is smooth because of the smoothness of f^i in Banach spaces. Moreover, for every i, j, k with $i \geq j \geq k$,

$$\rho_2^{jk} \circ R^j\left(\rho_1^{ij}(x^i)\right) = \rho_2^{jk} \circ Df^j\left(\rho_1^{ij}(x^i)\right)$$

[by the analog of (2.3.10)]
$$= Df^k\left(\rho_1^{jk}(\rho_1^{ij}(x^i))\right)$$
$$= R^k\left(\rho_1^{ik}(x^i)\right).$$

The maps F^i do take values in $\mathcal{H}^i(\mathbb{F}_1, \mathbb{F}_2)$, for all $i \in \mathbb{N}$. Indeed, for every

$x^j \in U^j$ and $j \geq i$, (2.3.6) implies that

$$\left(h^{ji} \circ F^j\right)(x^j) =$$
$$= h^{ji}\left(R^1\big(\rho_1^{j1}(x^j)\big), R^2\big(\rho_1^{j2}(x^j)\big), \ldots, R^j(x^j)\right)$$
$$= \left(R^1\big(\rho_1^{j1}(x^j)\big), R^2\big(\rho_1^{j2}(x^j)\big), \ldots, R^i\big(\rho_1^{ji}(x^j)\big)\right)$$
$$= \left(R^1\big(\rho^{i1}\big(\rho_1^{ji}(x^j)\big)\big), R^2\big(\rho_1^{i2}\big(\rho_1^{ji}(x^j)\big)\big), \ldots, R^i\big(\rho_1^{ii}\big(\rho_1^{ji}(x^j)\big)\big)\right)$$
$$= \left(F^i \circ \rho_1^{ji}\right)(x^j);$$

hence, the map

$$\varprojlim F^i : U = \varprojlim U^i \longrightarrow \varprojlim \mathcal{H}^i(\mathbb{F}_1, \mathbb{F}_2) = \mathcal{H}(\mathbb{F}_1, \mathbb{F}_2)$$

exists and is smooth, as a consequence of Theorem 2.3.10 and Proposition 2.3.11. Moreover, taking into account (2.3.4) and the identifications (2.3.9),(2.3.9′), we have that

$$\left(\varepsilon \circ \varprojlim F^i\right)(x) = \varepsilon \left(\varprojlim F^i(x^i)\right) =$$
$$= \varepsilon \left(R^1(x^1), \big(R^1(x^1), R^2(x^2)\big), \big(R^1(x^1), R^2(x^2), R^3(x^3)\big), \ldots\right)$$
$$\equiv \varepsilon \left(R^1(x^1), R^2(x^2), R^3(x^3), \ldots\right) = \varprojlim R^i(x^i)$$
$$= \varprojlim Df^i(x^i) = Df,$$

for every $x = \left(x^i\right)_{i \in \mathbb{N}}$. Applying the he smoothness of $\varepsilon \circ \varprojlim F^i$, we conclude the proof.   □

We close this section with a discussion on the *composition* and *evaluation* maps, whose use in the context of Fréchet spaces and their applications to Fréchet manifolds and bundles is often critical. However, their continuity is under question because spaces of continuous linear maps between Fréchet spaces drop out of the category of these spaces.

A way out of this difficulty is achieved by the two results proved below. For the first of them, let $\mathbb{F}$ and $\mathbb{G}$ be two Fréchet spaces whose topology is defined by the families of seminorms $\{p_n\}_{n \in \mathbb{N}}$ and $\{q_n\}_{n \in \mathbb{N}}$, respectively. We recall that the space of continuous linear maps $\mathcal{L}(\mathbb{F}, \mathbb{G})$ is a Hausdorff locally convex topological vector space, with topology defined by the non-countable family of seminorms $\{\,|\cdot|_{n,B}\}$ given by

$$|g|_{n,B} = \sup \{q_n(g(x)), \; x \in B\},$$

where $n \in \mathbb{N}$, and B is any bounded subset of $\mathbb{F}$ containing the zero element. $\mathcal{L}(\mathbb{F}, \mathbb{G})$ fails to be a Fréchet space since it is not metrizable.

On the other hand, the subspace

(2.3.11) $\mathcal{L}_I(\mathbb{F}) = \left\{ f \in \mathcal{L}(\mathbb{F}) : \sup\left\{ \dfrac{p_n(f(x))}{p_n(x)},\ p_n(x) \neq 0 \right\} < \infty \right\}$

of $\mathcal{L}(\mathbb{F}) = \mathcal{L}(\mathbb{F}, \mathbb{F})$ becomes a Fréchet space with corresponding semi-norms

$$|f|_n = \sup\left\{ \frac{p_n(f(x))}{p_n(x)},\ p_n(x) \neq 0 \right\}.$$

Setting, for the sake of simplicity [see also Proposition 2.3.7 and (2.3.3)],

(2.3.12) $\mathcal{H}(\mathbb{F}) := \mathcal{H}(\mathbb{F}, \mathbb{F}),$

we prove the following result, connecting the spaces (2.3.11) and (2.3.12):

Proposition 2.3.13   *Let $f \colon \mathbb{F} \to \mathbb{F}$ be a continuous linear map. Then $f \in \mathcal{L}_I(\mathbb{F})$ if and only if it is a projective limit of continuous linear maps between Banach spaces.*

Proof   Assume first that $f \in \mathcal{L}_I(\mathbb{F})$. If

$$C_i := \sup\left\{ \frac{p^i(f(x))}{p^i(x)}, p^i(x) \neq 0 \right\}; \qquad i \in \mathbb{N},$$

then, following the analysis of Theorem 2.3.8 leading to the realization of the Fréchet space $\mathbb{F}$ as a projective limit of Banach spaces, we define the maps

$$f^i \colon \mathbb{E}^i \longrightarrow \mathbb{E}^i \colon \left[x + \ker p^i \right] \mapsto \left[f(x) + \ker p^i \right], \qquad i \in \mathbb{N}.$$

Each, f^i is a continuous linear map, since $\left\| f^i(u) \right\|_{\mathbb{E}^i} \leq C_i \cdot \left\| u \right\|_{\mathbb{E}^i}$, for every $u \in \mathbb{E}^i$. Moreover, $\left(f^i \right)_{i \in \mathbb{N}}$ is a projective system, since

$$\left(\rho^{ji} \circ f^j \right)\left(\left[x + \ker p^j \right] \right) = \rho^{ji}\left(\left[f(x) + \ker p^j \right] \right) = \left[f(x) + \ker p^i \right]$$
$$= f^i\left(\left[x + \ker p^i \right] \right) = \left(f^i \circ \rho^{ji} \right)\left(\left[x + \ker p^j \right] \right),$$

whereas, each f^i ($i \in \mathbb{N}$) is the i-th projection of f (by the corresponding canonical map of the projective system), i.e.

$$\left(f^i \circ \rho^i \right)(x) = f^i\left(\left[x + \ker p^i \right] \right) = \left[f(x) + \ker p^i \right] = \rho^i(f(x)).$$

Therefore, $f = \varprojlim f^i$.

Conversely, assume that $f = \varprojlim f^i$, with $f^i \in \mathcal{L}(\mathbb{E}^i)$. Then, for every

$x = ([x + \ker p^i])_{i \in \mathbb{N}} \in \mathbb{F}$, we have that

$$p_i(f(x)) = \sum_{j=1}^{i} \left\| f^j \left([x + \ker p^i]\right) \right\|_{\mathbb{E}^j}$$

$$\leq \sum_{j=1}^{i} \left\| f^j \right\|_{\mathcal{L}(\mathbb{E}^j)} \cdot p^j(x) \leq \left(\sum_{j=1}^{i} \left\| f^j \right\| \right) \cdot p_i(x).$$

As a result, $f \in \mathcal{L}_I(\mathbb{F})$. $\square$

Based on (2.3.11), we prove the continuity of the following form of composition, needed in § 3.6:

Proposition 2.3.14   *The composition map*

$$\mathrm{comp} \colon \mathcal{L}(\mathbb{F}, \mathbb{G}) \times \mathcal{L}_I(\mathbb{F}) \longrightarrow \mathcal{L}(\mathbb{F}, \mathbb{G}) \colon (g, f) \mapsto f \circ g$$

is continuous.

Proof   Since $\mathcal{L}_I(\mathbb{F})$ is a Fréchet space, and therefore metrizable, it suffices to show that, for any sequence $(f_n)_{n \in \mathbb{N}}$ in $\mathcal{L}_I(\mathbb{F})$ and every net $(g_i)_{i \in I}$ in $\mathcal{L}(\mathbb{F}, \mathbb{G})$ converging to f and g, respectively, it follows that $\mathrm{comp}(g_i, f_n) = g_i \circ f_n$ converges to $g \circ f$. Before this, recall that

$$B \subseteq \mathbb{F} \quad \text{is bounded}$$

$$\Leftrightarrow \quad B \text{ is bounded with respect to all seminorms}$$

$$\Leftrightarrow \forall\, k \in \mathbb{N} \; \exists\, M_k > 0 \, : \, |x|_k \leq M_k, \; \forall\, x \in B.$$

Moreover, for $(f_n)_{n \in \mathbb{N}}$ as above, the set

$$D = \{ f_n(x) \,|\, n \in \mathbb{N}, \; x \in B \}$$

is also bounded, because, for every k-seminorm (for convenience the seminorms of all spaces involved are denoted by $|\cdot|_a$ with appropriate indices a), we have that

$$\left| f_n(x) \right|_k \leq \left| f_n \right|_k \cdot |x|_k \leq |f|_k \cdot M_k,$$

since $f_n \to f$ and $\left| f_n \right|_k \leq \left| f_n - f \right|_k + |f|_k \leq |f|_k$.

Now, as a first step to the proof of our claim, we consider the case $f = g = 0$. Then $g_i \circ f_n \to 0$. Indeed,

$$\left| g_i \circ f_n \right|_{n, B} = \sup \left\{ \left| g_i(f_n(x)) \right|_k, \; x \in B \right\}$$

$$\leq \sup \left\{ \left| g_i(y) \right|_k, \; y \in D \right\}$$

$$= \left| g_i \right|_{k, D} \longrightarrow 0.$$

In the general case, where $f_n \to f$ and $g_i \to g$, with f, g not necessarily zero, we see that

$$((g_i - g) \circ (f_n - f))(x) = g_i(f_n(x)) - g_i(f(x)) - g(f_n(x)) + g(f(x)) =$$
$$= (g_i(f_n(x)) - g(f(x))) + (g(f(x)) - g_i(f(x))) + (g(f(x)) - g(f_n(x))),$$

or

$$g_i(f_n(x)) - g(f(x)) = ((g_i - g) \circ (f_n - f))(x) + (g_i(f(x)) - g(f(x))$$
$$+ (g(f_n(x)) - g(f_n(x))),$$

thus

$$(2.3.13) \quad \begin{aligned} |(g_i \circ f_n) - (g \circ f)|_{k,B} &= \sup\left\{\left|g_i(f_n(x)) - g(f(x))\right|_k, \; x \in B\right\} \\ &\leq \sup\left\{\left|(g_i - g) \circ (f_n - f)\right|_k, \; x \in B\right\} \\ &+ \sup\left\{\left|(g_i - g)(f(x))\right|_k, \; x \in B\right\} \\ &+ \sup\left\{\left|g(f_n(x)) - f(x))\right|_k \; x \in B\right\}. \end{aligned}$$

Since $(g_i - g) \to 0$ and $(f_n - f) \to 0$, the first case implies that

$$(2.3.14) \quad \begin{aligned} \sup\left\{\left|(g_i - g) \circ (f_n - f)\right|_k, \; x \in B\right\} &= \\ \left|(g_i - g) \circ (f_n - f)\right|_{k,B} &\longrightarrow 0. \end{aligned}$$

Also, by the convergence of g,

$$(2.3.15) \quad \sup\left\{\left|(g_i - g)(f(x))\right|_k, \; x \in B\right\} = \left|g_i - g\right|_{k, f(B)} \longrightarrow 0,$$

whereas, by the convergence of f and the continuity of g,

$$f_n \longrightarrow f \;\Rightarrow\; f_n(x) \longrightarrow f(x) \;\Rightarrow\; g(f_n(x) - f(x)) \longrightarrow 0,$$

for all $x \in B$; hence,

$$(2.3.16) \quad \sup\left\{\left|g(f_n(x) - f(x))\right|_k \; x \in B\right\} \longrightarrow 0.$$

As a result, (2.3.13)–(2.3.16) imply that $\left|(g_i - f_n) \circ (g - f)\right|_{k,B} \to 0$, for any seminorm $|\cdot|_{k,B}$ of $\mathcal{L}(\mathbb{F}, \mathbb{G})$; therefore, $(g_i - f_n) \to (g - f)$, as desired. $\qquad\square$

Remarks 2.3.15 1) With similar arguments, we prove the continuity of the composition map

$$\mathrm{comp}\colon \mathcal{L}_I(\mathbb{F}) \times \mathcal{L}(\mathbb{F}, \mathbb{G}) \longrightarrow \mathcal{L}(\mathbb{F}, \mathbb{G})\colon (f, g) \mapsto f \circ g$$

For the sake of completeness, we sketch the proof: Taking seminorms, a sequence (f_n) and a net (g_i) as in the main proof, we consider again two cases:

i) $f = g = 0$. Then, for an arbitrary seminorm $|\cdot|_{k,B}$ of $\mathcal{L}(\mathbb{F}, \mathbb{G})$ (again for $B \subseteq \mathbb{F}$ bounded containing 0),

$$\left|(f_n \circ g_i)(x)\right|_k = \left|f_n(g_i(x))\right|_k \leq \left|f_n\right|_k \cdot \left|g_i(x)\right|_k, \qquad x \in B.$$

Consequently,

$$\sup \left\{\left|(f_n \circ g_i)(x)\right|_k, \ x \in B\right\} \leq \left|f_n\right|_k \cdot \sup \left\{\left|g_i(x)\right|_k, \ x \in B\right\}$$

or, equivalently, $\left|f_n \circ g_i\right|_{k,B} \leq \left|f\right|_n \cdot \left|g_i\right|_{k,B}$. Since f_n and g_i converge to zero, the same is true for their composition.

ii) $f_n \to f$ and $g_n \to g$, with f, g not necessarily zero. Then

$$\left|(f_n \circ g_i) - (f \circ g)\right|_{k,B} = \sup \left\{\left|f_n(g_i(x)) - f(g(x))\right|_k, \ x \in B\right\}$$
$$\leq \sup \left\{\left|(f_n - f)(g_i(x) - g(x))\right|_k, \ x \in B\right\}$$
$$+ \sup \left\{\left|(f_n - f)(g(x))\right|_k, \ x \in B\right\}$$
$$+ \sup \left\{\left|f\big(g_i(x) - g(x)\big)\right|_k, \ x \in B\right\}$$

Since $(f_n - f) \to 0$ and $(g_i - g) \to 0$, it follows that

$$\sup \left\{\left|(f_n - f)(g_i(x) - g(x))\right|_k \ x \in B\right\} =$$
$$= \left|(f_n - f) \circ (g_i - g)\right|_{n,B} \longrightarrow 0.$$

On the other hand, since g is continuous linear, $g(B)$ is also a bounded set containing the zero element, thus providing a seminorm $|\cdot|_{k,g(B)}$ of $\mathcal{L}(\mathbb{F}, \mathbb{G})$, for which

$$\sup \left\{\left|(f_n - f)(g(x))\right|_k \ x \in B\right\} = \left|f_n - f\right|_{k,g(B)} \longrightarrow 0$$

because $(f_n - f) \to 0$ in $\mathcal{L}_I(\mathbb{F}) \subseteq \mathcal{L}(\mathbb{F})$. Finally,

$$\sup \left\{\left|f(g_i(x) - g(x))\right|_k, \ x \in B\right\} \leq$$
$$\leq \left|f\right|_k \cdot \sup \left\{\left|g_i(x) - g(x)\right|_k, \ x \in B\right\}$$
$$= \left|f\right|_k \cdot \left|g_i - g\right|_{k,B} \longrightarrow 0.$$

As a result, $\left|(f_n \circ g_i) - (f \circ g)\right|_{k,B} \to 0$, for any seminorm $|\cdot|_{k,B}$ of $\mathcal{L}(\mathbb{F}, \mathbb{G})$; therefore, $f_n \circ g_i \to f \circ g$.

2) We note that, in virtue of (2.3.4), $\mathcal{L}_I(\mathbb{F}) = \varepsilon(\mathcal{H}(\mathbb{F}))$.

To prove the continuity of the evaluation map, we consider the Fréchet spaces $\mathbb{E}$ and $\mathbb{F}$, with topologies derived from the families of seminorms $\{p_n\}_{n \in \mathbb{N}}$ and $\{q_n\}_{n \in \mathbb{N}}$, respectively. As usual, $\mathcal{L}(E, \mathbb{F})$ is a locally convex space with topology based on the seminorms given by

$$\left|f\right|_{n,B} = \sup \left\{q_n(f(x)), \ x \in B\right\},$$

where $n \in \mathbb{N}$, and $B \subseteq \mathbb{E}$ is any bounded set containing 0.

Proposition 2.3.16 *The evaluation map*

$$\mathrm{ev}\colon \mathcal{L}(\mathbb{E}, \mathbb{F}) \times \mathbb{E} \longrightarrow \mathbb{F}\colon (f, a) \mapsto f(a)$$

is continuous.

Proof Let (f_i) be an arbitrary net in $\mathcal{L}(\mathbb{E}, \mathbb{F})$ converging to a continuous linear map f, and let (a_n) be sequence in $\mathbb{E}$ converging to $a \in \mathbb{E}$. Then, for any seminorm q_l of $\mathbb{F}$,

$$q_l\big(f_i(a_n) - f(a)\big) \le q_l\big(f_i(a_n) - f(a_n)\big) + q_l\big(f(a_n) - f(a)\big).$$

Since $f \in \mathcal{L}(\mathbb{E}, \mathbb{F})$, there exists a seminorm p_k of $\mathbb{E}$ and a positive constant M such that $q_l(f(u)) \le M \cdot p_k(u)$, for every $u \in \mathbb{E}$. Then,

$$q_l\big(f(a_n) - f(a)\big) = q_l\big(f(a_n - a)\big) \le M \cdot p_k(a_n - a) \xrightarrow[n]{} 0.$$

On the other hand, if we consider the bounded set

$$B = \{a_n - a \mid n \in \mathbb{N}\} \cup \{0, a\}$$

and the corresponding seminorm $|\cdot|_{l,B}$ of $\mathcal{L}(\mathbb{E}, \mathbb{F})$, we have that

$$\big|f_i - f\big|_{l,B} = \sup\big\{q_l\big((f_i - f)(x)\big),\ x \in B\big\} \xrightarrow[i]{} 0,$$

as well as

$$
\begin{aligned}
q_l\big(f_i(a_n) - f(a_n)\big) &\le q_l\big(f_i(a_n) - f_i(a) - f(a_n) + f(a)\big) \\
&\qquad + q_l(f_i(a) - f(a)) \\
&= q_l\big((f_i - f)(a_n - a)\big) + q_l\big((f_i - f)(a)\big) \\
&\le \big|f_i - f\big|_{l,B} + \big|f_i - f\big|_{l,B}.
\end{aligned}
$$

This completes the proof. $\qquad\qquad\square$

The following result is an obvious consequence of Propositions 2.3.14, 2.3.16 and Remark 2.2.4(2).

Corollary 2.3.17 *The preceding maps* comp *and* ev *are smooth.*

2.4 Differential equations in Fréchet spaces

One of the main problems in the study of infinite dimensional non-Banach locally convex topological vector spaces is the lack of a general solvability theory for differential equations (even the linear ones), analogous to the theory established in the Banach framework. This serious drawback directly impinges on many problems and applications of differential geometry. For example, Fréchet modelled differential manifolds and fibre bundles with applications (especially in theoretical physics) constantly increasing, appear to have very poor geometrical structures because of the aforementioned weakness.

The question of solvability has been approached by a number of authors in recent years by suggesting various ways of studying *concrete* classes of differential equations in infinite-dimensional spaces, particularly in Fréchet spaces (see, for instance, [Ham82], [Lem86], [Lob92] and [Pap80]). In this section, a new approach, based on the representation of Fréchet spaces by projective limits of Banach spaces, is proposed.

Our method exploits the compatibility of algebraic, topological and differential structures of Fréchet spaces with projective limits. We note that this approach not only addresses the solvability question of the equations at hand, but also provides a detailed description of the solutions, a fact not always achieved in other approaches. Clearly, this is essentially advantageous in many applications.

In what follows, $\mathbb{F}$ will denote a Fréchet space whose topology is generated by the countable family of seminorms $(p_i)_{i \in \mathbb{N}}$ and is realized as the limit of a projective system of Banach spaces $\{\mathbb{E}^i; \rho^{ji}\}_{i,j \in \mathbb{N}}$. In this framework, the notion of *Lipschitz (continuity) condition* is generalized as follows.

Definition 2.4.1 A map $\phi \colon \mathbb{F} \to \mathbb{F}$ will be called k-**Lipschitz**, where k is a positive real number, if

$$p_i(\phi(x_2) - \phi(x_1)) \leq k \cdot p_i(x_2 - x_1)$$

for every $x_1, x_2 \in \mathbb{F}$ and $i \in \mathbb{N}$.

In the case where ϕ is the projective limit of a projective system of maps $\{\phi^i \colon \mathbb{E}^i \to \mathbb{E}^i\}_{i \in \mathbb{N}}$, we check whether ϕ satisfies Definition 2.4.1 or not by the behaviour of its components. More precisely:

Proposition 2.4.2 *A map $\phi = \varprojlim \phi^i$ is k-Lipschitz if and only if each ϕ^i is k-Lipschitz in $\mathbb{E}^i$.*

Proof Let $x_i = \rho^i(x)$, $y_i = \rho^i(y)$ be two arbitrarily chosen points in $\mathbb{E}^i$. (Here, deviating from our usual notation, we set $(x_i)_{i \in \mathbb{N}}$ for typographical reasons which will be clear soon.) The relations between ϕ and

(ϕ^i) lead to

$$\left\| \phi^i\left(x_i\right) - \phi^i\left(y_i\right) \right\|_{\mathbb{E}^i} = p_i\big(\phi(x) - \phi(y)\big),$$

(recall that $\|\cdot\|_{\mathbb{E}^i}$ stands for the norm of $\mathbb{E}^i$). Therefore, a k-Lipschitz condition for ϕ in the generalized sense of Definition 2.4.1, implies the classical analog for each ϕ^i in $\mathbb{E}^i$, i.e.

$$\left\| \phi^i\left(x_i\right) - \phi^i\left(y_i\right) \right\|_{\mathbb{E}^i} \leq k \cdot p_i(x - y) = k \cdot \|x_i - y_i\|_{\mathbb{E}^i}.$$

Conversely, the k-Lipschitz conditions for all ϕ^i's (with the same constant k, of course) yield

$$p_i(\phi(x) - \phi(y)) \leq k \cdot \|x_i - y_i\|_{\mathbb{E}^i} = k \cdot p_i(x - y),$$

by which we conclude the proof. $\qquad\Box$

The previous generalization of the Lipschitz continuity allows one to approach a wide class of differential equations in $\mathbb{F}$ by using the techniques of projective limits, as the next result illustrates.

Theorem 2.4.3 *Let* $\phi = \varprojlim \phi^i : \mathbb{R} \times \mathbb{F} \to \mathbb{F}$ *be a projective limit k-Lipschitz map. If, for an initial point* $(t_0, x_0) \in \mathbb{R} \times \mathbb{F}$, *there exists a constant* $\tau \in \mathbb{R}$ *such that*

$$(*)\qquad M := \sup\big\{ p_i(\phi(t, x_0)); \ i \in \mathbb{N}, \ t \in [t_0 - \tau, t_0 + \tau] \big\} < +\infty,$$

then the differential equation

$$(2.4.1)\qquad\qquad x' = \phi(t, x)$$

admits a unique solution defined on the interval $I = [t_0 - a, t_0 + a]$, *satisfying the initial condition* $x(t_0) = x_0$. *Here,* $a = \inf\left\{ \tau, \frac{1}{M_1 + k} \right\}$.

Proof We define the system of ordinary differential equations in the Banach spaces $\mathbb{E}^i$

$$(2.4.2)\qquad\qquad x_i' = \phi^i(t, x_i).$$

According to Proposition 2.4.2, each ϕ^i is k-Lipschitz and, for every $i \in \mathbb{N}$ and $x \in \mathbb{E}^i$ with $\|x_i - \rho^i(x_0)\|_i \leq 1$, we see that

$$\begin{aligned}
\left\| \phi^i(t, x) \right\|_{\mathbb{E}^i} &\leq \left\| \phi^i(t, \rho^i(x_0)) \right\|_{\mathbb{E}^i} + \left\| \phi^i(t, x) - \phi^i(t, \rho^i(x_0)) \right\|_{\mathbb{E}^i} \\
&\leq \left\| \rho^i(\phi(t, x_0)) \right\|_{\mathbb{E}^i} + k \cdot \left\| x - \rho^i(x_0) \right\|_{\mathbb{E}^i} \\
&= p_i(\phi(t, x_0)) + k \leq M_1 + k.
\end{aligned}$$

Therefore (see [Car67(a), Corollary 1.7.2]), each Equation 2.4.1 admits a unique solution x_i on the interval $I = [t_0 - a, t_0 + a]$.

On the other hand, the previous solutions form a projective system. This is deduced from

$$\left(\rho^{ji} \circ x_j\right)'(t) = \rho^{ji}(x_j'(t)) = \rho^{ji}\left(\phi^j(t, x_j(t))\right) = \phi^i(t, (\rho^{ji} \circ x_j)(t)),$$

which means that $\rho^{ji} \circ x_j$ is a solution of (2.4.2) with the same initial condition $(t_0, \rho^i(x_0))$. As a result, the C^1-map $x = \varprojlim x_i$ can be defined on I and determines the desired solution of (2.4.1), since

$$x'(t) = \left(x_i'(t)\right) = \left(\phi^i(t, x_i(t))\right) =$$
$$= \left(\phi^i(t, \rho^i(x(t)))\right) = \left(\rho^i(\phi(t, x(t)))\right) = \phi(t, x(t)).$$

This is the unique solution satisfying the initial condition (t_0, x_0). Indeed, if $y \colon J \subseteq \mathbb{R} \to \mathbb{F}$ is another solution of (2.4.1) with $y(t_0) = x_0$, then y can be realized as the limit of the projective system $\{\rho^i \circ y^i : J \to \mathbb{E}^i\}_{i \in \mathbb{N}}$. Employing analogous computations as before, we check that $\rho^i \circ y$ coincides with x^i (for each $i \in \mathbb{N}$) as a solution of (2.4.2) with the same initial condition $(t_0, \rho^i(x_0))$. Thus

$$y = \varprojlim(\rho^i \circ y) = \varprojlim x^i = x,$$

as claimed.   □

Since a Lipschitz condition is fulfilled in the case of linear differential equations, taking into account (2.3.12), we prove the following:

Theorem 2.4.4 *The n-order linear differential equation*

$$x^{(n)} = A_0 \cdot x + A_1 \cdot x' + \cdots + A_{n-1} \cdot x^{(n-1)} + B,$$

where $A_i \colon [0,1] \to \mathcal{L}(\mathbb{F})$ and $B \colon [0,1] \to \mathbb{F}$ are continuous maps, admits a unique solution for a given initial condition, provided that each factor A_i decomposes to

$$A_i = \varepsilon \circ A_i^*,$$

where $A_i^* \colon [0,1] \to \mathcal{H}(\mathbb{F})$ is continuous and, as in (2.3.4),

$$\varepsilon \colon \mathcal{H}(\mathbb{F}) \longrightarrow \mathcal{L}(\mathbb{F}) \colon \left(f^i\right)_{i \in \mathbb{N}} \mapsto \varprojlim f^i.$$

Proof The assumptions imply that each factor A_i can be thought of as a projective limit of continuous maps; namely,

$$A_i(t) = \varepsilon(A_i^*(t)) = \varprojlim_{j} \left(A_i^j(t)\right),$$

where $A_i^j \colon [0,1] \to \mathcal{L}(\mathbb{E}^i)$ $(j \in \mathbb{N})$ are the components of A_i^*. In this way, the n-order differential equation in study is equivalent to

$$X' = \Phi(t, X),$$

where $X = \left(x, x', \dots, x^{(n-1)}\right)$ and $\Phi(t, X) = A(t) \cdot X + B$, with

$$A \colon [0,1] \longrightarrow \mathcal{H}(\mathbb{F}^n) \colon t \longmapsto \begin{pmatrix} 0 & 1_{\mathbb{F}} & 0 & \dots & 0 \\ 0 & 0 & 1_{\mathbb{F}} & \dots & 0 \\ \multicolumn{5}{c}{\dotfill} \\ 0 & 0 & 0 & \dots & 1_{\mathbb{F}} \\ A_0(t) & A_1(t) & A_2(t) & \dots & A_{n-1}(t) \end{pmatrix}.$$

Since $A_i = \varepsilon \circ A_i^*$, the operator A can be realized as the projective limit of the continuous operators A^j given by

$$A^j \colon [0,1] \longrightarrow \mathcal{H}^j(\mathbb{F}^n) \colon t \longmapsto \begin{pmatrix} 0 & 1_{\mathbb{E}^j} & 0 & \dots & 0 \\ 0 & 0 & 1_{\mathbb{E}^j} & \dots & 0 \\ \multicolumn{5}{c}{\dotfill} \\ 0 & 0 & 0 & \dots & 1_{\mathbb{E}^j} \\ A_0^j(t) & A_1^j(t) & A_2^j(t) & \dots & A_{n-1}^j(t) \end{pmatrix}.$$

The same is true for the map $B \colon [0,1] \to \mathbb{F}$, since $B = \varprojlim(\rho^j \circ B)$. As a result, Φ itself is the projective limit of the maps

$$\Phi^j(t, X) = A^j(t) \cdot X + B^j,$$

which are linear with respect to the second factor, hence Lipschitz. Applying now Theorem 2.4.3, we conclude the proof. $\qquad\square$

Remark 2.4.5 It is worth noticing at this point that the above approach generalizes the works of R. S. Hamilton [Ham82] and N. Papaghiuc [Pap80] on *linear* differential equations.

In particular, R. S. Hamilton has studied ordinary differential equations of first order $x' = \varphi(t, x)$, where the function φ decomposes into

$$\varphi = g \circ f, \quad \text{where} \quad g \colon \mathbb{B} \to \mathbb{F}, \ f \colon \mathbb{F} \to \mathbb{B},$$

$\mathbb{B}$ denoting a Banach space. Maps of this type are called *smooth-Banach functions*. However, it is proved (see [Gal97(a)]) that

a smooth-Banach linear operator can always be realized as a projective limit, falling thus within the framework of Theorem 2.4.4

On the other hand, the linear differential equations studied by N. Papaghiuc are restricted to the Fréchet space $\mathcal{L}_I(\mathbb{F}) \equiv \mathcal{H}(\mathbb{F})$ (see Proposition 2.3.13, thus the result of [Pap80] is again a special case of Theorem 2.4.4.

The projective limit approach, although not answering the general problem of solvability of differential equations in Fréchet spaces, nevertheless applies to every Fréchet space and not to particular classes as is the case in, e.g., [Lob92]. In fact, the latter is dealing with operators admitting converging exponential series.

3

Fréchet manifolds

The study of manifolds modelled on infinite-dimensional spaces is receiving an increasing interest in recent decades due to interactions and applications extending beyond the borders of (classical) differential geometry and associated problems in mathematical analysis and theoretical physics. For instance, fibrations and foliations, jet fields, connections, sprays, Lagrangians and Finsler structures ([EE67], [AA96], [AIM93], [DRP95], [GP05] and [Sau87]) are objects naturally listed in this framework. In particular, manifolds modelled on non-Banach locally convex spaces have been studied from different points of view as in [Omo70], [Omo74], [Omo78], [Omo97], [KM97], [AM99], [Nee06], [Vero74] and [Vero79]. However, several questions remain open as a result of the internal problems of the space models.

To be more precise, in conjunction with the problems encountered in the framework of Fréchet spaces or more general topological vector spaces and discussed at the end of §,2.1, two issues still remain critical in the study of infinite-dimensional manifolds: The lack of a general solvability theory for differential equations and the pathological structure of the general linear groups involved in this framework. Both issues seriously affect the study of topological, differential and geometrical aspects in the non-Banach framework.

To give a concrete example, let us take the tangent bundle TM of a smooth manifold M modelled on a Fréchet space $\mathbb{F}$. Even the existence of a vector bundle structure on TM cannot be ensured, since the general linear group $GL(\mathbb{F})$, serving as the structural group in the finite-dimensional case, does not admit a reasonable Lie group structure here.

On the other hand, the study of many geometric features on M, such as connections, parallel translations, holonomy groups, etc., reduce to appropriate differential equations in the models. The problems regarding the solvability of the latter raise questions about whether the corresponding results, already obtained for finite-dimensional or Banach manifolds, can be still transferred to the non-Banach case.

In this chapter we intend to propose a new way for addressing the above mentioned problems, by taking advantage of the realization of Fréchet spaces as projective limits of Banach spaces, already discussed in §.2.3. Exploiting the fact that projective limits are compatible with algebraic structures on the models, as well as with the differentiability tools adopted in Chapter 2, we are able to study a wide sub-category of infinite-dimensional non-Banach manifolds, namely those modelled on Fréchet spaces that can be viewed as projective limits of Banach manifolds.

3.1 Smooth structures on Fréchet manifolds

In Chapter 2, we have exploited the compatibility of projective limits with algebraic and topological data in order to reduce basic problems in Fréchet spaces to their Banach counterparts. However, the same reduction is not always successful when dealing with differential geometric objects, even at the lower level of the smooth structure of a manifold. For example, without appropriate restrictions, the domain of a projective limit of local charts may collapse to a singleton. The following definition provides the optimum conditions ensuring the smooth interaction between projective limits and manifolds.

Definition 3.1.1 Let $\{M^i; \mu^{ji}\}_{i,j\in\mathbb{N}}$ be a projective system of smooth manifolds modelled on the Banach spaces $\{\mathbb{E}^i\}_{i\in\mathbb{N}}$, respectively (with smooth connecting morphisms). A system $\{(U^i, \phi^i)\}_{i\in\mathbb{N}}$ of corresponding charts will be called a *(projective) limit chart* if and only if the limits $\varprojlim U^i$, $\varprojlim \phi^i$ can be defined, and the sets $\varprojlim U^i$, $\varprojlim \phi^i(\varprojlim U^i)$ are open in $\varprojlim M^i$, $\varprojlim \mathbb{E}^i$, respectively.

Limit charts will determine a smooth structure on the projective limit of Banach manifolds. Before proving this, we need the following:

Definition 3.1.2 With the notations of Definition 3.1.1, the space $M = \varprojlim\{M^i; \mu^{ji}\}_{i,j\in\mathbb{N}}$ is called a *projective limit of Banach manifolds*, or *plb-manifold* for short, provided that:

(1) The models $\{\mathbb{E}^i\}_{i\in\mathbb{N}}$ form a projective system with connecting morphisms $\{\rho^{ji}: \mathbb{E}^j \to \mathbb{E}^i;\ j \geq i\}$ and limit the Fréchet space $\mathbb{F} = \varprojlim \mathbb{E}^i$.

(2) M is covered by a family $\{(U_\alpha, \phi_a)\}_{\alpha \in I}$ of limit charts, where $U_\alpha = \varprojlim U_\alpha^i$ and $\phi_a = \varprojlim \phi_\alpha^i$, the limits taken with respect to $i \in \mathbb{N}$.

If there is no danger of confusion, we simply write $M = \varprojlim\{M^i\}$.

Proposition 3.1.3 *A plb-manifold $M = \varprojlim M^i$, as before, is a smooth manifold modelled on the Fréchet space $\mathbb{F}$.*

Proof The desired differential structure on M is naturally determined by the limit charts $\left(\varprojlim U_\alpha^i, \varprojlim \phi_\alpha^i\right)$. They obviously cover M. Their (smooth) compatibility is ensured by the compatibility of the corresponding factors and the fact that projective limit of differentiable maps remain also differentiable. More precisely, if

$$\left(U_\alpha = \varprojlim U_\alpha^i, \phi_\alpha = \varprojlim \phi_\alpha^i\right) \quad \text{and} \quad \left(U_\beta = \varprojlim U_\beta^i, \phi_\beta = \varprojlim \phi_\beta^i\right)$$

are two limit charts with $U_\alpha \cap U_\beta \neq \emptyset$, then the transition functions are the diffeomorphisms

$$\phi_\beta \circ \phi_\alpha^{-1} = \varprojlim \left(\phi_\beta^i \circ (\phi_\alpha^i)^{-1}\right) : \phi_\alpha(U_\alpha \cap U_\beta) \longrightarrow \phi_\beta(U_\alpha \cap U_\beta). \quad \square$$

Remarks 3.1.4 1) We recall that the smooth structure of Proposition 3.1.3 is based on the differentiability of J. A. Leslie ([Les67], [Les68]) discussed in §.2.2. Other methods of differentiability (for instance those of M. C. Abbati-A. Manià [AM99], A. Kriegl-P. Michor ([KM97]), M.E. Verona [Vero74], [Vero79]) can also be applied, because the projective limits are compatible with the algebraic and topological structures involved in the definitions. However, the respective manifold structures are not in general equivalent to that of the aforementioned proposition.

2) Let us elaborate on the requirement concerning the existence of limit charts covering the projective limit $M = \varprojlim M^i$: The assumption that $\{M^i; \mu^{ji}\}_{i,j \in \mathbb{N}}$ is a projective system of *smooth* manifolds already means, by definition, that the connecting morphisms $\mu^{ji} : M^j \to M^i$ ($j \geq i$) are compatible with the smooth structures of the manifolds M^i ($i \in \mathbb{N}$); that is, they are smooth in the ordinary sense. Therefore, for each $j \geq i$, there exist charts (U^i, ϕ^i) and (U^j, ϕ^j) of M^i and M^j, respectively, such that $\mu^{ji}(U^j) \subseteq U^i$ and the local representation

(3.1.1) $$\phi^i \circ \mu^{ji} \circ (\phi^j)^{-1} : \mathbb{E}^j \supseteq \phi^j(U^j) \longrightarrow \phi^i(U^i) \subseteq \mathbb{E}^i$$

is smooth. This fact does not necessarily lead to the construction of corresponding smooth charts on M, a gap that led a number of authors (see, e.g., [AM99], [Vero74], [Vero79]) to adopt rather algebraic approaches.

However, our definition of limit charts ensures that (3.1.1) not only is a smooth map but also coincides with the connecting morphisms of the models; that is,

$$(3.1.2) \qquad \phi^i \circ \mu^{ji} \circ (\phi^j)^{-1} = \rho^{ji} \quad j \geq i.$$

In other words, the charts under consideration connect the projective systems of the manifolds and the corresponding models involved.

As an application of the preceding constructions, we prove the following immediate result.

Proposition 3.1.5 *If $M = \varprojlim \{M^i; \mu^{ji}\}_{i,j \in \mathbb{N}}$ is a plb-manifold, then the canonical projections $\mu^i \colon M \to M^i$ ($i \in \mathbb{N}$) are smooth.*

Proof Let $(U = \varprojlim x^i, \phi = \varprojlim \phi^i)$ be any chart at an arbitrary point in $x \in M$. The local representation of μ^{ji} with respect to the charts (U, ϕ) of M and (x^i, ϕ^i) of M^i, coincides with the corresponding canonical projection on the model spaces, i.e.

$$\phi^i \circ \mu^i \circ \phi^{-1} = \rho^i,$$

thus proving the smoothness of μ^i at x. $\qquad\qquad\qquad\qquad\qquad\square$

The previous proposition, in conjunction with Remark 3.1.4(2), ensures the categorical consistency of the derived structures.

Examples 3.1.6

1. Every Banach manifold M can be trivially considered as a plb-manifold coinciding with the limit of the single element projective system $\{M, \mathrm{id}_M\}$.

2. Every Fréchet space $\mathbb{F}$, being always a projective limit of Banach spaces $\varprojlim \mathbb{E}^i$, fulfils also the assumptions of the Definition 3.1.2 with respect to the total chart $(\mathbb{F}, \mathrm{id}_{\mathbb{F}}) = \left(\varprojlim \mathbb{E}^i, \varprojlim \mathrm{id}_{\mathbb{E}^i} \right)$.

3. The group $C^0(\mathbb{R}, \mathbb{R}_+)$ of all continuous positive real-valued curves is a plb-manifold modelled on the Fréchet space $C^0(\mathbb{R}, \mathbb{R})$ via the isomorphism $g = \varprojlim g^n$, where

$$g^n : C^0(\mathbb{R}, \mathbb{R}_+) \longrightarrow C^0\left([-n.n], \mathbb{R}_+\right) : f \mapsto f\big|_{[-n,n]}; \qquad n \in \mathbb{N},$$

and corresponding limit charts $\left(C^0([-n, n], \mathbb{R}_+), \phi_n \right)$, where ϕ_n is given by $\phi_n(f) = \log \circ f$.

4. The group of all smooth maps $C^\infty(M, G)$ from a compact manifold M to a finite dimensional Lie group G is a Fréchet plb-manifold since

$$C^\infty(M, G) \cong \bigcap_{n \in \mathbb{N}} C^n(M, G)$$

with corresponding charts $\left(C^n(M, V), \phi_n\right)_{n \in \mathbb{N}}$, where V is an open subset of G over which the exponential map $\exp_G$ of G is a diffeomorphism, and

$$\phi_n(f) = \exp_G^{-1} \circ f.$$

5. H. Omori in [Omo70] introduced the notion of *inverse limit manifolds* as the intersection of a countable number of Banach manifolds forming a nested sequence

$$M^1 \supseteq M^2 \supseteq \cdots \supseteq M^n \supseteq M^{n+1} \supseteq \cdots$$

on which projective limits of charts can be defined. The space $M = \bigcap_{n \in \mathbb{N}} M^n$ *cannot* be always endowed with the structure of a smooth manifold, since the domains of the charts employed may collapse to single point sets. However, if the intersection of the domains are open sets, then Omori's manifolds are special cases of plb-manifolds with connecting morphisms being the natural embeddings.

6. The space of infinite jets $J^\infty(E)$ of a Banach vector bundle (E, π, B) is a plb manifold modelled on the Fréchet space

$$\mathbb{E} \times \mathcal{L}(\mathbb{B}, \mathbb{E}) \times \mathcal{L}_s^2(\mathbb{B}, \mathbb{E}) \times \mathcal{L}_s^3(\mathbb{B}, \mathbb{E}) \times \cdots,$$

where $\mathcal{L}_s^k(\mathbb{B}, \mathbb{E})$ is the space of continuous symmetric k-linear maps between the model spaces $\mathbb{B}$ and $\mathbb{E}$ of the base B and the total space E, respectively. In particular, $J^\infty(E)$ is isomorphic with the projective limit of the finite dimensional jets:

$$J^\infty(E) \cong \varprojlim J^n(E).$$

The charts of $J^\infty(E)$ are the pairs $(J^\infty(U), \phi_U)$, where

$$\phi_U \colon J^\infty(U) \longrightarrow U \times \mathcal{L}(\mathbb{B}, \mathbb{E}) \times \mathcal{L}_s^2(\mathbb{B}, \mathbb{E}) \times \cdots$$

is given by

$$\phi_U\left(j_x^\infty \xi\right) = \left(x, \xi(x), d\xi(x), d^2\xi(x), \dots\right),$$

if $j_x^\infty \xi$ denotes the infinite jet (: equivalent class) of a section ξ of E, and U is an open subset of B. Further details will be given in § 6.3.

7. M .E. Verona in [Vero74] and [Vero79] studied projective limits of manifolds adopting, however, a rather topological approach.

Smooth maps between plb-manifolds are defined in the usual way. However, there is a particular category of maps which deserves special attention.

Definition 3.1.7   Let $\{M^i; \mu^{ji}\}_{i,j\in\mathbb{N}}$ and $\{N^i; \nu^{ji}\}_{i,j\in\mathbb{N}}$ be two plb-manifolds and let $\{f^i: M^i \to N^i\}_{i\in\mathbb{N}}$ be a projective system of *smooth* maps. Then the limit

$$\varprojlim f^i : (x^i)_{i\in\mathbb{N}} \longmapsto (f^i(x^i))_{i\in\mathbb{N}}$$

is called a **projective limit of smooth maps** (**pls-map**, in short).

The two commutative diagrams of Definition 2.3.4, relating a projective system of maps and its limit with the connecting morphisms and the canonical projections of the projective systems of the domains and ranges of the maps, translated in the present situation read as follows:

$$(3.1.3) \qquad\qquad \nu^{ji} \circ f^j = f^i \circ \mu^{ji},$$

$$(3.1.4) \qquad\qquad \nu^i \circ f = f^i \circ \mu^i,$$

for all $i, j \in \mathbb{N}$ and $j \geq i$.

In the approaches of [AM99], [Pap80], [Vero74], [Vero79], projective limits of smooth maps between projective limits of smooth manifolds are taken, by definition, as the smooth maps between projective limits of smooth manifolds. This is a rather algebraic way of defining differentiability. In our framework, the differentiation defined in Section § 2.2) and the definition of plb-manifolds allow one to prove that pls-maps are rendered smooth in the standard way, i.e. they have differentiable local representations with respect to limit charts. We clarify this matter in the next proposition.

Proposition 3.1.8   *Every pls-map*

$$f = \varprojlim f^i : M = \varprojlim M^i \longrightarrow N = \varprojlim N^i$$

is a smooth map between plb-manifolds.

Proof   Let $x = (x^i)_{i\in\mathbb{N}}$ be an arbitrary point in M. We consider the limit charts $(U, \phi) = \left(\varprojlim x^i, \varprojlim \phi^i\right)$ and $(V, \psi) = (\varprojlim V^i, \varprojlim \psi^i)$ of M and N containing x and $f(x)$, respectively. Without loss of generality, we may assume that $f(U) \subseteq V$, otherwise we restrict ourselves to an open subset of $U \cap f^{-1}(V)$. Now, the local representation $\psi \circ f \circ \phi^{-1} \colon \phi(U) \to \psi(V)$ of f, with respect to the chosen charts, is given by

$$\psi \circ f \circ \phi^{-1} = \varprojlim \psi^i \circ \varprojlim f^i \circ \varprojlim (\phi^i)^{-1} = \varprojlim (\psi^i \circ f^i \circ \phi^i)^{-1}.$$

Therefore, f is smooth at x if and only if its local representation is smooth at $\phi(x)$; equivalently (in virtue of Proposition 2.3.11) if and only if each $\psi_i \circ f_i \circ \phi_i^{-1} \colon \phi_i(U_i) \to \psi_i(V_i)$ is smooth at x_i, for all $i \in \mathbb{N}$. $\quad\square$

Remarks 3.1.9 1) It is worth noting here that the previous result, although ensuring the smoothness of pls-maps, does not restrict the set of smooth maps between plb-manifolds only to pls-maps, as is the case of [AM99]), [Pap80], [Vero74] and [Vero79] mentioned earlier [see also Remark 3.1.4(1)]. Therefore, smooth maps that are not necessarily projective limits are also included in our framework.

2) In the special case of smooth curves $\mathcal{C}^\infty(\mathbb{R}, M)$, where M is a plb-manifold, the projective limit approach proves to be very convenient since every curve $\alpha \colon \mathbb{R} \to M$ can be naturally realized as the inverse limit of the corresponding projections to the factors, i.e.

$$\alpha = \varprojlim(\mu^i \circ \alpha),$$

where $\mu^i \colon M \to M^i$ $(i \in \mathbb{N})$ are the canonical projections of the projective limit to the factors. We conclude that:

A curve on a plb-manifold is smooth if and only if it is a projective limit of smooth curves. Consequently,

(3.1.5) $$\mathcal{C}^\infty(\mathbb{R}, M) \equiv \varprojlim \mathcal{C}^\infty(\mathbb{R}, M^i)$$

within a natural isomorphism.

3) The preceding identification of smooth curves might be used to find a way connecting the differentiability adopted here with that of [KM97]. This is an open question.

3.2 The tangent bundle of a plb-manifold

Using the tools of the previous section, we proceed to the study of the tangent spaces and bundles of projective limit manifolds. The identification (3.1.4) allows one to follow the classical pattern.

Definition 3.2.1 Let $M = \{M^i; \mu^{ji}\}_{i,j \in \mathbb{N}}$ be a plb manifold and $x = (x^i)_{i \in \mathbb{N}} \in M$. Then, the **tangent space of M at x** is the quotient space

$$T_x M = \mathcal{C}_x^\infty(\mathbb{R}, M)/\sim_x$$

with respect to the equivalence relation

(3.2.1) $\alpha \sim_x \beta \quad \Leftrightarrow \quad (\phi \circ \alpha)'(0) = (\phi \circ \beta)'(0)$

between smooth curves in M with $\alpha(0) = \beta(0) = x$, and for any chart (U, ϕ) of M with $x \in U$.

We note that (see also Proposition 2.3.11)

$$(\phi \circ \alpha)'(t) = (D(\phi \circ \alpha)(0))\,(1)$$

$$= \Big(D(\varprojlim \phi^i \circ \alpha)(0)\Big)\,(1)$$

(3.2.2)
$$= \Big(D(\varprojlim \phi^i \circ \varprojlim \alpha^i))(0)\Big)\,(1)$$

$$= \Big(D(\varprojlim(\phi^i \circ \mu^i \circ \alpha))(0)\Big)\,(1)$$

$$= \varprojlim \big(D(\phi^i \circ \mu^i \circ \alpha)(0)\big)\,(1)$$

As usual, (3.2.1) does not depend on the choice of the chart at x. Hence, it is sufficient to check this condition for a limit chart.

The equivalence class of a curve α passing through x is denoted by $[\alpha, x]$, instead of the customary $[(\alpha, x)]$. This is a minor deviation aiming to simplify expressions involving successive parentheses, as will be often the case below.

Obviously,

$$TM = \bigcup_{x \in M} T_x M$$

is the **tangent bundle** of M. Its projection is the map

$$\pi \colon TM \longrightarrow M \colon [\alpha, x] \mapsto x.$$

We shall show that the tangent bundle of a plb-manifold remains in the same category of manifolds, thus we obtain yet another example of a plb-manifold. First we prove the following:

Proposition 3.2.2   *Let $M = \varprojlim M^i$ be a plb-manifold. Then the tangent space $T_x M$ at an arbitrary point $x = (x^i)_{i \in \mathbb{N}} \in M$ is in bijective correspondence with $\varprojlim T_{x^i} M^i$. Therefore, we obtain the identification*

(3.2.3) $T_x M \equiv \varprojlim T_{x^i} M^i$

Proof   The differentials $T_{x^j} \mu^{ji} \colon T_{x^j} M^j \longrightarrow T_{x^i} M^i$ $(j \geq i)$ of the connecting morphisms of M determine the connecting morphisms of the projective system $\{T_{x^j} M^j, T_{x^j} \mu^{ji}\}$, since

$$T_{x^k} \mu^{kj} \circ T_{x^j} \mu^{ji} = T_{x^k}(\mu^{kj} \circ \mu^{ji}) = T_{x^k} \mu^{ki}, \quad k \geq j \geq i.$$

Then we relate $\varprojlim T_{x^i} M^i$ with the tangent space $T_x M$ by the map

(3.2.4) $\qquad R_x \colon T_x M \longrightarrow \varprojlim T_{x^i} M^i \colon [\alpha, x] \mapsto \left([\mu^i \circ \alpha, \mu^i(x)]^i \right)_{i \in \mathbb{N}},$

where the bracket $[\ ]^i$ stands for the equivalence class of curves in M^i.

R_x is an *injection*: Assume that $R_x([\alpha, x]) = R_x([\beta, x])$. If $(U, \phi) = \left(\varprojlim x^i, \varprojlim \phi^i \right)$ is any limit chart of M containing x, then, in virtue of (3.2.2), we see that

$$(\phi \circ \alpha)'(0) = (\phi \circ \beta)'(0)$$
$$\Leftrightarrow \quad \varprojlim \left(D(\phi^i \circ \mu^i \circ \alpha)(0) \right)(1) = \varprojlim \left(D(\phi^i \circ \mu^i \circ \beta)(0) \right)(1)$$
$$\Leftrightarrow \quad \left(D(\phi^i \circ \mu^i \circ \alpha)(0) \right)(1) = \left(D(\phi^i \circ \mu^i \circ \beta)(0) \right)(1)$$
$$\Leftrightarrow \quad [\mu^i \circ \alpha, x^i]^i = [\mu^i \circ \beta, x^i]^i$$
$$\Leftrightarrow \quad [\mu^i \circ \alpha, \mu^i(x)]^i = [\mu^i \circ \beta, \mu^i(x)]^i$$
$$\Leftrightarrow \quad R_x([\alpha, x]) = R_x([\beta, x])$$

from which we obtain the injectivity of R_x.

R_x is a *surjection*: Take any element $\left([\alpha^i, x^i]^i \right)_{i \in \mathbb{N}}$ in $\varprojlim T_{x^i} M^i$. Using again the limit chart $(U, \varphi) = \left(\varprojlim x^i, \varprojlim \varphi^i \right)$, we define the elements of $\mathbb{F} = \varprojlim \mathbb{E}_i$

$$u = (\phi^i(x^i))_{i \in \mathbb{N}} \quad \text{and} \quad v = \left((\phi^i \circ \alpha^i)'(0) \right)_{i \in \mathbb{N}},$$

(recall that $\mathbb{E}^i$ is the model of M^i, for each $i \in \mathbb{N}$), and the smooth curves

$$h \colon \mathbb{R} \longrightarrow \mathbb{F} \colon t \mapsto u + t \cdot v,$$
$$\alpha \colon J \longrightarrow M \colon t \mapsto \phi^{-1}(h(t)),$$

where J is an open interval such that $\alpha(J) \subset U$. We shall show that $R_x([\alpha, x]) := ([\mu^i \circ \alpha, \mu^i(x)]^i)_{i \in \mathbb{N}} = ([\alpha^i, x^i]^i)_{i \in \mathbb{N}}$.

To this end we recall that, by the definition of the projective limit of maps, the following diagram is commutative :

$$
\begin{array}{ccc}
M \supseteq \varprojlim U^i & \xrightarrow{\ \phi\ } & \varprojlim \phi^i(U^i) \subseteq \mathbb{F} \\[2mm]
{\scriptstyle \mu^i} \Big\downarrow & & \Big\downarrow {\scriptstyle \rho^i} \\[2mm]
M \supseteq U^i & \xrightarrow[\ \phi^i\ ]{} & \phi^i(U^i) \subseteq \mathbb{F}_i
\end{array}
$$

Therefore,

$$\begin{aligned}
(\mu^i \circ \alpha)(0) &= (\mu^i \circ \phi^{-1})(h(0)) \\
&= (\phi^i)^{-1} \left(\rho^i \left((\phi^i(x^i))_{i \in \mathbb{N}} \right) \right) \\
&= (\phi^i)^{-1}(\phi^i(x^i)) = x^i \\
&= \mu^i(x) = (\mu^i \circ \alpha)(0).
\end{aligned}$$

Similarly, since ρ^i is continuous linear for all $i \in \mathbb{N}$,

$$\begin{aligned}
(\phi^i \circ \mu^i \circ \alpha)'(0) &= (\rho^i \circ \phi \circ \alpha)'(0) = (\rho^i \circ h)'(0) \\
&= \left(D(\rho^i \circ h)(0) \right) (1) = \left(D\rho^i(h(0)) \circ Dh(0) \right) (1) \\
&= \rho^i(h'(0)) = \rho^i \left(((\phi^i \circ \alpha^i)'(0))_{i \in \mathbb{N}} \right) \\
&= (\phi^i \circ \alpha^i)'(0).
\end{aligned}$$

As a result, the curves $\mu^i \circ \alpha$ and α^i are equivalent at $x^i \in M^i$ ($i \in \mathbb{N}$), thus proving the desired surjectivity of R_x. □

Corollary 3.2.3 *By means of R_x, the structure of a Fréchet space on $\varprojlim T_{x^i} M^i$ is transferred to $T_x M$. Therefore, $T_x M$ is isomorphic to the Fréchet model $\mathbb{F}$, and (3.2.3) is completed as follows:*

(3.2.5) $$T_x M \equiv \varprojlim T_{x^i} M^i \cong \varprojlim \mathbb{E}^i = \mathbb{F},$$

for every $x = (x^i)_{i \in \mathbb{N}}$

Of course, the isomorphism $T_x M \cong \mathbb{F}$ can be established as in the case of finite-dimensional or Banach manifolds. However, (3.2.5) shows that the tangent space of a plb-manifold is a projective limit itself and provides a complete picture of the way the tangent spaces of a plb-manifold are related with the tangent spaces of the factors and their models.

The differential of a smooth map between plb-manifolds can be defined as in the Banach case.

Definition 3.2.4 Let $M = \varprojlim\{M^i; \mu^{ji}\}_{i,j \in \mathbb{N}}$, $N = \varprojlim\{N^i; \nu^{ji}\}_{i,j \in \mathbb{N}}$ be two plb-manifolds, and let $f \colon M \to N$ be a differentiable map at the point $x = (x^i)_{i \in \mathbb{N}} \in M$. Then the **differential of f at** x is the map

$$T_x f \colon T_x M \longrightarrow T_{f(x)} N \colon [\alpha, x] \mapsto [f \circ \alpha, f(x)].$$

Accordingly, the **(total) differential** of f is the map $Tf \colon TM \to TN$, given by $Tf|_{T_x M} = T_x f$.

Working as in the Banach framework, one easily verifies that $T_x f$ is a linear map. In particular, if f is a pls-map, then its differential can be realized as the projective limit of the differentials of the factor maps. As a matter of fact, we prove the following:

Proposition 3.2.5 *Let $f := \varprojlim f^i \colon \varprojlim M^i \to \varprojlim N^i$ be a pls-map as in Proposition 3.1.5. Then the differential of f at any point $x = (x^i)_{i \in \mathbb{N}} \in M$ coincides—up to isomorphism—with the projective limit of the differentials of the f^i's; that is,*

$$(3.2.6) \qquad T_x f \equiv \varprojlim T_{x^i} f^i.$$

Proof Differentiating (3.1.2) at every $x^i \in T_{x^j} M^j$, we obtain

$$(3.2.7) \qquad T_{f^j(x^j)} \nu^{ji} \circ T_{x^j} f^j = T_{\mu^{ji}(x^j)} f^i \circ T_{x^j} \mu^{ji}.$$

Taking into account that the differentials of the connecting morphisms $\{\mu^{ji}\}$ and $\{\nu^{ji}\}$ are themselves connecting morphisms for the projective systems of the tangent spaces (see the beginning of the proof of Proposition 3.2.2), we immediately see that $\varprojlim T_{x^i} f^i$ exists.

For the proof of (3.2.6) it suffices to verify that the diagram

$$
\begin{array}{ccc}
T_x M & \xrightarrow{\;\; T_x f \;\;} & T_{f(x)} N \\[4pt]
{\scriptstyle R^M_x} \Big\downarrow & & \Big\downarrow {\scriptstyle R^N_{f(x)}} \\[4pt]
\varprojlim T_{x^i} M^i & \xrightarrow[\;\; \varprojlim T_{x^i} f^i \;\;]{} & \varprojlim T_{f^i(x^i)} N^i
\end{array}
$$

is commutative, where the vertical maps are the corresponding isomorphisms (3.2.4). Indeed, for an arbitrary $[\alpha, x] \in T_x M$, we have that

$$\left(R^N_{f(x)} \circ T_x f \right)([\alpha, x]) = \left([\nu^i \circ f \circ \alpha, \nu^i(f(x))] \right)_{i \in \mathbb{N}},$$

whereas

$$\left(\varprojlim T_{x^i} f^i \circ R^M_x \right)([\alpha, x]) = \left([f^i \circ \mu^i \circ \alpha, f^i(\mu^i(x))] \right)_{i \in \mathbb{N}}.$$

The right-hand sides of the above equalities coincide in virtue of (3.1.4); hence, the diagram is commutative, as claimed. $\qquad\square$

Corollary 3.2.6 *The canonical projections $q^i \colon \varprojlim T_{x^i} M^i \to T_{x^i} M^i$ satisfy $T_x \mu^i = q^i \circ R_x$, $i \in \mathbb{N}$, where $T_x \mu^i \colon T_x M \to T_{x^i} M^i$ are the differentials of the canonical projections of $M = \varprojlim M^i$. Therefore, in virtue of (3.2.4), $T_x \mu^i = q^i$ up to isomorphism.*

Proof For every $[\alpha, x] \in T_x M$,

$$(q^i \circ R_x)([\alpha, x]) = q^i\left(\left([\mu^i \circ \alpha, \mu^i(x)]^i\right)_{i \in \mathbb{N}}\right)$$

$$= [\mu^i \circ \alpha, \mu^i(x)]^i = T_x \mu^i([\alpha, x]).$$

Schematically, the diagram

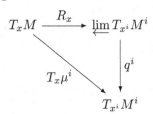

is commutative. This completes the proof, since R_x is a (linear) isomorphism. $\qquad \square$

Proposition 3.2.7 *If $M = \varprojlim M^i$ is a plb-manifold, then so is the limit space $\varprojlim T M^i$.*

Proof Since M is a plb-manifold there exist local charts $\left(\varprojlim U^i, \varprojlim \phi^i\right)$ covering M. Each chart (U^i, ϕ^i) of M^i induces the corresponding chart $\left(\pi_i^{-1}(U^i), \Phi^i\right)$ of $T M^i$, where $\pi_i \colon T M^i \to M^i$ is the projection and

$$\Phi^i \colon \pi_i^{-1}(U^i) \longrightarrow \phi^i(U^i) \times \mathbb{E}^i$$

is given by $\Phi^i(x^i) = \left(x^i, \overline{\phi^i}(x^i)\right)$, $x^i \in T_{x^i} M^i$. Recall that $\overline{\phi^i} \colon T_{x^i} M^i \to \mathbb{E}^i$ is the linear isomorphism with $\overline{\phi^i}(x^i) := (\phi^i \circ \alpha)'(0)$ if $x^i = [\alpha, x^i]$.

It is an easy exercise to verify that:

i) $\left\{T M^j ; T \mu^{ji} \colon T M^j \to T M^i\right\}_{i, j \in \mathbb{N}}$ is a projective system, thus the limit $\varprojlim T M^i$ exists.

ii) The pairs $\left(\pi_i^{-1}(U^i), \Phi^i\right)$, for all $i \in \mathbb{N}$, form a projective system of charts as in Definition 3.1.1.

We shall show that $\varprojlim \pi_i^{-1}(U^i)$ and $\varprojlim \Phi^i\left(\pi_i^{-1}(U^i)\right)$ are open subsets of $T M$ and $\mathbb{F} \times \mathbb{F}$, respectively. Indeed, by

$$\varprojlim \pi_i^{-1}(U^i) = \varprojlim \pi_i^{-1}\left(\varprojlim U^i\right) = \left(\varprojlim \pi_i\right)^{-1}\left(\varprojlim U^i\right)$$

we see that $\varprojlim \pi_i^{-1}(U^i) \subset T M$ is open, because $\varprojlim U^i$ is open in M and $\varprojlim \pi_i \colon \varprojlim T M^i \to \varprojlim M^i = M$ is a continuous map.

On the other hand,

$$\varprojlim \Phi^i\left(\pi_i^{-1}(U^i)\right) = \varprojlim \left(\phi^i(U^i) \times \mathbb{E}^i\right) =$$

$$= \varprojlim \phi^i(U^i) \times \varprojlim \mathbb{E}^i = \varprojlim \phi^i(U^i) \times \mathbb{F}.$$

Since $\phi^i(U^i) \subset \mathbb{E}^i$ is open, then $\varprojlim \phi^i(U^i)$ is open in $\mathbb{F}$, consequently the set $\varprojlim \Phi^i(\pi_i^{-1}(U^i))$ is an open subset of $\mathbb{F} \times \mathbb{F}$.

The previous assertions show that the pairs $(\pi_i^{-1}(U^i), \Phi^i)$, for all $i \in \mathbb{N}$, determine a projective limit chart (see Definition 3.1.1). Moreover, all $\left(\varprojlim \pi_i^{-1}(U^i), \varprojlim \Phi^i\right)$ cover $\varprojlim TM^i$; thus, in virtue of Definition 3.1.2, the latter becomes a plb-manifold (see also Proposition 3.1.3).   □

The tangent bundle TM of a plb-manifold $M = \varprojlim\{M^i; \mu^{ji}\}_{i,j\in\mathbb{N}}$ is endowed with a differential structure in the standard way of finite-dimensional or Banach manifolds: As in the previous proof, for every chart (U, ϕ) of M (a limit chart or not), we define the chart $(\pi^{-1}(U), \Phi)$ of TM, where $\pi\colon TM \to M$ is the projection of the bundle and

$$\Phi\colon \pi^{-1}(U) \longrightarrow \phi(U) \times \mathbb{F}\colon [\alpha, x] \mapsto \left(\phi(x), (\phi \circ \alpha)'(0)\right).$$

By routine checking, we see that the previous charts determine the structure of a Fréchet manifold on TM. However, the realization of the tangent spaces of TM as projective limits of the corresponding tangent spaces of the factors (see Proposition 3.2.2) allows us to show that TM is diffeomorphic with the projective limit of the corresponding tangent bundles of the factors. Therefore, as alluded to before Proposition 3.2.2, the tangent bundle of a plb-manifold remains—up to a diffeomorphism—within the category of plb-manifolds.

Theorem 3.2.8   *Within a diffeomorphism, the tangent bundle TM of a plb-manifold $M = \varprojlim M^i$ coincides with $\varprojlim TM^i$.*

Proof   We consider the map $R\colon TM \to \varprojlim TM^i$, given by [see (3.2.4)]

$$R\big|_{T_xM} = R_x\colon T_xM \longrightarrow \varprojlim T_{x^i}M^i.$$

The map R is well-defined because

$$\varprojlim TM^i = \bigcup_{x\in M} \varprojlim T_{x^i}M^i \equiv \bigsqcup_{x\in M} \varprojlim T_{x^i}M^i \quad \text{(disjoint union)}.$$

This is a consequence of the fact that the projective limits, involved in both sides of the equality, have the same connecting morphisms.

Obviously, R is a bijection by the analogous properties of the R_x's, established in the proof of Proposition 3.2.2.

To show that R is a diffeomorphism, it suffices to examine its local behaviour. To this aim let $u \in T_xM$ be an arbitrary tangent vector. For a limit chart $\left(U = \varprojlim x^i, \phi = \varprojlim \phi^i\right)$ of M at x, we construct the chart

$(\pi^{-1}(U), \Phi)$ of TM at u (see the comments preceding the present statement), and the chart $\left(\varprojlim \pi_i^{-1}(x^i), \varprojlim \Phi^i \right)$ of $\varprojlim TM^i$ (see the proof of Proposition 3.2.7).

We check immediately that $R\left((\pi^{-1}(U)) = \varprojlim \pi_i^{-1}(x^i)\right)$, thus the local representation, say, R_ϕ of R, with respect to the last two charts, is given by $R_\phi = \varprojlim \Phi^i \circ R \circ \Phi^{-1}$ as in the following diagram:

$$
\begin{array}{ccc}
\pi^{-1}(U) & \xrightarrow{\;\;R\;\;} & \varprojlim \pi_i^{-1}(x^i) \\[2mm]
\Big\downarrow{\scriptstyle\Phi} & & \Big\downarrow{\scriptstyle\varprojlim \Phi^i} \\[2mm]
\phi(U) \times \mathbb{F} & \xrightarrow[\;\;R_\phi\;\;]{} & \varprojlim \phi^i(x^i) \times \mathbb{F}
\end{array}
$$

But now $\phi(U) = \varprojlim \phi^i(x^i)$, and

$$
\begin{aligned}
(\varprojlim \Phi^i \circ R)([\alpha, x]) &= \varprojlim \Phi^i (R_x([\alpha, x])) \\
&= \left(\Phi^i([\mu^i \circ \alpha, \mu^i(x)]^i) \right)_{i \in \mathbb{N}} \\
&= \left(\phi^i(x^i), (\phi^i \circ \mu^i \circ \alpha)'(0) \right)_{i \in \mathbb{N}} \\
&= \left(\varprojlim \phi^i)(x), (\varprojlim(\phi^i \circ \alpha)')(0) \right) \\
&= (\phi(x), (\phi \circ \alpha)'(0)) = \Phi([\alpha, x]);
\end{aligned}
$$

in other words, $R_\phi = \mathrm{id}_{\phi(U) \times \mathbb{F}}$. This shows that R is a diffeomorphism at u and concludes the proof. $\qquad\square$

Corollary 3.2.9 *Let $f = \varprojlim f^i$ be a pls-map between the plb-manifolds $M = \varprojlim M^i$ and $N = \varprojlim N^i$. Then the total differential $Tf\colon TM \to TN$ is also a pls-map such that*

(3.2.8) $$Tf = \varprojlim Tf^i$$

within a diffeomorphism.

Proof The total differentiation of (3.1.3) yields [see also (3.2.7)]

$$T\nu^{ji} \circ Tf^j = Tf^i \circ T\mu^{ji};$$

hence, in virtue of Definition 2.3.4, $\varprojlim Tf^i$ exists. by

Now, denoting by $R^M\colon TM \to \varprojlim TM^i$ and $R^N\colon TN \to \varprojlim TN^i$ the isomorphism of Theorem 3.2.8, for M and N respectively, and taking into account the commutative diagram in the proof of Proposition 2.3.6, we have that

$$R^N \circ Tf\big|_{T_xM} = R^N_{f(x)} \circ T_xf = \varprojlim T_{x^i}f^i \circ R^M_x = \varprojlim Tf^i \circ R^M\big|_{T_xM},$$

for every $x \in M$. Therefore,

(3.2.9) $$R^N \circ Tf = \varprojlim Tf^i \circ R^M,$$

whence the result. □

For the sake of completeness, the diffeomorphisms of Theorem 3.2.8 and Corollary 3.2.9 can be written in the respective forms

$$T\left(\varprojlim M^i\right) \cong \varprojlim TM^i,$$

$$T\left(\varprojlim f\right) \cong \varprojlim Tf^i.$$

Consequently, up to isomorphism, $\varprojlim$ commutes with the tangent functor T in the category of plb-manifolds.

3.3 Vector fields

Having established the structure of a Fréchet manifold on plb-manifolds and their tangent bundles, we can now define vector fields. We are interested in projective limits of vector fields, in consistency with our framework.

Definition 3.3.1 Let $M = \{M^i; \mu^{ji}\}$ be a plb-manifold. A **projective system of vector fields** is a (countable) family of smooth vector fields on the factors

$$\{\xi^i \colon M^i \longrightarrow TM^i \mid i \in \mathbb{N}\}$$

satisfying the natural condition

(3.3.1) $$T\mu^{ji} \circ \xi^j = \xi^i \circ \mu^{ji}; \qquad j \geq i,$$

which means that ξ^j and ξ^i are μ^{ji}-related vector fields.

The previous definition, along with Proposition 3.1.8, implies that

$$\varprojlim \xi^i \colon M = \varprojlim M^i \longrightarrow \varprojlim TM^i$$

exists and is a pls-map, thus it is smooth. Moreover,

(3.3.2) $$\left(\varprojlim \pi_i\right) \circ \left(\varprojlim \xi_i\right) = \varprojlim (\pi_i \circ \xi) = \varprojlim id_M = id_M.$$

Proposition 3.3.2 *With the previous notations, $\varprojlim \xi^i$ identifies with a smooth vector field of $M = \varprojlim M^i$.*

Proof Let $R = R^M \colon TM \to \varprojlim TM^i$ be the diffeomorphism of Theorem 3.2.8. We set $\xi := R^{-1} \circ \varprojlim \xi^i$. Clearly, ξ is a smooth map. By routine computations, we see that

$$(3.3.3) \qquad\qquad \pi = \left(\varprojlim \pi_i\right) \circ R.$$

Therefore, (3.3.3) and (3.3.2) imply that $\pi \circ \xi = id_M$; hence ξ is a smooth vector field on M. $\qquad\qquad\qquad\qquad\qquad\qquad\qquad\qquad\qquad\qquad$ □

Henceforth, in virtue of the preceding proposition, projective limit of vector fields $\xi = \varprojlim \xi^i$ will be thought of as a smooth vector field of M.

We now proceed to the study of the integral curves of a vector field $\xi = \varprojlim \xi^i$ as before. As in the classical case [see § 1.1.11 and (1.1.15)], we are led to solving a local equation of the form

$$(3.3.4) \qquad\qquad \beta'(t) = \xi_\phi(\beta(t)), \qquad t \in J_\beta$$

where now $\xi_\phi \colon \phi(U) \to \mathbb{F}$, with $\mathbb{F}$ a Fréchet space. Therefore, we are confronted with the problems already discussed in §,2.4. The existence and uniqueness of the solutions depend on appropriate conditions.

The main result here is the following:

Theorem 3.3.3 *Let $M = \varprojlim\{M^i; \mu^{ji}\}_{i,j\in\mathbb{N}}$ be a plb-manifold modelled on a Fréchet space $\mathbb{F} = \varprojlim \mathbb{E}^i$. Assume that every M^i is a Hausdorff space, and $\xi = \varprojlim \xi^i$ is a vector field on M. If there exists a limit chart $\left(\varprojlim x^i, \varprojlim \varphi^i\right)$ at a given point $x_0 = \left(x_0^i\right) \in M$, such that condition $(*)$ of Theorem 2.4.3 is satisfied, then there is a unique integral curve α of ξ with initial condition $\alpha(0) = x_0$.*

Proof Equation (3.3.4) is equivalent to the system of countable equations

$$(3.3.5) \qquad\qquad \left(\beta^i\right)'(t) = \xi_{\phi_i}^i(\beta^i(t)); \qquad i \in \mathbb{N},$$

with $\beta^i(0) = \phi^i(x_0^i)$, where $\beta^i = \mu^i \circ \beta$. By the assumptions and Theorem 2.4.3, there exists a unique solution (β^i), with *all the curves* β^i *defined on the same interval*, say J. Clearly, each $\alpha^i = \left(\phi^i\right)^{-1} \circ \beta^i$ is an integral curve of ξ^i with initial condition x_0^i. Moreover, following the argument of the same Theorem 2.4.3, we obtain the limit $\beta = \varprojlim \beta^i$ which is a solution of (3.3.4) with $\beta(0) = \phi(x_0)$; hence, $\alpha = \phi^{-1} \circ \beta$ is an integrable curve of ξ with $\alpha(0) = x_0$. The uniqueness of α is obvious: If γ

is any other integral curve of ξ with $\gamma(0) = x_0$, then the first part of the proof implies that $\alpha^i = \gamma^i$; hence, $\alpha = \gamma$. The proof is now complete. $\square$

Remarks 3.3.4 1) Regarding the above statement, it should be noted that the assumption that condition $(\ast)$ of Theorem 2.4.3 is satisfied is not an additional requirement, imposed by the differential structure of the manifolds under consideration. It is a necessary condition in order to resolve the involved equations in the context of Fréchet spaces.

2) The same condition guarantees the existence of solutions of (3.3.4) whose domains do not collapse to a single point. Indeed, without it, the factor solutions could have different domains, thus the projective limit of the latter could be a single point.

3.4 Fréchet-Lie groups

Here we extend the mechanism of plb-manifolds to the case of Lie groups modelled on Fréchet spaces. Plb-groups defined below admit always an exponential map. Note that projective limits of Lie groups have been used in the study of the group of diffeomorphisms of a compact manifold (see [Les67], [Omo70]).

Let $\{G^i; g^{ji}\}_{i,j\in\mathbb{N}}$ be a projective system of groups. By definition, the connecting morphisms of the system are group homomorphisms. The limit $G = \varprojlim G^i$ is also a group whose multiplication and inversion are given, respectively, by

$$(x^i) \cdot (y^i) := (x^i \cdot y^i),$$
$$((x^i))^{-1} := ((x^i)^{-1}),$$

for every $x = (x^i)$ and $y = (y^i)$ in G, with $i \in \mathbb{N}$. We denote by

$$g^i : G = \varprojlim G^i \longrightarrow G^i$$

the canonical projection of G to G^i, which are group (homo)morphisms.

If $L_x : G \to G$ is the left translation of G by $x = (x^i) \in G$ and, analogously, $L_y^j : G^j \to G^j$ that of G^j by $y \in G^i$, it is immediate that

(3.4.1) $L_x = \varprojlim L_{x^i}^i$

(3.4.2) $g^i \circ L_x = L_{x^i}^i \circ g^i,$

(3.4.3) $g^{ji} \circ L_y^j = L_{g^{ji}(y)}^i \circ g^{ji}.$

In the remainder of this chapter we are dealing with projective systems of Banach-Lie groups; that is, projective systems of groups satisfying the conditions of Definition 3.1.2. The connecting morphisms $g^{ji} \colon G^j \to G^i$ are morphisms of Banach-Lie groups.

Proposition 3.4.1 *Let* $\{G^i; g^{ji}\}_{i,j\in\mathbb{N}}$ *be a projective system of Banach-Lie groups, where* G^i *is modelled on the Banach space* $\mathbb{G}^i$ *(*$i \in \mathbb{N}$*). Then the plb-manifold* $G = \varprojlim G^i$ *is a Lie group modelled on the Fréchet space* $\mathbb{G} = \varprojlim \mathbb{G}^i$.

Proof By Proposition 3.1.3, G is a Fréchet manifold. On the other hand, the comment at the beginning of the present section implies that the operations of multiplication and inversion of G coincide with the projective limits of their counterparts on the factor Banach-Lie groups. As a result, in virtue of Proposition 3.1.8, the former operations are smooth maps, and G is a Lie group.   □

The group G lb-group is called a **plb-group**. Note that the connecting morphisms $g^{ji} \colon G^j \to G^i$ ($j \geq i$) are, by definition, Lie group morphisms, and so are the canonical projections $g^i \colon G \to G^i$, in virtue of Proposition 3.1.5.

Examples 3.4.2

1. Every Banach-Lie group G is trivially a Fréchet-Lie group by setting $G = \varprojlim G^i$, with $G^i = G$ for all $i \in \mathbb{N}$.

2. Let $\mathbb{F} = \varprojlim \mathbb{E}^i$ be a Fréchet space represented by a projective limit of a countable family of Banach spaces (see Theorem 2.3.8). If $G^i := (\mathbb{E}^i, +)$, then $\mathbb{F} = \varprojlim G^i$ is an (abelian) Fréchet-Lie group.

3. The groups $C^0(\mathbb{R}, \mathbb{R}_+)$ and $C^\infty(M, G)$ (where M is a compact manifold and G a finite dimensional Lie group), defined in Examples 3.1.6, are also Fréchet-Lie groups.

Proposition 3.4.3 *Let* $G = \varprojlim G^i$ *be a Fréchet-Lie group as in the previous statement. If* $\xi = \varprojlim \xi^i$ *is a smooth vector field on* G, *then:*

i) ξ^j *and* ξ^i *are* g^{ji}-*related, for every* $j \geq i$.

ii) ξ *and* ξ^i *are* g^i-*related.*

Moreover, a vector field on G *is left invariant if and only if it is a projective limit of invariant vector fields on the factors, i.e.*

$$\xi \in \mathcal{L}(G) \quad \Leftrightarrow \quad \xi = \varprojlim \xi^i \; : \; \xi^i \in \mathcal{L}(G^i) \; \forall \; i \in \mathbb{N}.$$

Proof Property i) is a result of the analog of (3.3.1), whereas ii) is a consequence of the second diagram of Definition 2.3.4 adapted to the present case.

Take now any $\xi \in \mathcal{L}(G)$. Then ξ corresponds to a unique $v \in T_e G \equiv \varprojlim T_{e^i} G^i$ such that $\xi_e = v \equiv (v^i)$. Clearly, $e \equiv (e^i)$ is the identity element of G and e^i the identity of G^i. Also, in a reverse way, each v^i determines a unique $\xi^i \in \mathcal{L}(G^i)$, with $\xi^i_{e^i} = v^i$, for every $i \in \mathbb{N}$. We shall prove that the collection (ξ^i) is a projective system of vector fields. To this end it suffices to prove the analog of (3.3.1), namely

$$(3.4.4) \qquad Tg^{ji} \circ \xi^j = \xi^i \circ g^{ji}, \qquad j \geq i.$$

Before proving this, we recall that $T_{e^j} g^{ji}$ are the connecting morphisms and $T_e g^i$ the canonical projections of $\varprojlim T_{e^i} G^i$, thus $T_{e^j} g^{ji}(v^j) = v^i$ and $T_e g^i(v) = v^i$ $(i, j \in \mathbb{N}; j \geq i)$. Therefore, for every $x \in G^j$, (3.4.3) implies that

$$\begin{aligned}
T_x g^{ji}(\xi^i(x)) &= T_x g^{ji}\big(T_{e^j} L_x^j(\xi_{e^j}^j)\big) = T_{e^j}(g^{ji} \circ L_x^j)(v^j) \\
&= T_{e^j}\big(L_{g^{ji}(x)}^i \circ g^{ji}\big)(v^j) = T_{g^{ji}(e^j)} L_{g^{ji}(x)}^i\big(T_{e^j} g^{ji}(v^j)\big) \\
&= T_{e^i} L_{g^{ji}(x)}^i(v^i) = T_{e^i} L_{g^{ji}(x)}^i(\xi_{e^i}^i) = \xi(g^{ji}(x)),
\end{aligned}$$

which proves (3.4.4). Consequently, $\varprojlim \xi^i$ exists.

To prove that $\xi = \varprojlim \xi^i$, we need to show that

$$(3.4.5) \qquad Tg^i \circ \xi = \xi^i \circ g^i,$$

according to Proposition 2.3.5. Indeed, for every $x \in G$, working as in the proof of (3.4.4) and applying (3.4.2), we find:

$$\begin{aligned}
T_x g^i(\xi(x)) &= T_x g^i\big(T_e L_x(\xi_e)\big) = T_e(g^i \circ L_x)(\xi_e) \\
&= T_e\big(L_{x^i}^i \circ g^i\big)(v) = T_{g^i(e)} L_{x^i}^i\big(T_e g^i(v)\big) \\
&= T_{g^i(e)} L_{x^i}^i(v^i) = T_{g^i(e)} L_{x^i}^i(\xi_{e^i}^i) \\
&= \xi^i(x^i) = \xi^i(g^i(x)),
\end{aligned}$$

thus we obtain (3.4.5). In conclusion, we have proved that every $\xi \in \mathcal{L}(G)$ is the projective limit of $\{\xi^i \in \mathcal{L}(G^i)\}_{i \in \mathbb{N}}$.

Conversely, if ξ is a vector field of G such that $\xi = \varprojlim \xi^i$, where $\xi^i \in \mathcal{L}(G^i)$, for every $i \in \mathbb{N}$, then necessarily $\xi \in \mathcal{L}(G)$. This is so, for if $x \in G$, then

$$\begin{aligned}
T_e L_x(\xi(e)) &= T_e L_x\left(\varprojlim \xi^i(e)\right) = T_e L_x\left(\big(\xi^i(e^i)\big)_{i \in \mathbb{N}}\right) \\
&= \varprojlim T_{e^i} L_{x^i}^i\left(\big(\xi^i(e^i)\big)_{i \in \mathbb{N}}\right) = \big(T_{e^i} L_{x^i}^i\big(\xi^i(e^i)\big)\big)_{i \in \mathbb{N}} \\
&= \big(\xi^i(x^i)\big)_{i \in \mathbb{N}} = \left(\varprojlim \xi^i\right)(e) = \xi(x),
\end{aligned}$$

which completes the proof. $\qquad\qquad\square$

Theorem 3.4.4 *For $G = \varprojlim G^i$ as before, its Lie algebra $\mathcal{L}(G)$ can be also realized as a projective limit, i.e.*

$$\mathcal{L}(G) \equiv \varprojlim \mathcal{L}(G^i),$$

by an isomorphism of Lie algebras, where $\mathcal{L}(G^i)$ is the Lie algebra of G^i ($i \in \mathbb{N}$). Hence, $\mathcal{L}(G)$ is a Fréchet-Lie algebra.

Proof From the identification (1.2.3), equality (3.2.3), and the proof of Proposition 3.4.3, we obtain the linear isomorphisms

(3.4.6) $\mathcal{L}(G) \equiv T_e G \equiv \varprojlim T_{e^i} G^i \equiv \varprojlim \mathcal{L}(G^i),$

thus $\mathcal{L}(G)$ is a Fréchet space.

It remains to show that $\mathcal{L}(G)$ and $\varprojlim \mathcal{L}(G^i)$ are isomorphic Lie algebras. To this end we first show that $\mathfrak{g} := T_e G$ and $\varprojlim \mathfrak{g}^i := \varprojlim T_{e^i} G^i$ are isomorphic Lie algebras. Hence we need to define an appropriate bracket on $\varprojlim \mathfrak{g}^i$. As one may guess, the latter will be the projective limit of the brackets $[\,,\,]^i$ of the factor algebras $\mathfrak{g}^i$. To ensure the existence of such a limit of brackets, it suffices to prove that the following the diagram is commutative:

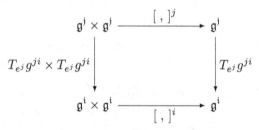

Indeed, assume that u^j, v^j are any elements of $\mathfrak{g}^j$, and ξ^j, η^j the corresponding (left invariant) vector fields in $\mathcal{L}(G^j)$. We denote by $\xi^i, \eta^i \in \mathcal{L}(G^i)$ the respective fields of the $T_{e^j} g^{ji}(u^j)$ and $T_{e^j} g^{ji}(v^j)$ in $\mathfrak{g}^i$. As in the proof of Proposition 3.4.3, we obtain equality (3.4.4), thus ξ^j and ξ^i are g^{ji}-related, and similarly for η^j, η^i. Therefore,

$$\left[T_{e^j} g^{ji}(u^j), T_{e^j} g^{ji}(v^j)\right]^i = \left[\xi^i, \eta^i\right]^i(e^i) =$$
$$= \left(\left[\xi^i, \eta^i\right]^i \circ g^{ji}\right)(e^j) = \left(Tg^{ji} \circ \left[\xi^j, \eta^j\right]^j\right)(e^j) = T_{e^j} g^{ji}\left(\left[u^j, v^j\right]^j\right).$$

This proves the desired commutativity and, consequently, the existence of $\varprojlim [\,,\,]^j$.

The next step is to show that $T_e g^i \colon \mathfrak{g} \to \mathfrak{g}^i$ is a morphism of Lie algebras with respect to the corresponding brackets $[\,,\,]$ and $[\,,\,]^i$: For arbitrary vectors $u, v \in \mathfrak{g}$, we consider the corresponding fields $\xi, \eta \in$

$\mathcal{L}(G)$. Also, we denote by $\xi^i, \eta^i \in \mathcal{L}(G^i)$ the fields corresponding to $T_e g^i(u), T_e g^i(u) \in \mathfrak{g}^i$. Then, working as in the proof of the commutativity of the previous diagram, we see that $[\xi, \eta]$ and $[\xi^i, \eta^i]^i$ are g^i-related. Therefore,

$$[T_e g^i(u), T_e g^i(u)] = [\xi^i, \eta^i]^i (e) =$$

$$= \left([\xi^i, \eta^i]^i \circ g^i\right)(e) = T_e g\left([\xi^i, \eta^i](e)\right) = T_e g\left([\xi^i_e, \eta^i_e]\right) = T_e g([u, v]);$$

that is, $T_e g^i$ is a morphism of Lie algebras. Taking now the limits, we have that

$$\varprojlim T_e g^i \colon T_e G \longrightarrow \varprojlim T_{e^i} G^i$$

is also a morphism of Lie algebras, which gives an isomorphism after the identification $T_e G \equiv \varprojlim T_{e^i} G^i$. As a matter of fact,

$$[\,,\,] \equiv \varprojlim [\,,\,]^i.$$

Finally, (3.4.6) extends the previous Lie algebra isomorphism to one between $\mathcal{L}(G)$ and $\varprojlim \mathcal{L}(G^i)$.   □

We come now to the question of the existence of an exponential map. Naturally, this is related to the solution of appropriate differential equations which determine the integral curves of the left invariant vector fields of the Lie groups at hand. As we have already mentioned in §,2.4, we lack a general solvability theory of differential equations in Fréchet or, more generally, non-Banach locally convex spaces. This deficiency has led many authors to propose various approaches to define a *kind* of exponential map (in this respect see the instructive notes by K. H. Neeb [Nee06]). Here, following the main methodology of this work, we prove the existence of an exponential map in the classical sense.

Theorem 3.4.5 *Let $G = \varprojlim G^i$ be a Fréchet-Lie group as in Proposition 3.4.1. Then G admits an exponential map $\exp_G \colon T_e G \to G$ such that*

$$\exp_G \equiv \varprojlim \exp_{G^i},$$

if $\exp_{G^i}$ $(i \in \mathbb{N})$ are the exponential maps of the factor groups.

Before the proof we need the following auxiliary result:

Lemma 3.4.6 *Let $G = \varprojlim G^i$ be a Fréchet-Lie group and $\xi \in \mathcal{L}(G)$, thus (by Proposition 3.4.3) $\xi = \varprojlim \xi^i$, with $\xi^i \in \mathcal{L}(G^i)$. Let also an arbitrary point $x_0 \equiv \left(x_0^i\right) \in G$. If, for every $i \in \mathbb{N}$, $\alpha^i \colon \mathbb{R} \to G^i$ is the integral curve of ξ^i, with initial condition $\alpha^i(0) = x_0^i$, then $\alpha := \varprojlim \alpha^i$*

exists and $\alpha \colon \mathbb{R} \to G$ *is the integral curve of* ξ, *with initial condition* $\alpha(0) = x_0$.

Proof For every $j \geq i$,

$$(g^{ji} \circ \alpha^j)(0) = g^{ji}(x_0^j) = x_0^i.$$

On the other hand, the velocity vector of $g^{ji} \circ \alpha^j \colon \mathbb{R} \to G^i$ at any $t \in \mathbb{R}$ yields [see also equality (1.1.11) and the ensuing comments]:

$$\left(g^{ji} \circ \alpha^j\right)^{\cdot}(t) = T_e(g^{ji} \circ \alpha^j)\left(\frac{d}{dt}\Big|_t\right)$$

$$= T_e g^{ji}\left((\alpha^j)^{\cdot}(t)\right) = T_e g^{ji}\left(\xi^j(\alpha^j(t))\right)$$

or, in virtue of (3.4.4),

$$= \left(T_e g^{ji} \circ \xi^j\right)(\alpha^j(t)) = \left(\xi^i \circ g^{ji}\right)(\alpha^j(t))$$

$$= \xi^i\left((g^{ji} \circ \alpha^j)(t)\right).$$

Therefore, $g^{ji} \circ \alpha^j$ is an integral curve of ξ^i with initial condition x_0^i; hence, $g^{ji} \circ \alpha^j = \alpha^i$, which means that $\left(\alpha^i\right)_{i \in \mathbb{N}}$ is a projective system. We set $\alpha := \varprojlim \alpha^i$.

We claim that α is the integral curve of ξ with initial condition $\alpha(0) = x_0$. Indeed,

$$\alpha(0) = \left(\varprojlim \alpha^i\right)(0) = \left(\alpha^i(0)\right)_{i \in \mathbb{N}} = \left(x_0^i\right)_{i \in \mathbb{N}} = x_0.$$

Since both $\dot{\alpha}(t)$ and $\xi(\alpha(t))$ are elements of $T_{\alpha(t)}G \equiv \varprojlim T_{\alpha^i(t)}G^i$, equality $\dot{\alpha}(t) = \xi(\alpha(t))$ holds if and only if the previous vectors have the same projections (components). Here, by Corollary 3.2.6, the canonical projections are $T_{\alpha(t)}g^i \colon T_{\alpha(t)}G \to T_{\alpha^i(t)}G^i$. Therefore,

$$\dot{\alpha}(t) = \xi(\alpha(t))$$

$$\Leftrightarrow \quad T_{\alpha(t)}g^i(\dot{\alpha}(t)) = T_{\alpha(t)}g^i(\xi(\alpha(t)))$$

$$\Leftrightarrow \quad T_{\alpha(t)}g^i\left(T_{\alpha(t)}\left(\frac{d}{dt}\Big|_t\right)\right) = T_{\alpha(t)}g^i\left(\varprojlim \xi^i(\alpha(t))\right)$$

$$\Leftrightarrow \quad T_t(g^i \circ \alpha)\left(\frac{d}{dt}\Big|_t\right) = T_{\alpha(t)}g^i\left(\left(\xi^i(\alpha^i(t))\right)_{i \in \mathbb{N}}\right)$$

$$\Leftrightarrow \quad (g^i \circ \alpha)^{\cdot}(t) = \left(\xi^i \circ \alpha^i\right)(t)$$

$$\Leftrightarrow \quad (\alpha^i)^{\cdot}(t) = (\xi^i \circ \alpha^i)(t),$$

which is true because α^i is an integral curve of ξ. $\qquad\qquad\square$

We are now in a position to give the

Proof of Theorem 3.4.5. Let an arbitrary vector $v \equiv (v^i) \in T_e G \equiv$

$\varprojlim T_e^i G^i$ [recall that $e = (e^i)$ is the identity of G and e^i the one of G^i $(i \in \mathbb{N})$]. If $\xi \in \mathcal{L}(G)$ is the left invariant vector field of G corresponding to v, then $\xi = (\xi^i)_{i \in \mathbb{N}}$, where $\xi^i \in \mathcal{L}(G^i)$ (see Proposition 3.4.3). By the preceding lemma, we obtain

$$\exp_G(v) = \alpha(1) = \left(\varprojlim \alpha^i \right)(1) =$$

$$= (\alpha^i(1))_{i \in \mathbb{N}} = (\exp_{G^i}(v^i))_{i \in \mathbb{N}} = \left(\varprojlim \exp_{G^i} \right)(v),$$

thus concluding the proof. □

Remark 3.4.7 It should be noted that $\exp_G$ is not necessarily a local diffeomorphism at $0 \in T_e G$ although each $\exp_G^i$ is, for every $i \in \mathbb{N}$. In fact, for each $i \in \mathbb{N}$, there are open neighborhoods U^i of $0 \in T_{e^i} G^i$, and N^i of $e^i \in G^i$, such that $\exp_G^i \colon U^i \to N^i$ is a diffeomorphism. Then $\exp_G |_{\varprojlim U^i} \colon \varprojlim U^i \to \varprojlim N^i$ is defined, but $\varprojlim U^i$ is not always an open neighborhood of $0 \in T_e G$. However, the existence of an exponential map, which is a local diffeomorphism at the identity of the group, characterizes *commutative* Fréchet-Lie groups in the sense of the next result.

Theorem 3.4.8 ([Gal96]) *Let G be a commutative Fréchet-Lie group, and assume that there is a smooth map $f \colon T_e G \to G$ satisfying the properties:*

i) $f(u + v) = f(u) \cdot f(v)$, for every $u, v \in T_e G$;

ii) There are open neighborhoods $V_0 \subseteq T_e G$, $N_e \subseteq G$ of 0 and e, respectively, such that $f \colon V_0 \to N_e$ is a diffeomorphism.

Then G is a projective limit of Banach-Lie groups.

3.5 Equations with Maurer-Cartan differential

We briefly discuss the analog of equation (1.2.7) in the context of Fréchet manifolds and Fréchet-Lie groups.

Let B be *Banach* space and $G = \varprojlim G^i$ a Fréchet-Lie group with Lie algebra $\mathfrak{g} = \varprojlim \mathfrak{g}^i$. As in § 1.4.4 [constructions (c) and (e)], we consider the the linear map bundle $L(B, \mathfrak{g})$. Anticipating the general theory of projective limit vector bundles studied in Chapter 5, we prove:

Proposition 3.5.1 *The following properties hold:*

i) $\{(L(B, \mathfrak{g}^i); p^{ji})\}_{i,j \in \mathbb{N}}$ is a projective system of Banach vector bundles, where $p^{ji}(f) := T_{e^j} g^{ji} \circ f$.

ii) $\varprojlim L(B, \mathfrak{g}^i) \equiv L(B, \mathfrak{g})$, *and the canonical projections are*

$$p_i \colon L(B, \mathfrak{g}) \longrightarrow L(B, \mathfrak{g}^i) \colon f \mapsto p_i(f) = T_e g_i \circ f.$$

Regarding now differential forms with values in $\mathfrak{g}$, we have:

Proposition 3.5.2 *If $\theta \in \Lambda^1(B, \mathfrak{g})$, there are $\theta_i \in \Lambda^1(B, \mathfrak{g}^i)$, such that $\theta = \varprojlim \theta^i$. Moreover, for every vector fields $X, Y \in \mathcal{X}(B)$, we have:*

(i) $\theta(X) = \varprojlim \left(\theta^i(X) \right)$,

(ii) $Y(\theta(X)) = \varprojlim \left(Y(\theta^i(X)) \right)$,

(iii) $d\theta(X, Y) = \varprojlim \left(d\theta^i(X, Y) \right)$,

(iv) $[\theta, \theta](X, Y) = \varprojlim \left([\theta^i, \theta^i]^i(X, Y) \right)$.

within appropriate isomorphisms.

Proof For every $i \in \mathbb{N}$, we define the map

$$\theta^i \colon B \to L(B, \mathfrak{g}^i) \colon x \mapsto T_e g^i \circ \theta_x,$$

where $g^i \colon G \to G^i$ ($i \in \mathbb{N}$) are the canonical projections of $G = \varprojlim G^i$. It is clear that $\theta^i \in \Lambda^1(B, \mathfrak{g}^i)$, for every $i \in \mathbb{N}$.

Easy computations show that the diagrams (for all $i, j \in \mathbb{N}$, $j \geq i$)

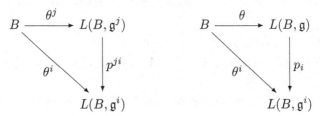

are commutative. The left of them implies that the limit $\varprojlim \theta^i$ exists, while the right one and Proposition 2.3.5 imply that $\theta = \varprojlim \theta^i$.

To prove (i), we check that

$$\left(T_{e^j} g^{ji} \circ \theta^j(X) \right)(x) = T_{e^j} g^{ji} \left(\theta^j_x(X_x) \right) = \theta^i(X)(x), \qquad x \in B.$$

This implies the existence of the $\varprojlim \left(\theta^i(X) \right)$. Similarly, $T_e g_i \circ \theta(X) = \theta^i(X)$, thus (again by Proposition 2.3.5) $\theta(X) = \varprojlim \left(\theta^i(X) \right)$.

For the proof of (ii) we note that

$$(3.5.1) \qquad \left(T_{e^j} g^{ji} \circ Y(\theta^j(X)) \right)(x) = \left(T_{e^j} g^{ji} \circ T_x(\theta^j(X)) \right)(Y_x),$$

for every $x \in B$. Also, from (i) and Proposition 3.2.2,

$$T_x(\theta(X)) = T_x \left(\varprojlim \theta^i(X) \right) = \varprojlim T_x \left(\theta^i(X) \right),$$

thus $T_x\left(\theta^i(X)\right) = T_{e^j}g^{ji}\left(T_x\theta^j(X)\right)$. As a result, (3.5.1) transforms into

$$\left(T_{e^j}g^{ji} \circ Y(\theta^j(X))\right)(x) = T_x\left(\theta^i(X)\right)(Y_x) = Y(\theta^i(X))(x);$$

that is, $T_{e^j}g^{ji} \circ Y(\theta^j(X) = Y(\theta^i(X))$, for all $i,j \in \mathbb{N}$ with $j \geq i$. This ensures the existence of $\varprojlim Y(\theta^i(X))$. On the other hand, by similar arguments, $T_e g^i \circ Y(\theta(X) = Y(\theta^i(X))$, for every $i \in \mathbb{N}$. Hence, we obtain property (ii).

Property (iii) is a direct consequence of (i):

$$\begin{aligned}
d\theta(X,Y) &= X(\theta(Y)) - Y(\theta(X)) - \theta([X,Y]) \\
&= \varprojlim X(\theta^i(Y)) - \varprojlim Y(\theta^i(X)) - \varprojlim\left(\theta^i([X,Y])\right) \\
&= \varprojlim\left(d\theta^i(X,Y)\right).
\end{aligned}$$

For (iv) we apply (i), along with Theorem 3.4.4, as follows:

$$\begin{aligned}
[\theta,\theta](X,Y) &= [\theta(X),\theta(Y)] = \left[\varprojlim\theta(X),\varprojlim\theta(Y)\right] \\
&= \varprojlim[\theta^i(X),\theta^i(Y)]^i = \varprojlim\left([\theta^i,\theta^i]^i(X,Y)\right). \qquad \square
\end{aligned}$$

For simplicity, in what follows, we denote by $D\colon \mathcal{C}^\infty(B,G) \to \Lambda^1(B,\mathfrak{g})$ the operator induced by the *right* Maurer-Cartan differential (cf. § 1.2.6). We intend to study the analog of (1.2.7) in the present setting; namely, equation

$$(3.5.2) \qquad\qquad Dx = \theta,$$

where $\theta \in \Lambda^1(B,\mathfrak{g})$.

Lemma 3.5.3 *A differential form $\theta \in \Lambda^1(B,\mathfrak{g})$ is integrable; that is, $d\theta = \frac{1}{2}[\theta,\theta]$, if and only if every θ^i $(i \in \mathbb{N})$ is integrable.*

Proof Immediate consequence of Proposition 3.5.2. $\qquad\square$

Theorem 3.5.4 *Let an arbitrary $(x_0,g_0) \in B \times G$. Equation (3.5.2) has a unique solution $f\colon U \to G$ (U open neighborhood of x_0) with $f(x_0) = g_0$ if and only if θ is integrable.*

Proof We first prove the statement for B simply connected. Assuming that θ is integrable, Lemma 3.5.3 implies that each equation (in the Banach framework)

$$(3.5.3) \qquad\qquad Dx^i = \theta^i; \qquad i \in \mathbb{N},$$

has a unique (global) solution $f^i\colon B \to G^i$ with initial condition $f^i(x_0) =$

$g^i(g_0) := g_0^i$. Because the connecting morphisms $g^{ji}: G^j \to G^i$ are group homomorphisms (see the beginning of § 3.4, we easily check that

$$D\left(g^{ji} \circ f^j\right)(x)) = T_{e^j} g^{ji} \circ Df^j(x) = T_{e^j} g^{ji} \circ \theta_x^j = \theta_x^i. \qquad x \in B.$$

Since $(g^{ji} \circ f^j)(x_0) = g^i(g_0)$, it follows that $g^{ji} \circ f^j$ is also a solution of (3.5.3) with the same initial condition, thus $g^{ji} \circ f^j = f^i$, for every $j \geq i$. As a result $f := \varprojlim f^i: B \to G$ exists and determines a smooth map in the sense of our framework (see also Proposition 3.1.8). On the other hand, if R_g $(g \in G)$ denotes the right translation of G by g, and $R_{g^i}^i$ $(g^i \in G^i)$, denotes the right translation of G^i by g^i, then the analog of (3.4.1) implies

$$\begin{aligned} Df(x) &= T_{f(x)} R_{f(x)^{-1}} \circ T_x f \\ &= \varprojlim \left(T_{f^i(x)} R_{f^i(x)^{-1}}^i \circ T_x f^i \right) \\ &= \varprojlim Df^i(x) = \varprojlim \theta^i(x) = \theta(x), \end{aligned}$$

for every $x \in B$. Hence, f is a solution of (3.5.2) such that

$$f(x_0) = \left(f^i(x_0)\right)_{i \in \mathbb{N}} = \left(g^i(g_0)\right)_{i \in \mathbb{N}} = g_0.$$

We shall show that f is the unique solution of (3.5.2) with the given initial condition. Indeed, assume that $\varphi: B \to G$ is another solution such that $\varphi(x_0) = g_0$. Then, $\varphi = \varprojlim \varphi^i$, where $\varphi^i = g^i \circ \varphi: B \to G^i$. We check that $D\varphi^i = \theta^i$ as follows: Since $D\varphi = \theta$, we have that

$$T_{e^i} g^i \circ T_{\varphi(x)} R_{\varphi(x)^{-1}} \circ T_x \varphi = T_{e^i} g^i \circ \theta_x,$$

or, by Corollary 3.2.6 applied to the case of G,

$$T_{\varphi^i(x)} R_{\varphi^i(x)^{-1}}^i \circ T_x \varphi^i = \theta_x^i; \qquad x \in B,$$

thus $D\varphi^i = \theta^i$. In addition, $\varphi^i(x_0) = g^i(g_0)$ holds for all $i \in \mathbb{N}$. Consequently, by the uniqueness of the solutions of (3.5.3), $\varphi^i = f^i$, for all $i \in \mathbb{N}$; hence, $\varphi = f$ as claimed.

Conversely, assume that (3.5.2) has a unique solution f with $f(x_0) = g_0$. Setting $f^i := g^i \circ f$ $(i \in \mathbb{N})$, we have that $f = \varprojlim f^i$. Using similar arguments as before, we check that $Df^i = \theta^i$, $i \in \mathbb{N}$. Therefore, θ^i is integrable for every $i \in \mathbb{N}$, which, by Lemma 3.5.3, implies the integrability of θ.

We consider now the case where B is not necessarily simply connected. Denoting by $(\widetilde{B}, \widetilde{\pi}, B)$ the universal covering space of B, we set $\widetilde{\theta} := \widetilde{\pi}^* \theta$. It is immediate that $\widetilde{\theta} = \varprojlim \widetilde{\theta}^i$. If θ is integrable, so are the forms

θ^i and $\widetilde{\theta^i}$, for every $i \in \mathbb{N}$. Thus each lifted equation $Dz^i = \widetilde{\theta^i}$ [see (1.2.8)] has a unique (global) solution $\widetilde{f^i} \colon \widetilde{B} \to G^i$ with $\widetilde{f^i}(\widetilde{x}_0) = g^i(g_0)$, where $\widetilde{x}_0$ is an arbitrarily chosen point of $\widetilde{B}$ with $\widetilde{p}(\widetilde{x}_0) = x_0$. By the structure of the universal covering, there is an open neighborhood U of x_0 and a connected component V of $\widetilde{\pi}^{-1}(U)$ such that $\widetilde{\pi} \colon V \to U$ is a diffeomorphism. Setting, $f^i := \widetilde{f^i} \circ \widetilde{\pi}^{-1}|_U$, $i \in \mathbb{N}$, we check that $Df^i = \theta^i$ and $f^i(x_0) = g^i(g_0)$. As in the first case of the proof, $f := \varprojlim f^i \colon U \to G$ exists and is the unique solution of (3.5.2) with $f(x_0) = g_0$.

For the converse we proceed as in its counterpart in the first case. $\square$

In conjunction with the terminology of fundamental solution (of the lifted equation on the universal covering) and the monodromy homomorphism of the original equation, induced in the end of § 1.2.6, we obtain the following byproduct of the proof of Theorem 3.5.4:

Corollary 3.5.5 *Let $\theta \in \Lambda^1(B, \mathfrak{g})$ be an integrable form such that $\theta = \varprojlim \theta^i$, with $\theta^i \in \Lambda^1(B, \mathfrak{g}^i)$, $i \in \mathbb{N}$. Then:*

(i) $F_\theta = \varprojlim F_{\theta^i}$,

(ii) $\theta^\# = \varprojlim \theta^{i\#}$.

3.6 Differential forms

Since differential forms are smooth sections of map bundles, arbitrary forms on Fréchet manifolds, with values in Fréchet spaces, are not necessarily represented as projective limits of forms on Banach manifolds. Nevertheless, as we explain below (Proposition 3.6.3), the converse is partially true: 'point-wise' projective limits of ordinary vector valued differential forms on Banach manifolds yield differential forms on Fréchet manifolds. On the other hand, connection forms on limit principal bundles (treated in § 4.2) provide an important example of differential forms that can always be represented—in a point-wise fashion—as projective limits of (connection) forms on Banach principal bundles.

In view of concrete applications in Chapter 4, throughout this section we consider:

- A Fréchet manifold $M = \varprojlim M^i$, derived from the projective system of Banach manifolds $\{M^i; \mu^{ji}\}_{i,j \in \mathbb{N}}$, as in § 3.1. We recall that M is modelled on $\mathbb{F} = \varprojlim \mathbb{E}^i$, where $\mathbb{E}^i$ is the model of M^i ($i \in \mathbb{N}$), and $\{\mathbb{E}^i; \rho^{ji}\}_{i,j \in \mathbb{N}}$ is the projective system generating $\mathbb{F}$.

- A Fréchet-Lie group $G = \varprojlim G^i$, where $\{G^i; g^{ji}\}_{i,j \in \mathbb{N}}$ is a projective system of Banach-Lie groups. We already know (see Theorem 3.4.4) that the Lie algebras $\mathfrak{g}^i \equiv \mathcal{L}(G^i)$ of G^i ($i \in \mathbb{N}$) determine a projective system such that $\mathfrak{g} \equiv \mathcal{L}(G) = \varprojlim \mathfrak{g}^i$. The connecting morphisms of the latter system are the Lie algebra morphisms $\bar{g}^{ji} \equiv T_{e^j} g^j : \mathfrak{g}^j \to \mathfrak{g}^i$, $(j \geq i)$, induced by g^{ji} after the identification (1.2.3).

Then we define the set

$$L(TM, \mathfrak{g}) := \bigcup_{x \in M} \mathcal{L}(T_x M, \mathfrak{g})$$

and the natural projection

$$L : L(TM, \mathfrak{g}) \longrightarrow M : f \mapsto L(f) := x, \text{ if } f \in \mathcal{L}(T_x M, \mathfrak{g}).$$

Proposition 3.6.1 $L(TM, \mathfrak{g})$ *is a smooth manifold modelled on the locally convex space* $\mathcal{L}(\mathbb{F}, \mathfrak{g})$.

Proof Let (U, ϕ) be a chart of M and $(\pi^{-1}(U), \Phi)$ the induced chart of the tangent bundle (TM, M, π). We define the chart $(L^{-1}(U), \widehat{\Phi})$ of $L(TM, \mathfrak{g})$, where the map

$$\widehat{\Phi} : L^{-1}(U) \longrightarrow \phi(U) \times \mathcal{L}(\mathbb{F}, \mathfrak{g})$$

is given by

$$\widehat{\Phi}(f) = \left(\phi(L(f)), f \circ \overline{\phi}^{-1}_{L(f)} \right) = \left(\phi(x), f \circ \overline{\phi}^{-1}_x \right),$$

if $f \in \mathcal{L}(T_x M, \mathfrak{g})$. Recall that $\overline{\phi}_x^{-1} : T_x M \to \mathbb{F}$ is the linear isomorphism induced by ϕ.

Let (V, ψ) be another chart of M, compatible with (U, Φ), $U \cap V \neq \emptyset$, and let $(L^{-1}(V), \widehat{\Psi})$ the corresponding chart of $L(TM, \mathfrak{g})$. To verify the compatibility of $(L^{-1}(U), \widehat{\Phi})$ and $(L^{-1}(V), \widehat{\Psi})$, we evaluate

$$\widehat{\Psi} \circ \widehat{\Phi}^{-1} : \phi(U \cap V) \times \mathcal{L}(\mathbb{F}, \mathfrak{g}) \longrightarrow \psi(U \cap V) \times \mathcal{L}(\mathbb{F}, \mathfrak{g})$$

at any point (a, g) of the domain:

$$\left(\widehat{\Psi} \circ \widehat{\Phi}^{-1} \right)(a, g) = \widehat{\Psi} \left(\phi^{-1}(a), g \circ \overline{\phi}_{\phi^{-1}(a)} \right)$$

$$= \left((\psi \circ \phi^{-1})(a), g \circ \overline{\phi}_{\phi^{-1}(a)} \circ \overline{\psi}^{-1}_{\phi^{-1}(a)} \right)$$

[see §1.1.4] $$= \left((\psi \circ \phi^{-1})(a), g \circ D(\phi \circ \psi^{-1})((\psi \circ \phi^{-1})(a)) \right),$$

or, setting $F := \phi \circ \psi^{-1}$,

$$\left(\widehat{\Psi} \circ \widehat{\Phi}^{-1} \right)(a, g) = \left(F^{-1}(a), g \circ DF(F^{-1}(a)) \right).$$

Hence, for the desired compatibility, it suffices to show that

(3.6.1) $\quad \chi \colon \phi(U \cap V) \times \mathcal{L}(\mathbb{F}, \mathfrak{g}) \longrightarrow \mathcal{L}(\mathbb{F}, \mathfrak{g}) \colon (a, g) \mapsto g \circ DF\left(F^{-1}(a)\right)$

is a smooth map. To this end, we exploit the structure of plb-manifolds. In this regard, we can choose charts of M such that

$$(U, \phi) = \left(\varprojlim U^i, \varprojlim \phi^i\right).$$

Then $\overline{\phi}_x = \varprojlim \overline{\phi}^i_{x^i}$, for every $x = (x^i) \in M$, $\widehat{\Phi} = \varprojlim \widehat{\Phi}^i$, where $(x^i, \widehat{\Phi}^i)$ is the analogous chart of the Banach bundle $L(TM^i, \mathfrak{g}^i)$ [see §1.4.4(c)], and $F = \varprojlim F^i$, with

$$F^i := \phi^i \circ (\psi^i)^{-1} \colon \mathbb{E}^i \supseteq \psi^i(U^i \cap V^i) \longrightarrow \phi^i(U^i \cap V^i) \subseteq \mathbb{E}^i.$$

Consequently, since F is a diffeomorphism, so is the map

$$\phi(U \cap V) \ni a \longmapsto DF(F^{-1}(a)) \in \mathcal{L}is(\mathbb{F}) \subset \mathcal{L}(\mathbb{F})$$

in virtue of Proposition 2.3.12. The commutative diagram below clarifies the latter argument.

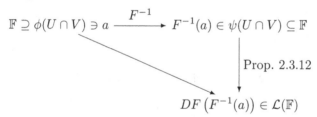

Now, because $DF\left(F^{-1}(a)\right)$ is a projective limit of corresponding differentials, it follows in particular that $DF\left(F^{-1}(a)\right) \in \mathcal{L}_I(\mathbb{F})$ [see (2.3.11) and Proposition 2.3.13]. Therefore, applying the composition map

$$\mathrm{comp} \colon \mathcal{L}(\mathbb{F}, \mathfrak{g}) \times \mathcal{L}_I(\mathbb{F}) \longrightarrow \mathcal{L}(\mathbb{F}, \mathfrak{g}) \colon (g, h) \mapsto g \circ h,$$

which is continuous bilinear, hence smooth [see Proposition 2.3.14 and Remark 2.2.4(2)], we conclude that (3.6.1) is smooth, by which we prove the compatibility of the aforementioned charts and the existence of a smooth structure on $L(TM, \mathfrak{g})$. □

Corollary 3.6.2 *The triplet $(L(TM, \mathfrak{g}), M, L)$ is a locally trivial fibration of fibre type $\mathcal{L}(\mathbb{F}, \mathfrak{g})$*

As usual, the smooth sections of $\mathcal{L}(TM, \mathfrak{g})$ are the $\mathfrak{g}$-valued 1-forms on M and their set is denoted by $\Lambda^1(M, \mathfrak{g})$.

In preparation of the connection forms, treated in §4.2, we prove the following.

Proposition 3.6.3   *Let $\theta^i \in \Lambda^1(M^i, \mathfrak{g}^i)$, $i \in \mathbb{N}$, be differential forms such that the projective limit $\varprojlim \theta^i(x^i)$ exists, for every $x = (x^i) \in M$. Then the map $\theta \colon M \to L(TM, \mathfrak{g})$, given by $\theta(x) \equiv \theta_x := \varprojlim \theta^i(x^i)$, is a $\mathfrak{g}$-valued 1-form on M.*

Proof   In virtue of Proposition 3.2.2, $L \circ \theta = \mathrm{id}_M$; thus we need only to show the smoothness of θ. For this purpose, we modify the pattern of the proof of Proposition 2.3.12. More precisely, using the limit charts (in the proof) of Proposition 3.6.1, we consider the composite map $\widehat{\Phi} \circ \theta|_U$, shown also in the diagram:

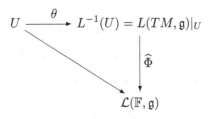

Hence, for every $x = (x^i) \in U = \varprojlim U^i$

$$
\begin{aligned}
(\widehat{\Phi} \circ \theta)(x) = \widehat{\Phi}(\theta_x) &= \widehat{\Phi}\left(\varprojlim \theta^i(x^i) \right) \\
&= \left(\phi(x), \left(\varprojlim \theta^i(x^i) \right) \circ \overline{\phi}_x^{-1} \right) \\
&= \left(\phi(x), \varprojlim \left(\theta^i(x^i) \circ \overline{\phi}_x^{-1} \right) \right).
\end{aligned}
$$
(3.6.2)

The first component of the last equality in (3.6.2) is smooth. For the smoothness of the second one we define the maps

$$
R^i \colon U^i \longrightarrow \mathcal{L}(\mathbb{E}^i, \mathfrak{g}^i) \colon x^i \mapsto \theta^i(x^i) \circ \left(\overline{\phi}_{x^i}^i \right)^{-1},
$$
(3.6.3)

and [see also (2.3.5)]

$$
Q^i \colon U^i \longrightarrow \mathcal{H}^i(\mathbb{F}, \mathfrak{g}) \colon
$$

$$
x^i \longmapsto \left(R^1\big(\mu^{i1}(x^i) \big), R^2\big(\mu^{i2}(x^i) \big), \dots, R^i(\mu^i) \right).
$$
(3.6.4)

The map (3.6.3) is smooth because $\theta^i \in \Lambda^1(M^i, \mathfrak{g}^i)$, thus $\theta^i(x^i) \circ \left(\overline{\phi}_{x^i}^i \right)^{-1}$ coincides with the second component of $\Phi^i \circ \theta^i$. Also, each Q^i takes values in $\mathcal{H}^i(\mathbb{F}, \mathfrak{g})$ since the diagram

$$
\begin{array}{ccc}
\mathbb{E}^j & \xrightarrow{\;R^j(\mu^{ij}(x^i))\;} & \mathfrak{g}^j \\
{\scriptstyle \rho^{jk}}\downarrow & & \downarrow{\scriptstyle \bar{g}^{jk}} \\
\mathbb{E}^i & \xrightarrow[\;R^k(\mu^{ik}(x^i))\;]{} & \mathfrak{g}^k
\end{array}
$$

commutes, for every $i, j, k \in \mathbb{N}$ with (caution !) $i \geq j \geq k$. Indeed, the existence of $\varprojlim \theta^i(x^i)$ implies that

$$\bar{g}^{ij} \circ \theta^j(\mu^{ij}(x^i)) = \theta^k(\mu^{ik}(x^i)) \circ T_{\mu^{ij}(x^i)}\mu^{jk};$$

therefore,

$$
\begin{aligned}
\bar{g}^{jk} \circ R^j\big(\mu^{ij}(x^i)\big) &= \bar{g}^{jk} \circ \theta^j\big(\mu^{ij}(x^i)\big) \circ \big(\overline{\phi^j}_{\mu^{ij}(x^i)}\big)^{-1} \\
&= \theta^k\big(\mu^{ik}(x^i)\big) \circ T_{\mu^{ij}(x^i)}\mu^{jk} \circ \big(\overline{\phi^j}_{\mu^{ij}(x^i)}\big)^{-1} \\
&= \theta^k\big(\mu^{ik}(x^i)\big) \circ \big(\overline{\phi^k}_{\mu^{ik}(x^i)}\big)^{-1} \circ \rho^{jk} \\
&= R^k\big(\mu^{ik}(x^i)\big) \circ \rho^{jk}.
\end{aligned}
$$

On the other hand, if

$$h^{ji}\colon \mathcal{H}^j(\mathbb{F}, \mathfrak{g}) \to \mathcal{H}^i(\mathbb{F}, \mathfrak{g})\colon (f^1, f^2, \dots, f^j) \mapsto (f^1, f^2, ..., f^i); \qquad j \geq i,$$

it follows that

$$
\begin{aligned}
(h^{ji} \circ Q^j)(x^j) &= h^{ji}\Big(R^1\big(\mu^{j1}(x^j)\big), R^2\big(\mu^{j2}(x^j)\big), \dots, R^j(x^j)\Big) \\
&= \Big(R^1\big(\mu^{j1}(x^j)\big), R^2\big(\mu^{j2}(x^j)\big), \dots, R^i\big(\mu^{ji}(x^j)\big)\Big) \\
&= \Big(R^1\big(\mu^{i1}\big(\mu^{ji}(x^j)\big)\big), R^2\big(\mu^{i2}\big(\mu^{ji}(x^j)\big)\big), \dots, R^i\big(\mu^{ji}(x^j)\big)\Big) \\
&= \big(Q^i \circ \mu^{ji}\big)(x^j);
\end{aligned}
$$

hence, the map

$$Q = \varprojlim Q^i\colon U = \varprojlim U^i \longrightarrow \mathcal{H}(\mathbb{F}, \mathfrak{g}) = \varprojlim \mathcal{H}^i(\mathbb{F}, \mathfrak{g})$$

exists and is smooth.

Finally, applying the continuous linear embedding

$$\varepsilon\colon \mathcal{H}(\mathbb{F}, \mathfrak{g}) \longrightarrow \mathcal{L}(\mathbb{F}, \mathfrak{g})\colon (f^i) \mapsto \varprojlim f^i,$$

along with the identifications of Example 2.3.3(2), we see that

$$(\varepsilon \circ Q)((x^i)) = \left(\varepsilon \circ \varprojlim Q^i\right)((x^i)) =$$

$$= \varepsilon\left(R^1(x^1), (R^1(x^1), R^2(x^2)), (R^1(x^1), R^2(x^2), R^3(x^3)), \dots\right)$$

$$\equiv \varepsilon(R^1(x^1), R^2(x^2), R^3(x^3), \dots)$$

$$= \varprojlim R^i(x^i) = \varprojlim \left(\theta^i(x^i) \circ \left(\overline{\phi}_{x^i}^{-1}\right)\right).$$

This means that the last limit is a smooth map (since both Q and ε are smooth), and so are (3.6.2) and θ.   □

Remark 3.6.4   The previous result clarifies the terminology applied in the introduction of the present section: $\theta \in \Lambda^1(M, \mathfrak{g})$ is determined *point-wise* by the limits of $(\theta^i)_{i \in \mathbb{N}}$, i.e. $\theta(x) = \varprojlim \theta^i(x^i)$, $x = (x^i)$.

To define differential forms of higher degree, we consider the space of continuous k-linear maps $\mathcal{L}_k(\mathbb{F}, \mathfrak{g}) = \mathcal{L}(\mathbb{F}, \dots, \mathbb{F}; \mathfrak{g})$ (k factors), equipped with an appropriate topology (e.g. the bornologification, [KM97]), so that the composition map is continuous, and the space of continuous k-alternating (skew-symmetric) maps $\mathcal{A}_k(\mathbb{F}, \mathfrak{g})$ is a closed subspace of $\mathcal{L}_k(\mathbb{F}, \mathfrak{g})$, thus a locally convex space itself. Accordingly, we construct the triplet $(A_k(TM, \mathfrak{g}), M, L_k)$, where

$$A_k(TM, \mathfrak{g}) := \bigcup_{x \in M} \mathcal{A}_k(T_x M, \mathfrak{g}),$$

$$L_k \colon A_k(TM, \mathfrak{g}) \longrightarrow M \colon f \mapsto L_k(f) := x, \text{ if } f \in \mathcal{A}_k(T_x M, \mathfrak{g}).$$

The analogs of Proposition 3.6.1 and Corollary 3.6.2 are stated in the following.

Proposition 3.6.5 $(A_k(TM, \mathfrak{g}), M, L_k)$ *is a locally trivial fibration of fibre type* $\mathcal{A}_k(\mathbb{F}, \mathfrak{g})$.

Proof   We sketch its main steps since it is a simple extension of that of Proposition 3.6.1. From the charts (U, ϕ) and $(\pi^{-1}(U), \Phi)$ of M and TM, respectively, we define the chart $\left(L_k^{-1}(U), \widehat{\Phi}_k\right)$ of $A_k(TM, \mathfrak{g})$, where

$$\widehat{\Phi}_k \colon L_k^{-1}(U) \longrightarrow \phi(U) \times \mathcal{A}_k(\mathbb{F}, \mathfrak{g})$$

is given by

$$\widehat{\Phi}_k(f) := \left(\phi(x), f \circ \left(\overline{\phi}_x^{-1} \times \cdots \times \overline{\phi}_x^{-1}\right)\right)$$

if $f \in \mathcal{A}(T_x M, \mathfrak{g})$.

For the compatibility of two charts $\left(L_k^{-1}(U), \widehat{\Phi}_k\right)$ and $\left(L_k^{-1}(V), \widehat{\Psi}_k\right)$, with $U \cap V \neq \emptyset$, we check that

$$\widehat{\Psi}_k \circ \widehat{\Phi}_k^{-1} \colon \phi(U \cap V) \times \mathcal{A}_k(\mathbb{F}, \mathfrak{g}) \longrightarrow \psi(U \cap V) \times \mathcal{A}_k(\mathbb{F}, \mathfrak{g})$$

has the form

$$\left(\widehat{\Psi}_k \circ \widehat{\Phi}_k^{-1}\right)(a, h) = \left(F^{-1}(a), g \circ \left(DF(F^{-1}(a)) \times \cdots \times DF(F^{-1}(a))\right)\right),$$

for every $(a, h) \in \phi(U \cap V) \times \mathcal{A}_k(\mathbb{F}, \mathfrak{g})$, where $F := \phi \circ \psi^{-1}$. Its smoothness is shown as in Proposition 3.6.1, taking into account that the involved composition map $\mathrm{comp} \colon \mathcal{A}_k(\mathbb{F}, \mathfrak{g}) \times \mathcal{L}_I(\mathbb{F}) \times \cdots \times \mathcal{L}_I(\mathbb{F}) \to \mathcal{A}_k(\mathbb{F}, \mathfrak{g})$ is given by $\mathrm{comp}\,(h; f_1, \ldots, f_k) = h \circ (f_1 \times \cdots \times f_k)$. $\qquad \square$

We close by adding that the exterior differentials

$$d \colon \Lambda^k(M, \mathfrak{g}) \to \Lambda^{k+1}(M, \mathfrak{g}); \qquad k \geq 0,$$

are defined in the usual way; that is, by differentiating the local principal parts of the sections representing the given differential forms. The ordinary properties hold true also in the present context.

4

Projective systems
of principal bundles

In the previous chapter, taking advantage of the compatibility between Fréchet structures and projective limits, we set up an appropriate background for the study of certain manifolds and Lie groups modelled on Fréchet spaces. The results obtained allowed us to handle a wide variety of manifolds that cannot be modelled on Banach spaces, addressing the difficulty that prevents the transfer of classical tools from the finite-dimensional and Banach cases.

The same background will be exploited in order to study more complicated geometric structures. More precisely, in this chapter, we shall focus on principal bundles and their geometric properties within the Fréchet framework. It will be shown that the projective limit approach gives a way out of a number of significant difficulties emerging from the peculiarities of the space models. For example, in the case of an arbitrary Fréchet principal bundle, connections may not have parallel displacements, the standard proof of Cartan's (second) structural equation fails as based on the existence of 1-parameter subgroups of transformations, and so on. All the above issues can be addressed, up to a point, by using, whenever possible, the mechanism of projective limits.

4.1 Projective systems and Fréchet principal bundles

For the definition of projective limits of principal bundles, we follow the method adopted in the previous sections: All the spaces and local charts involved should form projective systems.

Definition 4.1.1 A countable family

$$\left\{\ell^i = (P^i, G^i, B, \pi^i); \; F^{ji} = \left(p^{ji}, g^{ji}, \mathrm{id}_B\right)\right\}_{i,j\in\mathbb{N}}$$

of Banach principal bundles and corresponding morphisms will be called
a **projective system of principal bundles** if the following conditions
are fulfilled:

(PLPB. 1) $\{P^i; p^{ji}\}_{i,j\in\mathbb{N}}$ is a projective system of Banach smooth manifolds converging to the plb-manifold $P = \varprojlim P^i$

(PLPB. 2) $\{G^i; g^{ji}\}_{i,j\in\mathbb{N}}$ is a projective system of Banach Lie groups
converging to the plb-group $G = \varprojlim G^i$.

(PLPB. 3) For every $j \geq i$, each

$$F^{ji} = (p^{ji}, g^{ji}, \mathrm{id}_B)\colon \ell^j = (P^j, G^j, B, \mathbb{P}i^j) \longrightarrow \ell^i = (P^i, G^i, B, \pi^i)$$

is a principal bundle morphism.

(PLPB. 4) There exists a trivializing cover $\{(U_\alpha, \Phi_\alpha)\}_{\alpha\in I}$ of P, whose
elements are obtained as projective limits of corresponding trivializations
of the factor bundles P^i; that is,

$$(4.1.1) \qquad (U_\alpha, \Phi_\alpha) = \left(\varprojlim_{i\in\mathbb{N}} U^i_\alpha, \; \varprojlim_{i\in\mathbb{N}} \Phi^i_\alpha\right), \qquad \alpha \in I.$$

As a matter of notation, if $\mathbb{B}$ is the model of B and the group G is
modelled on the Fréchet space $\mathbb{G} = \varprojlim\{\mathbb{G}^i, \zeta^{ji}\}_{i,j\in\mathbb{N}}$, then P is modelled
on $\mathbb{P} = \varprojlim\{\mathbb{P}^i, \rho^{ji} = \mathrm{id}_\mathbb{B} \times\zeta^{ji}\}_{i,j\in}$, where $\mathbb{P}^i = \mathbb{B} \times \mathbb{G}^i$.

It is important to note that all the factor bundles ℓ^i ($i \in \mathbb{N}$) in the
previous definition have the *same* base space. This assumption simpli-
fies many technical details of subsequent computations, *without any loss
of generality*. Alternatively, one could consider base spaces B^i ($i \in \mathbb{N}$),
assuming, at the same time, that they converge to a plb-manifold. However, this would lead to undue complications without any reasonably
significant gain. Thus, regarding (PLPB 4), in many cases it will be
sufficient to take $U^i_\alpha = U_\alpha$ for all $i \in \mathbb{N}$.

The conditions of Definition 4.1.1 ensure the existence of the projective
limit of the system $\{\ell_i; F^{ji}\}_{i,j\in\mathbb{N}}$, namely

$$\ell = \varprojlim \ell^i = \left(\varprojlim P^i, \varprojlim G^i, B, \varprojlim \pi^i\right),$$

called a **projective limit of (Banach) principal bundles** (or **plb-
principal bundle**, for short). We shall denote by $p^i\colon P = \varprojlim P^i \to P^i$
the canonical projection of the limit to the i-th factor bundle. The base

manifold B is assumed to be a Hausdorff space and admits smooth partitions of unity. The first assumption implies the uniqueness of solutions of differential equations (with given initial conditions), while the second ensures the existence of connections on the factor bundles.

Theorem 4.1.2 *Every plb-principal bundle is a Fréchet principal bundle.*

Proof In virtue of Propositions 3.1.3 and 3.4.1, in conjunction with Definition 4.1.1, the total space $P = \varprojlim P^i$ is a Fréchet manifold, and the structure group $G = \varprojlim G^i$ is a Fréchet-Lie group.

Denoting by $\pi^i \colon P^i \to B$ the projections of the factor bundles ℓ^i ($i \in \mathbb{N}$), and by $\delta^i \colon P^i \times G^i \to P^i$ the actions of G^i respectively on the total spaces, we see that (PLPB. 3) implies the equalities

$$(4.1.2) \qquad \pi^i \circ p^{ji} = \pi^j,$$

$$(4.1.3) \qquad p^{ji} \circ \delta^j = \delta^i \circ (p^{ji} \times g^{ji}), \quad j \geq i.$$

As a result, the pls-maps

$$\pi := \varprojlim \pi^i \colon P \to B, \quad \text{and} \quad \delta := \varprojlim \delta^i \colon P \times G \to P$$

exist and are smooth in virtue of Proposition 3.1.8. Moreover, the compatibility of the projective limits with any algebraic structure ensures that δ is a smooth action. Indeed, for every $u = (u^i) \in P$ and every $g_1 = (g_1^i), g_2 = (g_2^i) \in G$, we check that

$$\delta(u, e) = \varprojlim \delta^i \left((u^i), (e^i) \right) = \left(\delta^i(u^i, e^i) \right)_{i \in \mathbb{N}} = (u^i) = u,$$

as well as

$$\begin{aligned}
\delta(\delta(u, g_1), g_2) &= \varprojlim \delta^i \left(\varprojlim \delta^i \left((u^i), (g_1^i) \right), (g_2^i) \right) \\
&= \varprojlim \delta^i \left(\left(\delta^i(u^i, g_1^i), (g_2^i) \right)_{i \in \mathbb{N}} \right) \\
&= \left(\delta^i \left(\delta^i(u^i, g_1^i), (g_2^i) \right) \right)_{i \in \mathbb{N}} \\
&= \left(\delta^i \left(u^i, g_1^i g_2^i \right) \right)_{i \in \mathbb{N}} \\
&= \varprojlim \delta^i \left((u^i), (g_1^i g_2^i) \right) \\
&= \delta(u, g_1 g_2).
\end{aligned}$$

Customarily, we simplify the notations by writing $u \cdot g$ instead of $\delta(u, g)$, and analogously for the actions δ^i. This convention will be systematically applied throughout this chapter.

The trivializing cover for the desired principal bundle structure on P

is provided by the charts (U_α, Φ_α) given by (4.1.1), because, for every $u = (u^i) \in \pi^{-1}(U)$ and $g = (g^i) \in G$,

$$\left(\mathrm{pr}_1 \circ \Phi_\alpha\right)(u) = \mathrm{pr}_1\left(\Phi_\alpha(u)\right) = \mathrm{pr}_1\left(\Phi_\alpha((u^i))\right) =$$
$$= \left(\mathrm{pr}_1(\Phi_\alpha^i(u^i))\right)_{i \in \mathbb{N}} = \left(\pi^i(u^i)\right)_{i \in \mathbb{N}} = \pi(u),$$

while

$$\Phi_\alpha(u \cdot g) = \varprojlim \Phi_\alpha^i\left((u^i) \cdot (g^i)\right) = \left(\Phi_\alpha^i(u^i \cdot g^i)\right)_{i \in \mathbb{N}}$$
$$= \left(\pi^i(u^i \cdot g^i), \Phi_{\alpha,2}^i(u^i \cdot g^i)\right)_{i \in \mathbb{N}} = \left(\pi^i(u^i), \Phi_{\alpha,2}^i(u^i) \cdot (g^i)\right)_{i \in \mathbb{N}}$$
$$= \left(\pi(u), \Phi_{\alpha,2}(u) \cdot g)\right) = \left(\pi(u), \Phi_{\alpha,2}(u)\right) \cdot g$$
$$= \Phi_\alpha(u) \cdot g,$$

where $\Phi_{\alpha,2} \colon \pi^{-1}(U_\alpha) \to G$ denotes the projection of Φ_α to the second factor. $\qquad\square$

The previous bundle structure allows to transfer a number of classical results to the Fréchet framework.

Proposition 4.1.3 *Let $\left(P = \varprojlim P^i, G = \varprojlim G^i, B, \pi = \varprojlim \pi^i\right)$ be a plb-principal bundle. Then, for every $b \in B$, the fibre $\pi^{-1}(b)$ is a plb-manifold such that*

$$\pi^{-1}(b) = \varprojlim \left((\pi^i)^{-1}(b)\right).$$

Proof By (PLPB. 4) of Definition 4.1.1, we can find a trivialization, say, $(U, \Phi) = (\varprojlim U^i, \varprojlim \Phi^i)$ of P, with $b \in U$ (for simplicity we omit the index α), where (U^i, Φ^i) are trivializations of the bundles P^i ($i \in \mathbb{N}$). We already know (see §1.6.1) that, for every $i \in \mathbb{N}$, the fibre $(\pi^i)^{-1}(b)$ is a Banach manifold diffeomorphic with the structure group G^i via the map

$$(4.1.4) \qquad \Phi_b^i := \mathrm{pr}_2 \circ \Phi^i\big|_{(\pi^i)^{-1}(b)} \colon (\pi^i)^{-1}(b) \xrightarrow{\simeq} G^i.$$

The fact that $F^{ji} = (p^{ji}, \mathrm{id}_B, g^{ji})$ is a principal bundle morphism implies that $\pi^i \circ p^{ji} = \pi^j$ and

$$p^{ji}\left((\pi^j)^{-1}(b)\right) \subseteq (\pi^i)^{-1}(b), \qquad j \geq i.$$

On the other hand, the restrictions

$$p^{ji}\big|_{(\pi^j)^{-1}(b)} \colon (\pi^j)^{-1}(b) \longrightarrow (\pi^i)^{-1}(b)$$

are smooth, since each $(\pi^i)^{-1}(b)$ is a canonical submanifold of the corresponding total space P^i ($i \in \mathbb{N}$). Therefore, we obtain the projective system of smooth manifolds

$$\left\{(\pi^i)^{-1}(b); p^{ji}\right\}_{i,j\in\mathbb{N}}.$$

We shall show that the conditions of Definition 3.1.1 are satisfied, therefore, $\varprojlim\left((\pi^i)^{-1}(b)\right) = \pi^{-1}(b)$ is a plb-manifold.

Indeed, the models of $(\pi^i)^{-1}(b)$, identified with $\{\mathbb{G}^i\}_{i\in\mathbb{N}}$, form a projective system with connecting morphisms identified with the connecting morphisms $\zeta^{ji}\colon \mathbb{G}^j \to \mathbb{G}^i$ ($j \geq i$). Also, the fact that (U, Φ) is the projective limit of (U^i, Φ^i) implies that

(4.1.5) $$(\mathrm{id}_B \times g^{ji}) \circ \Phi^j = \Phi^i \circ p^{ji},$$

which, together with (4.1.4), gives for every $u = (u^i) \in \varprojlim\left((\pi^i)^{-1}(b)\right)$:

(4.1.6) $$g^{ji}\big(\Phi_b^j(u^j)\big) = \Phi_b^i(p^{ji}(u^j)) = \Phi_b^i(u^i).$$

Therefore, $\big(\Phi_b^i(u^i)\big) \in G = \varprojlim G^i$ and we can find a limit chart

$$\left(V = \varprojlim V^i, \Psi = \varprojlim \Psi^i\right)$$

of G so that $\big(\Phi_b^i(u^i)\big) \in V$. Setting now

$$W^i = (\Phi_b^i)^{-1}(V^i) \quad \text{and} \quad \chi^i = \Psi^i \circ \Phi_b^i,$$

we obtain a chart (W^i, χ^i) of $(\pi^i)^{-1}(b)$. Then, by (4.1.6),

$$p^{ji}(W^j) = p^{ji}\big((\Phi_b^j)^{-1}(V^j)\big) = (\Phi_b^i)^{-1}\big(g^{ji}(V^j)\big) = (\Phi_b^i)^{-1}(V^i) = W^i;$$

in other words, the domains of the charts (W^i, χ^i), $i \in \mathbb{N}$, determine a projective system. The same holds true for the corresponding maps, since

$$\zeta^{ji} \circ \chi^j = \zeta^{ji} \circ \Psi^j \circ \Phi_b^j = \Psi^i \circ g^{ji} \circ \Phi_b^j = \Psi^i \circ \Phi_b^i \circ p^{ji} = \chi^i \circ p^{ji},$$

for all $j \geq i$. In addition, the projective limits of the domains and the images of the same maps are open sets, because

$$\varprojlim \chi^i(W^i) = \varprojlim(\Psi^i(V^i)),$$

$$\varprojlim W^i = \varprojlim\left((\Phi_b^i)^{-1}(V^i)\right) = \left(\varprojlim \Phi_b^i\right)^{-1}\left(\varprojlim V^i\right).$$

Summarizing, we have seen that a limit chart, in the sense of Definition 3.1.1, can be defined at every point of $\varprojlim((\pi^i)^{-1}(b))$; hence, the latter is a plb-manifold as claimed. $\square$

Remark 4.1.4 It is useful to mention here another approach to the submanifold structure on the fibres $\pi^{-1}(b)$ of P: In the presence of the trivializations $\Phi = \varprojlim \Phi^i$, each $\Phi_b := \mathrm{pr}_2 \circ \Phi|_{\pi^{-1}(b)} \colon \pi^{-1}(b) \to G$ $(b \in B)$ coincides with $\varprojlim \Phi_b^i = \varprojlim \left(\mathrm{pr}_2 \circ \Phi^i|_{(\pi^i)^{-1}(b)}\right)$, a fact ensuring that the Fréchet manifold structure, determined as in the Banach case (see the list of properties of a Banach principal bundle in § 1.6.1), coincides with the structure obtained by the projective limit approach given here.

Proposition 4.1.5 *Let* $\left(P = \varprojlim P^i, G = \varprojlim G^i, B, \pi = \varprojlim \pi^i\right)$ *be a plb-principal bundle. The canonical projections* $p^i \colon P \to P^i$ *and* $g^i \colon G \to G^i$ *of the total space and the structure group, respectively, determine the principal bundle morphism*

$$F^i = (p^i, g^i, \mathrm{id}_B) \colon (P, G, B, \pi) \longrightarrow (P^i, G^i, B, \pi^i).$$

Proof Since $P = \varprojlim P^i$ is a plb-manifold and $G = \varprojlim G^i$ a plb-group, it follows that $p^i \colon P \to P^i$ is a smooth map and $g^i \colon G \to G^i$ a Lie group morphism, for every index $i \in \mathbb{N}$ (see Propositions 3.1.5 and 3.4.1). On the other hand, $\pi = \varprojlim \pi^i$ implies that

$$\pi^i \circ p^i = \mathrm{id}_B \circ \pi, \qquad i \in \mathbb{N}.$$

Moreover, p^i is equivariant with respect to G and G^i:

$$p^i(u \cdot a) = p^i \left((u^i), (a^i)\right) = p^i \left((u^i \cdot a^i)\right) = u^i \cdot a^i = p^i(u) \cdot g^i(a),$$

for every $(u, a) \in P \times G$. $\qquad\qquad\qquad\qquad\qquad\qquad\qquad\qquad$ $\square$

Concerning the local sections of a plb-principal bundle we obtain:

Proposition 4.1.6 *Every local section* $s \colon U \subseteq B \to P$ *of a plb-principal bundle* $\left(P = \varprojlim P^i, G = \varprojlim G^i, B, \pi = \varprojlim \pi^i\right)$ *coincides with a projective limit of local sections* $s^i \colon U \to P^i$ *on the factor bundles; that is,* $s = \varprojlim s^i$.

Proof The desired components of s are obtained by projecting it via the canonical projections of P to the factor bundles P^i. Indeed, each

(4.1.7) $\qquad\qquad\qquad s^i := p^i \circ s \colon U \subseteq B \to P^i$

is a smooth map (as a composite of smooth maps). It is also a section of the Banach bundle (P^i, G^i, B, π^i) since

$$\pi^i \circ s^i = \pi^i \circ p^i \circ s = \pi \circ s = \mathrm{id}_U.$$

Finally, the equalities

$$p^{ji} \circ s^j = p^{ji} \circ p^j \circ s = p^i \circ s = s^i; \qquad j \geq i,$$

ensure that $\varprojlim s^i$ exists and coincides with s, in virtue of the definition of the components (s^i) and Proposition 2.3.5. □

In Definition 4.1.1 we required the existence of limit trivializations in the structure of a projective limit principal bundle. However, by preceding proposition, and the close relationship between the local sections and the trivializations of a principal bundle, every trivialization is a limit. More precisely:

Theorem 4.1.7   *Every trivialization of a plb-principal bundle* $\big(P = \varprojlim P^i, G = \varprojlim G^i, B, \pi = \varprojlim \pi^i\big)$ *can be realized as the projective limit of trivializations of the factor bundles.*

Proof   Let (U, Φ) be an arbitrary trivialization of P. Then the map

$$s\colon U \longrightarrow \pi^{-1}(U)\colon x \mapsto \Phi^{-1}(x, e),$$

where e denotes the unit of G, is a smooth (local) section of P. As in the previous proposition, $s = \varprojlim s^i$, with s^i given by (4.1.7). It is a typical result of the theory of (Banach) principal bundles [see §1.6.3 and equality (1.6.5)] that each s^i induces the local trivialization of P^i

$$\Phi^i\colon (\pi^i)^{-1}(U) \longrightarrow U \times G^i\colon u \mapsto \big(\pi^i(u), \underline{k}^i(u)\big),$$

where $\underline{k}^i\colon (\pi^i)^{-1}(U) \to G^i$ is the smooth map, uniquely determined by

(4.1.8) $u = s^i(\pi^i(u)) \cdot \underline{k}^i(u), \qquad u \in (\pi^i)^{-1}(U).$

[Observe that $\underline{k}^i = k^i \circ \big(s^i \circ \pi^i, \mathrm{id}_{(\pi^i)^{-1}(U)}\big)$, where k^i is the analog of 1.6.6.] We see that the family $\{\underline{k}^i\colon (\pi^i)^{-1}(U) \to G^i\}_{i \in \mathbb{N}}$ is a projective system of smooth maps, since, for every $j \geq i$ and $u \in (\pi^i)^{-1}(U)$, (4.1.8) leads to

$$p^{ji}(u) = p^{ji}(s^j(\pi^j(u))) \cdot g^{ji}(\underline{k}^j(u)) = s^i(\pi^i(p^{ji}(u))) \cdot g^{ji}(\underline{k}^j(u));$$

hence, again by (4.1.8),

$$\underline{k}^i(p^{ji}(u)) = g^{ji}(\underline{k}^j(u)), \qquad u \in (\pi^i)^{-1}(U);$$

equivalently,

(4.1.9) $\underline{k}^i \circ p^{ji} = g^{ji} \circ \underline{k}^j.$

As a result, the pls-map $\varprojlim \underline{k}^i\colon U \to G$ is defined.

Analogously, the section s defined by (U, Φ) in the beginning of the proof, determines the map $\underline{k}\colon \pi^{-1}(U) \to G$, given by

(4.1.10) $u = s(\pi(u)) \cdot \underline{k}(u), \qquad u \in \pi^{-1}(U).$

Now, for every $u \in \pi^{-1}(U)$, (4.1.10) yields:

$$p^i(u) = p^i(s(\pi(u))) \cdot g^i(\underline{k}(u)) = s^i(\pi^i(p^i(u))) \cdot g^i(\underline{k}(u)),$$

from which follows that $\underline{k}^i(p^i(u)) = g^i(\underline{k}(u))$; that is,

$$(4.1.11) \qquad\qquad \underline{k}^i \circ p^i = g^i \circ \underline{k}, \qquad i \in \mathbb{N}.$$

Therefore, in virtue of Proposition 2.3.5, $\underline{k} = \varprojlim \underline{k}^i$.

The above equalities lead to the desired relationship between the trivializations (U, Φ) and (U, Φ^i). First observe that (4.1.9) yields

$$
\begin{aligned}
(\mathrm{id}_U \times g^{ji}) \circ \Phi^j &= (\mathrm{id}_U \times g^{ji}) \circ (\pi^j, \underline{k}^j) = (\pi^j, g^{ji} \circ \underline{k}^j) \\
(4.1.12) \qquad &= (\pi^i \circ p^{ji}, \underline{k}^i \circ p^{ji}) = (\pi^i, \underline{k}^i) \circ p^{ji} \\
&= \Phi^i \circ p^{ji},
\end{aligned}
$$

thus $\varprojlim \Phi^i$ exists. Moreover, in virtue of (4.1.11),

$$
\begin{aligned}
(\mathrm{id}_U \times g^i) \circ \Phi &= (\mathrm{id}_U \times g^i) \circ (\pi, \underline{k}) = (\pi, g^i \circ \underline{k}) \\
(4.1.13) \qquad &= (\pi^i \circ p^i, \underline{k}^i \circ p^i) = (\pi^i, \underline{k}^i) \circ p^i \\
&= \Phi^i \circ p^i.
\end{aligned}
$$

This implies $\varprojlim \Phi^i = \Phi$, by which we complete the proof. $\qquad \square$

In the same vein, the transition functions of a limit bundle remain also in the category of projective limits.

Proposition 4.1.8 *The transition functions of a plb-principal bundle $\left(P = \varprojlim P^i, G = \varprojlim G^i, B, \pi = \varprojlim \pi^i\right)$ are the pls-maps $\{t_{\alpha\beta} = \varprojlim t^i_{\alpha\beta}\}_{\alpha,\beta \in I}$, where $\{t^i_{\alpha\beta}\}_{\alpha,\beta \in I}$ are the transition functions of P^i, for every $i \in \mathbb{N}$.*

Proof According to Definition 4.1.1, there is a limit trivialization

$$\left\{ (U_\alpha, \Phi_\alpha) = (\varprojlim U^i_\alpha, \varprojlim \Phi^i_\alpha) \right\}_{\alpha \in I}$$

of P. Let

$$t_{\alpha\beta} \colon U_{\alpha\beta} \longrightarrow G \colon x \mapsto \left(\Phi_{\alpha,x} \circ \Phi^{-1}_{\beta,x} \right)(e); \qquad \alpha, \beta \in I,$$

be the corresponding transition functions, where [see also the analog of equality (4.1.4)]

$$\Phi_{\alpha,x} := \mathrm{pr}_2 \circ \Phi_\alpha|_{\pi^{-1}(x)} \colon \pi^{-1}(x) \xrightarrow{\;\simeq\;} G.$$

Analogously, the transition functions of the (P^i, G^i, B, π^i), defined by the trivializations $\{(U_\alpha^i, \Phi_\alpha^i)\}_{\alpha \in I}$ $(i \in \mathbb{N})$, are

$$t_{\alpha\beta}^i \colon U_{\alpha\beta}^i \longrightarrow G^i \colon x \mapsto \left(\Phi_{\alpha,x}^i \circ (\Phi_{\beta,x}^i)^{-1}\right)(e^i).$$

Recall that e and e^i denote the unit elements of G and G^i, respectively.

Since $\Phi_\alpha = \varprojlim \Phi_\alpha^i$, for every $\alpha \in I$, we obtain the analogs of (4.1.12) and (4.1.13); namely,

$$(\mathrm{id}_B \times g^{ji}) \circ \Phi_\alpha^j = \Phi_\alpha^i \circ p^{ji},$$

$$(\mathrm{id}_B \times g^i) \circ \Phi_\alpha = \Phi_\alpha^i \circ p^i,$$

for every $i, j \in \mathbb{N}$ with $j \geq i$.

Consequently, by restriction to the fibres, we obtain:

$$
\begin{aligned}
(g^{ji} \circ t_{\alpha\beta}^j)(x) &= g^{ji}\big(\Phi_{\alpha,x}^j\big((\Phi_{\beta,x}^j)^{-1}(e^j)\big)\big) = \Phi_{\alpha,x}^i\big(p^{ji}\big((\Phi_{\beta,x}^j)^{-1}(e^j)\big)\big) \\
&= \Phi_{\alpha,x}^i\big((\Phi_{\beta,x}^i)^{-1}(g^{ji}(e^j))\big) = \big(\Phi_{\alpha,x}^i \circ (\Phi_{\beta,x}^i)^{-1}\big)(e^i) \\
&= t_{\alpha\beta}^i(x),
\end{aligned}
$$

for every $\alpha, \beta \in I$, $j \geq i$ and every $x \in U_{\alpha\beta}$. This proves that the limit

$$\varprojlim t_{\alpha\beta}^i \colon U_{\alpha\beta} \longrightarrow G$$

exists. Moreover, by the same token,

$$
\begin{aligned}
(g^i \circ t_{\alpha\beta})(x) &= g^i\big(\Phi_{\alpha,x}\big((\Phi_{\beta,x})^{-1}(e)\big)\big) = \Phi_{\alpha,x}^i\big(p^i\big((\Phi_{\beta,x})^{-1}(e)\big)\big) \\
&= \Phi_{\alpha,x}^i\big((\Phi_{\beta,x}^i)^{-1}(g^i(e))\big) = \big(\Phi_{\alpha,x}^i \circ (\Phi_{\beta,x}^i)^{-1}\big)(e^i) \\
&= t_{\alpha\beta}^i(x),
\end{aligned}
$$

which implies that $t_{\alpha\beta} = \varprojlim t_{\alpha\beta}^i$, for all $\alpha, \beta \in I$. $\qquad\square$

The previous proposition, in conjunction with the bijective correspondence between principal bundles and cocycles (see § 1.6.3), can be exploited to obtain the following fundamental result in our framework.

Theorem 4.1.9 *Every Fréchet principal bundle $\ell = (P, G, B, \pi)$ with a Banach base and structure group a plb-group coincides, up to isomorphism, with a projective limit of Banach principal bundles.*

Proof Let $\{G^i; g^{ji}\}_{i,j \in \mathbb{N}}$ be the projective system of Banach-Lie groups such that $G = \varprojlim G^i$, the latter being modelled on the Fréchet space $\mathbb{G} = \varprojlim\{\mathbb{G}^i, \rho^{ji}\}$. Also, take $\mathbb{B}$ as the model of B. If $\{(U_\alpha, \Phi_\alpha)\}_{\alpha \in I}$ is a trivializing cover of P and $\{t_{\alpha\beta}\}$ the corresponding transition functions, then, for every $i \in \mathbb{N}$, the smooth maps

$$t_{\alpha\beta}^i := g^i \circ t_{\alpha\beta} \colon U_{\alpha\beta} \longrightarrow G^i; \qquad i \in \mathbb{N},$$

($: g^i \colon G \to G^i$ the canonical projections of G) determine a projective system, because

$$g^{ji} \circ t^j_{\alpha\beta} = g^{ji} \circ g^j \circ t_{\alpha\beta} = g^i \circ t_{\alpha\beta} = t^i_{\alpha\beta}.$$

Thus, by the definition of the factors, $t_{\alpha\beta} = \varprojlim t^i_{\alpha\beta}$, for all $\alpha, \beta \in I$.

On the other hand, for every $i \in \mathbb{N}$, $\left(t^i_{\alpha\beta}\right)_{\alpha,\beta \in I}$ is a (G^i-valued) smooth 1-cocycle, since, for every $x \in U_{\alpha\beta\gamma}$, we have:

$$t^i_{\alpha\beta}(x) \cdot t^i_{\beta\gamma}(x) = g^i(t_{\alpha\beta}(x)) \cdot g^i(t_{\beta\gamma}(x)) =$$
$$= g^i(t_{\alpha\beta}(x) \cdot t_{\beta\gamma}(x)) = g^i(t_{\alpha\gamma}(x)),$$

for all $\alpha, \beta, \gamma \in I$. Therefore, a unique Banach principal bundle $\ell^i = (P^i, G^i, B, \pi^i)$ can be defined with transition functions $\left(t^i_{\alpha\beta}\right)_{\alpha,\beta \in I}$. More precisely, following the construction described in the last part of § 1.6.3, $P^i = S^i / \sim_i$, where

$$S^i = \bigcup_{\alpha \in I} \left(\{\alpha\} \times U_\alpha \times G^i\right)$$

and the equivalence relation is defined by

$$(\alpha, x, g) \sim_i (\beta, x', g') \Leftrightarrow x = x' \text{ and } g' = t^i_{\beta\alpha}(x) \cdot g.$$

The natural projection of P^i is

$$\pi^i \colon P^i \longrightarrow B \colon [(\alpha, x, g)]_i \mapsto x,$$

where $[(\alpha, x, s)]_i$ denotes the equivalence class of (α, x, s), while the action of G^i on the (right of) P^i is given by

$$[(\alpha, x, g)]_i \cdot g' := [(\alpha, x, g \cdot g')]_i.$$

Finally, a (local) trivializing cover $(U_\alpha, \Phi^i_\alpha)$ of P^i is obtained by taking

$$\Phi^i_\alpha \colon (\pi^i)^{-1}(U_\alpha) \longrightarrow U_\alpha \times G^i \colon [(\gamma, y, h)]_i \mapsto (y, t^i_{\alpha\gamma}(y) \cdot h).$$

The bundles ℓ^i ($i \in \mathbb{N}$) are interconnected by the maps

$$p^{ji} \colon P^j \longrightarrow P^i \colon [(\alpha, x, g)]_j \mapsto [(\alpha, x, g^{ji}(g))]_i.$$

They are well-defined, since $[(\alpha, x, g)]_j = [(\beta, x, g')]_j$ implies that $g' = t^j_{\beta\alpha}(x) \cdot g$. Moreover, since the connecting morphisms $g^{ji} \colon G^j \to G^i$ ($j \geq i$) are group morphisms, and the transition functions $t^i_{\alpha\beta}$ form a projective systems, it follows that

$$g^{ji}(g') = g^{ji}(t^j_{\beta\alpha}(x) \cdot g) = t^i_{\beta\alpha}(x) \cdot g^{ji}(g),$$

which shows that $[(\alpha, x, g^{ji}(g))]_i = [(\alpha, x, g^{ji}(g'))]_i$.

On the other hand, the maps p^{ji} $(j \geq i)$ first commute with the natural projections of the bundles:

$$(\pi^i \circ p^{ji})\left([(\alpha, x, g)]_j\right) = \pi^i\left([(\alpha, x, g^{ji}(g))]_i\right) = x = \pi^j([(\alpha, x, g)]_j),$$

and, secondly, they respect the bundle actions:

$$p^{ji}\left([(\alpha, x, g)]_j \cdot g'\right) = p^{ji}\left([(\alpha, x, g \cdot g')]_j\right) =$$
$$= \left[(\alpha, x, g^{ji}(g \cdot g'))\right]_i = [(\alpha, x, g^{ji}(g) \cdot g^{ji}(g'))]_i =$$
$$= [(\alpha, x, g^{ji}(g))]_i \cdot g^{ji}(g') = p^{ji}([(\alpha, x, g)]_j) \cdot g^{ji}(g'),$$

where $[(a, x, g)]_j$ and g' are arbitrarily chosen elements of P^j and G^j respectively. In addition, p^{ji} are smooth maps since they are locally projected, via the trivializing maps, to $\mathrm{id}_{U_\alpha} \times g^{ji}$, as it follows from

$$\left(\Phi_\alpha^i \circ p^{ji}\right)\left([(\alpha, x, g)]_j\right) = \Phi_\alpha^i\left([(\alpha, x, g^{ji}(g))]_i\right) =$$
$$= \left(x, t_{\alpha\beta}^i(x) \cdot g^{ji}(g)\right) = \left(x, g^{ji}(t_{\alpha\beta}^j(x) \cdot g)\right) =$$
$$= (\mathrm{id}_{U_\alpha} \times g^{ji})\left(\Phi_\alpha^j([(\alpha, x, g)]_j)\right).$$

The last two arguments prove that the triplets

$$F^{ji} = (p^{ji}, g^{ji}, \mathrm{id}_B) \colon \ell^j \longrightarrow \ell^i; \qquad j \geq i,$$

are principal bundle morphisms. Moreover, the family $(\ell^i, F^{ji})_{i,j \in \mathbb{N}}$ is a projective system, because

$$(p^{ik} \circ p^{ji})([(\alpha, x, g)]_j) = [(\alpha, x, g^{ik}(g^{ji}(g)))]_k =$$
$$= [(\alpha, x, g^{jk}(g))]_k = p^{jk}([(\alpha, x, g)]_j),$$

for every $j \geq i \geq k$. This system fulfils all the requirements of Definition 4.1.1. As a matter of fact, taking into account that the trivializations

$$\left\{\left(U_\alpha, \varprojlim \Phi_\alpha^i\right)\right\}_{\alpha \in I}$$

satisfy condition (PLPB. 4), the only thing that remains to be checked is that $\{P^i; p^{ji}\}_{i,j \in \mathbb{N}}$ is a projective system of Banach manifolds converging to the plb-manifold $\varprojlim P^i$. The necessary plb-charts are obtained by the previously defined trivializations $(U_\alpha, \Phi_\alpha^i)$ of the principal bundles, the corresponding charts $\left\{\left(V_\alpha = \varprojlim V_\alpha^i, \Psi_\alpha = \varprojlim \Psi_\alpha^i\right)\right\}$ of the plb-group G, and the atlas $\{(U_\alpha, \varphi_\alpha, \mathbb{B})\}_{\alpha \in I}$ of B. More precisely, for each $i \in \mathbb{N}$, we define the local chart with map

$$\mu_\alpha^i := (\varphi_\alpha \times \Psi_\alpha^i) \circ \Phi_\alpha^i \colon (\Phi_\alpha^i)^{-1}(U_\alpha \times V_\alpha^i) \longrightarrow \varphi_\alpha(U_\alpha) \times \Psi_\alpha^i(V_\alpha^i),$$

where $\varphi_\alpha(U_\alpha)$, $\Psi_\alpha^i(V_\alpha^i)$ and $(\Phi_\alpha^i)^{-1}(U_\alpha \times V_\alpha^i)$ are open subsets of $\mathbb{B}$, $\mathbb{G}$ and $\pi^{-1}(U_\alpha)$, respectively.

These charts commute with the connecting morphisms of the involved projective limits. Indeed, recalling that $\zeta^{ji} \colon \mathbb{G}^j \to \mathbb{G}^i$ are the connecting morphisms of the projective system of the models of G^i,

$$(\mathrm{id}_\mathbb{B} \times \zeta^{ji}) \circ \mu_\alpha^j = (\mathrm{id}_\mathbb{B} \times \zeta^{ji}) \circ (\varphi_\alpha \times \Psi_\alpha^j) \circ \Phi_\alpha^j =$$
$$= \left(\varphi_\alpha \times (\zeta^{ji} \circ \Psi_\alpha^j)\right) \circ \Phi_\alpha^j = \left(\varphi_\alpha \times (\Psi_\alpha^i \circ g^{ji})\right) \circ \Phi_\alpha^j =$$
$$= (\varphi_\alpha \times \Psi_\alpha^i) \circ (\mathrm{id}_\mathbb{B} \times g^{ji}) \circ \Phi_\alpha^j = (\varphi_\alpha \times \Psi_\alpha^i) \circ (\Phi_\alpha^i \circ p^{ji}) = \mu_\alpha^i \circ p^{ji}.$$

Therefore, the charts

$$\left(\varprojlim \left((\Phi_\alpha^i)^{-1}(U_\alpha \times V_\alpha^i)\right), \varprojlim \mu_\alpha^i \right)$$

can be defined. The fact that $\{(V_\alpha, \Psi_\alpha)\}_{\alpha \in I}$ is a family of plb-charts on G ensures that the domains and ranges of the previous limit charts are open sets, because

$$\varprojlim \left((\Phi_\alpha^i)^{-1}(U_\alpha \times V_\alpha^i)\right) = \left(\varprojlim \Phi_\alpha^i\right)^{-1} \left(U_\alpha \times \varprojlim V_\alpha^i\right),$$

$$\varprojlim \left(\mu_\alpha^i \left((\Phi_\alpha^i)^{-1}(U_\alpha \times V_\alpha^i)\right)\right) =$$
$$= \varprojlim \left(\varphi_\alpha(U_\alpha) \times \Psi_\alpha^i(V_\alpha^i)\right) = \varphi_\alpha(U_\alpha) \times \varprojlim \left(\Psi_\alpha^i(V_\alpha^i)\right).$$

Summarizing, the plb-(and therefore Fréchet) principal bundle

$$\varprojlim \ell^i = \varprojlim (P^i, G^i, B, \pi^i)$$

is now completely defined. Its identification with the initial bundle $\ell = (P, G, B, \pi)$ follows from the fact that both bundles have the same transition functions $\{t_{\alpha\beta} = \varprojlim t_{\alpha\beta}^i\}$, as a consequence of their local structure. □

Obviously, the family $\{P^i; p^{ji}\}_{i,j \in \mathbb{N}}$, constructed in the previous proof, is an example of a projective system of principal bundles as in Definition 4.1.1.

4.2 Connections on limit principal bundles

This section focuses on the notion of connections on Fréchet principal bundles in the context of the preceding § 4.1. Although the bundles in question are projective limits of Banach principal bundles, a direct definition of connections as projective limits is not possible (see also the introductory discussion in § 3.6). For this reason, we propose a generalized approach to projective systems of connections resulting in their convergence to connections on projective limit principal bundles, so as to satisfy the usual properties. This generalization provides a characterization of connections on such limit bundles.

The most convenient way to handle connections, in the present context, is by using global and local connection forms. The latter play a key role in our approach and allow us to transfer to the Fréchet framework many important geometric features and results of finite-dimensional and Banach principal bundles. For instance, we mention the existence of parallel translations (in spite of the problems arising in solving differential equations in Fréchet spaces), holonomy groups, and flat connections. Of course, splittings of appropriate exact sequences (in the sense of § 1.7.1) can also be used. However, they are quite cumbersome and there is no significant gain from their use.

Throughout this section, $\ell = (P, G, B, \pi)$ is a Fréchet principal bundle over a Banach manifold, with structure group a Fréchet-Lie group G represented by a projective limit of Banach Lie groups, i.e. $G = \varprojlim\{G^i; g^{ji}\}_{i,j\in\mathbb{N}}$. Then, by Theorem 4.1.9, ℓ coincides, up to isomorphism, with a projective limit of Banach principal bundles,

$$\ell \equiv \varprojlim \ell^i = \varprojlim(P^i, G^i, B, \pi^i).$$

This means that the total space and the projection of the bundle are also projective limits, i.e. $P \equiv \varprojlim\{P^i, p^{ji}\}$ and $\pi \equiv \varprojlim \pi^i$.

Assume now that each bundle ℓ^i admits a connection whose corresponding connection form is denoted by $\omega^i \in \Lambda^1(P^i, g^i)$, $i \in \mathbb{N}$ (see § 1.7.2 and § 3.6). Recall that g^i is the Lie algebra of G^i. Then, an unconditional existence of a projective limit for the family $\{\omega^i\}_{i\in N}$ cannot be expected, since the manifolds $\{L(TP^i, g^i)\}_{i,j\in\mathbb{N}}$ fail to form a projective system. A way out of this problem is to require that the previous connections are properly related.

Before giving a precise definition, we note that the value of a differential form $\omega^i \in \Lambda^1(P^i, g^i)$ at a point $u^i \in P^i$, customarily denoted by $\omega^i_{u^i}$, for convenience will be written as $\omega^i(u^i)$. In this respect, it is understood that $\omega^i(u^i)(V)$ denotes the value of the linear map $\omega^i_{u^i} = \omega^i(u^i)$ at a vector $V \in T_{u^i}P^i$.

Definition 4.2.1 By a ***projective system of connections*** on ℓ we mean a countable family $\{\omega^i\}_{i\in\mathbb{N}}$ of connections (forms) on $P^i{}_{i\in\mathbb{N}}$, respectively, such that, for every pair of indices (i, j) with $j \geq i$, the connections ω^j and ω^i are $(p^{ji}, g^{ji}, \mathrm{id}_B)$-related in the sense of § 1.7.5 [see also (1.7.19)]. Then the map

$$\omega \colon P \longrightarrow L(TP, \mathfrak{g}) \colon u = (u^i)_{i \in \mathbb{N}} \mapsto \omega(u) = \varprojlim(\omega^i(u^i))$$

is said to be the **projective limit** of $\{\omega^i\}_{i \in \mathbb{N}}$ and will be simply denoted by $\omega = \varprojlim \omega^i$.

> The preceding notation is used only for simplicity, in spite of the aforementioned remarks about the non existence of limit of differential forms. Thus, its real meaning is the aforementioned *pointwise* convergence; that is to say, $\omega(u) \equiv \omega_u = \varprojlim \omega^i(u^i)$, for every $u = (u^i) \in P$.

From the above definition it becomes clear that *related* connections are of particular importance to our approach, because they ensure the existence of the projective limit involved therein. Indeed, according to the explicit formula following the equivalent conditions (1.7.19), the requirement that ω^j and ω^i are $(p^{ji}, g^{ji}, \mathrm{id}_B)$-related $(j \geq i)$ implies that

$$T_{e^j} g^{ji} \circ \omega^j(u^j) = \omega^i(u^i) \circ T_{u^j} p^{ji},$$

for every $u^j \in P^j$, $u^i = p^{ji}(u^j)$. This is exactly the condition implying the existence of $\varprojlim \omega^i(u^i)$, according to Definition 2.3.4. Note that $T_{e^j} g^{ji} = \bar{g}^{ji}$ in the notations of (1.7.19).

Our immediate goal is to prove that ω is a connection form in the usual sense (see § 1.7.2). To this end, we need a number preparatory lemmas.

Lemma 4.2.2 *The map $\omega \colon P \to L(TP, \mathfrak{g})$, as in Definition 4.2.1, determines a $\mathfrak{g}$-valued differential 1-form on P.*

Proof Direct consequence of Proposition 3.6.3.   □

Lemma 4.2.3 *If X is a left invariant vector field of the structure group G and X_* the corresponding fundamental (Killing) vector field on P, then $\omega(X_*) = X_e \equiv X$.*

The reason for changing the customary notation X^* of the fundamental vector field (see § 1.3.2) to X_* will be clear in the course of the proof.

Proof By Theorem 3.4.4, every $X \in \mathcal{L}(G) \equiv \mathfrak{g}$ identifies with $\varprojlim X^i$, where $X^i \in \mathfrak{g}^i$. We denote by X_*^i the fundamental vector fields on P^i, corresponding to X^i $(i \in \mathbb{N})$. If $u^j \in P^j$ then, with respect to the action of G^j on (the right) of P^j, we define the map

$$\widetilde{u}^j : G^j \longrightarrow P^j : g \mapsto u^j \cdot g.$$

Hence, for any choice of indices i, j with $j \geq i$, we see that:

$$(Tp^{ji} \circ X^j_*)(u^j) = T_{u^j} p^{ji} \big(T_{e^j} \widetilde{u}^j \big(X^j(e^j)\big)\big) = T_{e^j} \big(p^{ji} \circ \widetilde{u}^j\big)\big(X^j(e^j)\big).$$

On the other hand, since $(p^{ij}, g^{ji}, \mathrm{id}_B) : \ell^j \to \ell^i$ is a principal bundle morphism,

$$(p^{ji} \circ \widetilde{u}^j)(g) = p^{ji}(u^j \cdot g) = p^{ji}(u^j) \cdot g^{ji}(g) = \Big(\widetilde{(p^{ji}(u^j))} \circ g^{ji}\Big)(g),$$

for every $g \in G^i$. Combining the preceding equalities, we obtain

$$(Tp^{ji} \circ X^j_*)(u^j) = T_{e^j} \Big(\widetilde{p^{ji}(u^j)} \circ g^{ji}\Big)\big(X^j(e^j)\big)$$
$$= T_{e^i} \widetilde{p^{ji}(u^j)} \big(T_{e^j} g^{ji}\big(X^j(e^j)\big)\big)$$
$$= T_{e^i} \widetilde{p^{ji}(u^j)}\big(X^i(e^i)\big)$$
$$= (X^i_* \circ p^{ji})(u^j);$$

hence, $\{X^i_*\}_{i \in \mathbb{N}}$ is a projective system and $\varprojlim X^i_*$ exists.

Working similarly for the fields X_* and X^i_* on P and P^i, respectively, and taking into account that the canonical projections of P and G induce the principal bundle morphisms $(P^i, g^j, \mathrm{id}_B) : \ell \to \ell^i$, we check that

$$Tp^i \circ X_* = X^i_* \circ p^i, \qquad i \in \mathbb{N}.$$

As a result, $X_* = \varprojlim X^i_*$, in virtue of Proposition 2.3.5. Therefore, for every $u = (u^i) \in P$, condition $(\omega.\,2)$ of §1.7.2 implies

$$\omega_u(X_*(u)) = \Big(\varprojlim \omega^i(u^i)\Big)\big(X^i_*(u^i)\big)_{i \in \mathbb{N}} =$$
$$= \big(\omega^i(u^i)\big(X^i_*(u^i)\big)\big)_{i \in \mathbb{N}} = \big(X^i(e^i)\big)_{i \in \mathbb{N}} = X_e,$$

which concludes the proof. $\qquad\qquad\qquad\qquad\qquad\qquad\qquad\qquad\square$

The last requirement needed is the interplay of ω with the adjoint representation Ad of G.

Lemma 4.2.4 *Let* $R_g : P \to P : u \mapsto u \cdot g$ *be the right translation of* P *by* $g \in G$. *Then, in analogy to* $(\omega.\,1)$ *of* §1.7.2,

$$R^*_g \omega = \mathrm{Ad}(g^{-1})\omega.$$

Proof The adjoint representation of $G = \varprojlim G^i$ takes a projective limit form when applied to any element $g \in G$:

$$\mathrm{Ad}(g^{-1}) = \varprojlim \left(\mathrm{Ad}^i(g^i)^{-1}\right),$$

as a consequence of Proposition 3.2.5. Similarly, $R_g = \varprojlim R^i_{g^i}$. Then, for every $u = (u^i) \in P$, $g = (g^i) \in G$ and $\mathrm{w} = (\mathrm{w}^i) \in T_u P$,

$$
\begin{aligned}
(R_g^* \omega)_u(\mathrm{w}) &= \omega_{u \cdot g}(T_u R_g(\mathrm{w})) \\
&= \left(\varprojlim \omega^i(u^i \cdot g^i)\right)\left(\varprojlim T_{u^i} R^i_{g^i}(\mathrm{w}^i)\right) \\
&= \left(\omega^i(u^i \cdot g^i)(T_{u^i} R^i_{g^i}(\mathrm{w}^i))\right)_{i \in \mathbb{N}} \\
&= \left((R^{i*}_{g^i}\omega^i)(u^i)(\mathrm{w}^i)\right)_{i \in \mathbb{N}} \\
&= \left(\mathrm{Ad}^i((g^i)^{-1})(\omega^i(u^i)(\mathrm{w}^i))\right)_{i \in \mathbb{N}} \\
&= \varprojlim\left(\mathrm{Ad}^i(g^i)^{-1}\right)\left(\varprojlim(\omega^i(u^i)(\mathrm{w}^i))\right) \\
&= \mathrm{Ad}(g^{-1})(\omega_u(\mathrm{w})). \qquad\qquad \square
\end{aligned}
$$

The previous lemmata now prove the following main result.

Theorem 4.2.5 *Let $(\omega^i)_{i \in \mathbb{N}}$ be a projective system of connections on $\ell \equiv \varprojlim \ell^i = \varprojlim(P^i, G^i, B, \pi^i)$. Then the differential form $\omega = \varprojlim \omega^i \in \Lambda^1(P, \mathfrak{g})$ is a connection on P.*

Corollary 4.2.6 *The connections ω and ω^i are $(p^i, g^i, \mathrm{id}_B)$-related.*

Proof We recall that $p^i \colon P \to P^i$ and $g^i \colon G \to G^i$ are the canonical projections. Thus, for every $u = (u^i) \in P$ and $\mathrm{w} = (\mathrm{w}^i) \in T_u P$, Corollary 3.2.6 implies that

$$(4.2.1) \qquad \left((p^i)^* \omega\right)_u(\mathrm{w}) = \omega^i(u^i)\left(T_u p^i((\mathrm{w}^i))\right) = \omega^i(u^i)(\mathrm{w}^i).$$

On the other hand,

$$\omega_u(\mathrm{w}) = \left(\varprojlim \omega^i(u^i)\right)((\mathrm{w}^i)) = \varprojlim\left(\omega^i(u^i)(\mathrm{w}^i)\right) \in \varprojlim \mathfrak{g}^i = \mathfrak{g},$$

or, applying $\bar{g}^i := T_{e^i} g^i$

$$(4.2.2) \qquad\qquad \bar{g}^i(\omega_u(\mathrm{w})) = \omega^i(u^i)(\mathrm{w}^i).$$

From (4.2.4) and (4.2.5) it follows that $(p^i)^* \omega = \bar{g}^i \omega$, which proves the statement. $\qquad \square$

The horizontal and vertical subspaces of TP are naturally determined by the corresponding spaces on the factors.

Proposition 4.2.7 *Let* $\omega = \varprojlim \omega^i$ *be a connection on the limit bundle* $(P, G, B, \pi) = \varprojlim(P^i, G^i, B, \pi^i)$. *Then, for every* $u = (u^i)_{i \in \mathbb{N}} \in P$, *the respective horizontal and vertical subspaces* $H_u P$ *and* $V_u P$ *of* $T_u P$, *determined by* ω, *coincide with the projective limits of their Banach counterparts; in other words,*

$$H_u P = \varprojlim H_{u^i} P^i, \quad V_u P = \varprojlim V_{u^i} P^i.$$

Proof In virtue of Proposition 3.2.2 and Corollary 3.2.6, the tangent spaces $\{T_{u^i} P^i\}_{i \in \mathbb{N}}$ form a projective system with connecting morphisms and canonical projections the differentials of the connecting morphisms and projections, respectively, of $P = \varprojlim P^i$. Since the connections ω^j and ω^i are $(p^{ji}, g^{ji}, \mathrm{id}_B)$-related, for every $i, j \in \mathbb{N}$ with $j \geq i$, it follows that $T_{u^j} p^j(H_{u^j} P^j) \subseteq H_{u^i} P^i$, thus $\varprojlim H_{u^i} P^i$ is defined. In addition,

$$\mathrm{w} = (\mathrm{w}^i) \in H_u P \Leftrightarrow \omega(u)(\mathrm{w}) = 0$$
$$\Leftrightarrow \varprojlim \left(\omega^i(u^i) \right)((\mathrm{w}^i)) = 0$$
$$\Leftrightarrow \omega^i(u^i)(\mathrm{w}^i) = 0, \ i \in \mathbb{N}$$
$$\Leftrightarrow \mathrm{w}^i \in H_{u^i} P \Leftrightarrow \mathrm{w} \in \varprojlim H_{u^i} P^i,$$

which proves the statement for the horizontal spaces.

On the other hand,

$$V_u P = T_u P - H_u P = \varprojlim T_{u^i} P^i - \varprojlim H_{u^i} P^i$$
$$= \varprojlim \left(T_{u^i} P^i - H_{u^i} P^i \right) = \varprojlim V_{u^i} P^i. \qquad \square$$

We turn now to the local connection forms of a limit connection. This is an advantageous approach within the framework of limit bundles, since they provide an equivalent way of studying limit connections, while their domains remain in the same fixed base. The first result in this context is rather expected.

Proposition 4.2.8 *Let* $\omega = \varprojlim \omega^i$ *be a connection on* $(P, G, B, \pi) = \varprojlim(P^i, G^i, B, \pi^i)$. *Let also* $\{s_\alpha \colon U_\alpha \to P\}_{\alpha \in I}$ *be the family of natural local sections of* P *over an open cover* $\{U_\alpha\}_{\alpha \in I}$ *of* B. *Then the local connection forms* $\{\omega_\alpha\}_{\alpha \in I}$ *of* ω *are given by*

$$(4.2.3) \qquad \omega_\alpha = \varprojlim_{i \in \mathbb{N}} \omega_\alpha^i; \qquad \alpha \in I,$$

where $\{\omega_\alpha^i\}_{\alpha \in I}$ *are the local connection forms of the factor connection* ω^i, *for every* $i \in \mathbb{N}$.

Before the proof we notice that the limit (4.2.3) has the ordinary sense and not the point-wise sense of Definition 4.2.1 concerning connection forms.

Proof By definition [see also (1.7.5)] and Proposition 4.1.6,

$$\omega_\alpha = s_\alpha^* \omega \colon U_\alpha \longrightarrow L(TU_\alpha, \mathfrak{g}),$$
$$\omega_\alpha^i = (s_\alpha^i)^* \omega^i \colon U_\alpha \longrightarrow L(TU_\alpha, \mathfrak{g}^i),$$

where $s_\alpha^i := p^i \circ s_\alpha$ and $p^i \colon P = \varprojlim P^i \to P^i$ $(i \in \mathbb{N})$ the canonical projections. Because TU_α is the same for all indices $i \in \mathbb{N}$, as we explain in detail in § 6.2, the linear map bundle $L(TU_\alpha, \mathfrak{g})$ is a projective limit of vector bundles. More precisely, $L(TU_\alpha, \mathfrak{g}) = \varprojlim L(TU_\alpha, \mathfrak{g}^i)$, with connecting morphisms

$$(4.2.4) \quad \lambda^{ji} \colon L(TU_\alpha, \mathfrak{g}^j) \longrightarrow L(TU_\alpha, \mathfrak{g}^i) \colon f \mapsto T_{e^j} g^{ji} \circ f; \qquad j \geq i,$$

and canonical projections

$$(4.2.5) \quad \lambda^i \colon L(TU_\alpha, \mathfrak{g}) \longrightarrow L(TU_\alpha, \mathfrak{g}^i) \colon f \mapsto T_e g^i \circ f; \qquad i \in \mathbb{N},$$

both induced by the corresponding elements of the group G.

Now, for every $x \in U_\alpha$ and every pair of indices (j, i) with $j \geq i$, we see that

$$\begin{aligned}
(\lambda^{ji} \circ \omega_\alpha^j)(x) &= \left(T_{e^j} g^{ji} \circ \omega_\alpha^j\right)(x) \\
&= T_{e^j} g^{ji} \circ \omega^j(s_\alpha^j(x)) \circ T_x s_\alpha^j \\
&= \omega^i \left(p^{ji}(p^j(s_\alpha(x)))\right) \circ T_x(p^{ji} \circ p^j \circ s_\alpha) \\
&= \omega^i \left(p^i(s_\alpha(x))\right) \circ T_x(p^i \circ s_\alpha) \\
&= \omega^i(s_\alpha^i(x)) \circ T_x s_\alpha^i = \omega_\alpha^i(x),
\end{aligned}$$

i.e. $\lambda^{ji} \circ \omega_\alpha^j = \omega_\alpha^i$, which means that $\{\omega_\alpha^i\}_{i \in \mathbb{N}}$ exists, for each $\alpha \in I$. Analogous computations for the canonical projections show that $\omega_\alpha^i = \lambda^i \circ \omega_\alpha$. Therefore, we obtain (4.2.3). $\qquad\square$

The use of the local connection forms yields a much more important result characterizing *all* the connections of a limit principal bundle. More precisely:

Theorem 4.2.9 *Every connection on a projective limit principal bundle $(P, G, B, \pi) = \varprojlim(P^i, G^i, B, \pi^i)$ is the limit of a projective system of connections.*

The proof relies on a series of auxiliary results. To this end, we recall the following facts needed below: Let $\{U_\alpha\}_{\alpha \in I}$ be an open cover of the base space B, over which we define the local trivializations of the bundles involved. We denote by $\{s_\alpha \colon U_\alpha \to P\}_{\alpha \in I}$ the family of natural local sections of the bundle P. By projecting to each factor bundle, we obtain the local sections

(4.2.6) $$\left\{ s_\alpha^i := p^i \circ s_\alpha \colon U_\alpha \to P^i \right\}_{\alpha \in I},$$

Similarly, the local connection forms $\{\omega_\alpha = s_\alpha^* \omega\}_{\alpha \in I}$ of ω, induce the local forms [see also (4.2.5)]

(4.2.7) $$\omega_\alpha^i := \lambda^i \circ \omega_\alpha \colon U_\alpha \to L(TU_\alpha, \mathfrak{g}^i), \qquad i \in \mathbb{N}.$$

on the limit factors. With these notation, we obtain:

Lemma 4.2.10 *For each $i \in \mathbb{N}$, the following compatibility condition holds for every $\alpha, \beta \in I$:*

$$\omega_\beta^i = \mathrm{Ad}^i\!\left((g_{\alpha\beta}^i)^{-1}\right)\omega_\alpha^i + \left(g_{\alpha\beta}^i\right)^{-1} dg_{\alpha\beta}^i.$$

More explicitly, for every $x \in U_{\alpha\beta}$ and $v \in T_x B$,

$$\omega_{\beta,x}^i(v) = \mathrm{Ad}^i\!\left((g_{\alpha\beta}^i(x))^{-1}\right).\omega_{\alpha,x}^i(v) + T_x\!\left(L_{(g_{\alpha\beta}^i(x))^{-1}}^i \circ g_{\alpha\beta}^i\right)(v),$$

where Ad^i denotes the adjoint representation of G^i, and $L_{g^i}^i \colon G^i \to G^i$ is the left translation of G^i by any $g^i \in G^i$.

Proof As in the ordinary case, the local connection forms of ω satisfy the compatibility condition

$$\omega_{\beta,x}(v) = \mathrm{Ad}\!\left((g_{\alpha\beta}(x))^{-1}\right)(\omega_{\alpha,x}(v)) + T_x\!\left(L_{(g_{\alpha\beta}(x))^{-1}} \circ g_{\alpha\beta}\right)(v),$$

for every $x \in U_{\alpha\beta}$, $v \in T_x B$ and $\alpha, \beta \in I$ (L_g denotes the left translation of G). On the other hand, both the adjoint representation of G and the left translations can be realized as projective limits of their Banach counterparts [see (3.4.1) and Proposition 3.2.5], i.e.

$$\mathrm{Ad}(g) = \varprojlim\left(\mathrm{Ad}^i(g^i)\right); \quad L_g = \varprojlim\left(L_{g^i}^i\right), \ \forall \ g = (g^i) \in G.$$

Therefore, in virtue of (4.2.5),

$$
\begin{aligned}
\omega^i_{\beta,x}(v) &= \big(\lambda^i(\omega_{\beta,x})\big)(v) = \big(T_e g^i \circ \omega_{\beta,x}\big)(v)\\
&= T_e g^i \big(\mathrm{Ad}(g_{\alpha\beta}(x)^{-1}).\omega_{\alpha,x}(v)\big)\\
&\quad + T_e g^i\big(T_x\big(L_{(g_{\alpha\beta}(x))^{-1}} \circ g_{\alpha\beta}\big)(v)\big)\\
&= \big(T_e g^i \circ \varprojlim \big(\mathrm{Ad}^i\big(g^i_{\alpha\beta}(x))^{-1}\big)\big).\omega_{\alpha,x}(v)\\
&\quad + T_x\Big(g^i \circ L_{(g_{\alpha\beta}(x))^{-1}} \circ g_{\alpha\beta}\Big)(v)\\
&= \big(\mathrm{Ad}^i\big((g^i_{\alpha\beta}(x))^{-1}\big) \circ T_e g^i\big).\omega_{\alpha,x}(v)\big)\\
&\quad + T_x\Big(L^i_{(g^i_{\alpha\beta}(x))^{-1}} \circ g^i \circ g_{\alpha\beta}\Big)(v)\\
&= \mathrm{Ad}^i\big(g^i_{\alpha\beta}(x))^{-1}\big).\lambda^i(\omega_{\alpha,x}))(v)\\
&\quad + T_x\Big(L^i_{(g^i_{\alpha\beta}(x))^{-1}} \circ g^i_{\alpha\beta}\Big)(v)\\
&= \mathrm{Ad}^i\big((g^i_{\alpha\beta}(x))^{-1}\big).\omega^i_{\alpha,x}(v)\\
&\quad + T_x\Big(L^i_{(g^i_{\alpha\beta}(x))^{-1}} \circ g^i_{\alpha\beta}\Big)(v). \qquad \square
\end{aligned}
$$

A direct consequence of the the preceding is:

Corollary 4.2.11 *An arbitrary connection $\omega \equiv \{\omega_\alpha\}$ on $P = \varprojlim P^i$ induces on each factor bundle P^i a connection ω^i, whose local connection forms $\{\omega^i_\alpha\}$ are given by equality (4.2.7).*

As the reader may guess, the previous connections $\{\omega^i\}_{i\in\mathbb{N}}$ will converge to ω. But, to be able to verify Definition 4.2.1, we need also the following:

Lemma 4.2.12 *For every $i, j \in \mathbb{N}$ with $j \geq i$, the connections ω^j and ω^i of Corollary 4.2.11 are $(p^{ji}, g^{ji}, \mathrm{id}_B)$-related.*

Proof By Proposition 1.7.1, it suffices to prove the analog of (1.7.20)

$$(4.2.8) \qquad \bar{g}^{ji}\omega^j_\alpha = \mathrm{Ad}(h_\alpha^{-1})\omega^i_\alpha + h_\alpha^{-1}dh_\alpha,$$

for every $\alpha, \beta \in I$, where the smooth maps $h_\alpha \colon U_\alpha \to G^i$ are determined by the equalities $p^{ji}(s^j_\alpha(x)) = s^i_\alpha(x) \cdot h_\alpha(x)$, for every $x \in U_\alpha$. Since, in virtue of (4.2.6),

$$p^{ji} \circ s^j_\alpha = p^{ji} \circ p^j \circ s_\alpha = p^i \circ s_\alpha = s^i_\alpha,$$

it follows that $h_\alpha(x) = e^i$, for all $x \in U_\alpha$. Therefore, (4.2.8) reduces to

$$(4.2.8') \qquad \bar{g}^{ji}\omega^j_\alpha = \omega^i_\alpha.$$

This is true, because, for every x as before, (4.2.7) and (4.2.5) yield:

$$
\begin{aligned}
\bar{g}^{ji} \circ \omega^j_{\alpha,x} &= T_{e^j} g^{ji} \circ \lambda^j(\omega_{\alpha,x}) \\
[e = (e^i)] &= T_{e^j} g^{ji} \circ T_e g^j \circ \omega_{\alpha,x} \\
&= T_e(g^{ji} \circ g^j) \circ \omega_{\alpha,x} = T_{e^i} g^i(\omega_{\alpha,x}) \\
&= \lambda^i(\omega_{\alpha,x}) = \omega^i_{\alpha,x},
\end{aligned}
$$

which proves (4.2.8$'$).   □

We are now in a position to give the

Proof of Theorem 4.2.9. Let ω be an arbitrary connection on $P = \varprojlim P^i$. By Corollary 4.2.11, we obtain the projective system of connections $\{\omega^i\}_{i\in\mathbb{N}}$. Lemma 4.2.12 implies that this is a projective system of connections in the sense of Definition 4.2.1; therefore, the latter system determines a limit connection, say, $\bar{\omega} = \varprojlim \omega^i$. Then $\omega = \bar{\omega}$, since both have the same local connection forms. Indeed, for every $x \in U_\alpha$ and $v \in T_x B$, we obtain:

$$
\begin{aligned}
\bar{\omega}_{\alpha,x}(v) &= (s_\alpha^* \bar{\omega})_x(v) = \bar{\omega}(s_\alpha(x))\big(T_x s_\alpha(v)\big) \\
&= \left(\varprojlim \omega^i\big(s_\alpha^i(x)\big)\right)\left(\varprojlim T_x s_\alpha^i(v)\right) \\
&= \varprojlim \left(\omega^i\big(s_\alpha^i(x)\big)\big(T_x s_\alpha^i(v)\big)\right) \\
&= \varprojlim \left(\big((s_\alpha^i)^* \omega^i\big)_x(v)\right) \\
&= \varprojlim \left(\omega_\alpha^i(x)(v)\right) = \omega_{\alpha,x}(v).
\end{aligned}
$$

This concludes the proof of the theorem.   □

In virtue of Theorem 4.2.9, it is reasonable to call the connections on a plb-principal bundle **plb-connections**.

Remark 4.2.13 Let (Q, G, B, π_Q) be an arbitrary Fréchet principal bundle with structure group $G = \varprojlim G^i$. Assume that θ is a connection on Q. Then, by Theorem 4.1.9, there is an isomorphism $(f, \mathrm{id}_G, \mathrm{id}_B)$ of (Q, G, B, π_Q) onto a plb-bundle $\big(\varprojlim P^i, \varprojlim G^i, B, \varprojlim \pi^i\big)$. Obviously, the isomorphism f determines a unique connection ω on $\varprojlim P^i$ such that $\theta = f^*\omega$. On the other hand, Theorem 4.2.9 implies that $\omega = \varprojlim \omega^i$, where ω^i is a connection on P^i, for every $i \in \mathbb{N}$. Therefore, θ is $(f, \mathrm{id}_G, \mathrm{id}_B)$-related with $\varprojlim \omega^i$. Roughly speaking, one may say that (Q, θ) coincides—*up to isomorphism*—with the projective limit of pairs (P^i, ω^i).

4.3 Parallel translations and holonomy groups

The aim of this section is to study the geometric objects in the title. The problems in solving differential equations on the Fréchet models make impossible the direct application, to our framework, of the classical pattern that so successfully ensures the existence of parallel translations (along curves of the base space), as well as the holonomy groups, for finite-dimensional or Banach bundles. The key approach giving a way out is based again on the use of projective limits, by means of which we realize the previous notions, without solving the corresponding differential equations. However, some interesting deviations from the classical case emerge in the study of holonomy groups, which seem to drop out of the projective limit category, as explicitly shown in Theorem 4.3.5 below.

Throughout this section we consider a plb-principal bundle

$$\ell \equiv (P, G, B, \pi) = \varprojlim(P^i, G^i, B, \pi^i) \equiv \varprojlim \ell^i,$$

endowed with a connection $\omega = \varprojlim \omega^i$, as in Definition 4.2.1.

Referring to §1.1.9 [see also (1.9.1)], we denote by $\hat{\alpha} \colon I \to P$ the horizontal lift of a smooth curve curve $\alpha \colon I \to B$. We recall that

$$\dot{\hat{\alpha}}(t) \in H_{\hat{\alpha}(t)}P \quad \text{or, equivalently,} \quad \omega(\hat{\alpha}(t))\big(\dot{\hat{\alpha}}(t)\big) = 0.$$

Proposition 4.3.1 *Let $\alpha \colon [0, 1] \to B$ be a smooth curve with $\alpha(0) = b$, and take any $u = (u^i) \in \pi^{-1}(b)$. Then there exists a unique horizontal lift $\hat{\alpha} \colon [0, 1] \to P$ of α such that $\hat{\alpha}(0) = u$. In particular, $\hat{\alpha}$ coincides with the projective limit of horizontal lifts of α on the factor bundles ℓ^i.*

Proof The idea here is to exploit the fact that the Banach bundle analog holds true. Thus, the horizontal lift $\hat{\alpha}^j \colon [0, 1] \to P^j$ of α on P^j, with initial condition $\hat{\alpha}^j(0) = u^j$, exists. Now, composing $\hat{\alpha}^j$ with the connecting morphism $p^{ji} \colon P^j \to P^i$ $(j \geq i)$, we see that:

i) $p^{ji} \circ \hat{\alpha}^j$ projects to α:

$$\pi^i \circ (p^{ji} \circ \hat{\alpha}^j) = \pi^j \circ \hat{\alpha}^j = \alpha,$$

ii) $p^{ji} \circ \hat{\alpha}^j$ is horizontal with respect to ω^i. Indeed, setting $\partial_t = \frac{d}{dt}\big|_t$ and applying (1.1.11),

$$\omega^i\big((p^{ji} \circ \hat{\alpha}^j)(t)\big)\big(\overline{(p^{ji} \circ \hat{\alpha}^j)}(t)\big) = \omega^i\big(p^{ji}(\hat{\alpha}^j(t))\big)\big(T_t(p^{ji} \circ \hat{\alpha}^j)(\partial_t)\big)$$

$$= \omega^i\big(p^{ji}(\hat{\alpha}^j(t))\big)\big(T_{\hat{\alpha}^j(t)}p^{ji}\big(\dot{\hat{\alpha}}^j(t)\big)\big)$$

or, since ω^j and ω^j are $(p^{ji}, g^{ji}, \mathrm{id}_B)$-related,

$$\omega^i\big((p^{ji} \circ \widehat{\alpha}^j)(t)\big)\big(\overbrace{(p^{ji} \circ \widehat{\alpha}^j)}(t)\big) = T_{e^j}g^{ji}\big(\omega^j(\alpha^j(t))\big)\big(\dot{\widehat{\alpha}}^j(t)\big)$$
$$= T_{e^j}g^{ji}(0) = 0,$$

iii) $p^{ji} \circ \widehat{\alpha}^j$ and the horizontal lift $\widehat{\alpha}^i \colon [0,1] \to P^i$ of α on P^i have the same initial condition:

$$\big(p^{ji} \circ \widehat{\alpha}^j\big)(0) = p^{ji}(u^j) = u^i.$$

Therefore, $p^{ji} \circ \widehat{\alpha}^j = \widehat{\alpha}^i$, $(j \geq i)$, implying that the smooth curve

$$\widehat{\alpha} := \varprojlim \widehat{\alpha}^i \colon [0,1] \longrightarrow P$$

can be defined. This is a horizontal curve with respect to $\omega = \varprojlim \omega^i$, because

$$\omega\big(\widehat{\alpha}(t)\big)\big(\dot{\widehat{\alpha}}(t)\big) = \varprojlim \omega^i(\widehat{\alpha}^i(t))\big(\dot{\widehat{\alpha}}^i(t)\big) = 0.$$

It is also a lift of α,

$$(\pi \circ \widehat{\alpha})(t) = \big(\pi^i\big(p^i(\widehat{\alpha}(t))\big)\big)_{i \in \mathbb{N}} = \big(\pi^i\big(\widehat{\alpha}^i(t)\big)\big)_{i \in \mathbb{N}} (\alpha(t))_{i \in \mathbb{N}} = \alpha(t),$$

satisfying the initial condition

$$\widehat{\alpha}(0) = \big(\widehat{\alpha}^i(0)\big)_{i \in \mathbb{N}} = \big(u^i\big)_{i \in \mathbb{N}}) = u.$$

As a result, $\widehat{\alpha} := \varprojlim \widehat{\alpha}^i$ is the desired horizontal lift of α. The uniqueness of $\widehat{\alpha}$ is checked using similar arguments, since any other curve, with the same properties, coincides with $\widehat{\alpha}$ on every factor bundle P^i. □

The previous result allows us to define the parallel translation of fibres along any smooth curve in the base of a limit (Fréchet) principal bundle.

Proposition 4.3.2 Let $\ell \equiv \varprojlim \ell^i$ be a plb-principal bundle as in the beginning of this section and $\alpha \colon I = [0,1] \to B$ a smooth curve such that $\alpha(0) = b_0$ and $\alpha(1) = b_1$. If $\widehat{\alpha}_u \colon I \to P$ denotes the horizontal lift of α with $\widehat{\alpha}_u(0) = u$, then the following assertions hold:
  i) The **parallel translation** or **displacement along** α

$$\tau_\alpha \colon \pi^{-1}(b_0) \longrightarrow \pi^{-1}(b_1) \colon u \mapsto \widehat{\alpha}_u(1),$$

is defined.
  ii) $\tau_\alpha = \varprojlim \tau_\alpha^i$, where τ_α^i are the corresponding parallel translations on the factor bundles.

Proof The first assertion is an immediate consequence of Proposition 4.3.1. The second one is essentially based on the following fact (see [Vas82], equality (1.7.20) and the notations of § 1.7.5):

> *If two connections ω and ω' defined on the (Banach) principal bundles (P, G, B, π) and (P', G', B', π'), respectively, are (f, φ, h)-related, then*
>
> $$f \circ \tau_\alpha = \tau'_{\alpha'} \circ f\big|_{\pi^{-1}(\alpha(0))},$$
>
> *where τ_α is the parallel displacement along $\alpha \colon I \to B$, and $\tau'_{\alpha'}$ that along $\alpha' = h \circ \alpha$.*

Therefore, since ω^j and ω^i are $(p^{ji}, g^{ji}, \mathrm{id}_B)$-related, for every $j, i \in \mathbb{N}$ with $j \geq i$, it follows that

$$p^{ji} \circ \tau_\alpha^j = \tau_\alpha^i \circ p^{ji}\big|_{(\pi^j)^{-1}(\alpha(0))},$$

which implies that the projective limit

$$\varprojlim \tau_\alpha^i \colon \varprojlim \left((\pi^i)^{-1}(b_0)\right) = \pi^{-1}(b_0) \longrightarrow \pi^{-1}(b_1) = \varprojlim \left((\pi^i)^{-1}(b_1)\right)$$

is defined. The latter coincides with the parallel translation along α. Indeed, for every $i \in \mathbb{N}$ and $u = (u^i) \in \pi^{-1}(b_0)$, composition with the canonical projection $p^i \colon P \to P^i$ yields:

$$(p^i \circ \tau_\alpha)(u) = p^i\big(\widehat{\alpha}_u(1)\big) = p^i\big(\varprojlim \widehat{\alpha}^i_{u^i}(1)\big) =$$

$$= p^i\left(\big(\widehat{\alpha}^i_{u^i}(1)\big)_{i \in \mathbb{N}}\right) = \widehat{\alpha}^i_{u^i}(1) = \tau_\alpha^i(u^i) = (\tau_\alpha^i \circ p^i)(u).$$

These equalities, along with Proposition 2.3.5, prove the last claim and conclude the proof. □

Corollary 4.3.3 *With the notations of Proposition 4.3.2, the following assertions are true:*

i) For every $g \in G$ and $u \in P$, $\tau_\alpha(u \cdot g) = \tau_\alpha(u) \cdot g$; consequently, τ_α is a G-equivariant diffeomorphism.

ii) $\tau_{\alpha^{-1}} = \tau_\alpha^{-1}$, if α^{-1} is the inverse (or reverse) curve of α.

iii) $\tau_{\beta * \alpha} = \tau_\beta \circ \tau_\alpha$, if $\beta * \alpha$ is the (appropriate) composition of α followed by β.

Proof An easy extension of the analogous results on ordinary (Banach or finite-dimensional) bundles in the vein of Proposition 4.3.2. □

Remark 4.3.4 Obviously, all the previous results, concerning horizontal lifts and parallel displacements, are valid for *piecewise smooth curves*, under the obvious modifications.

The fact that every connection on a plb-principal bundle is the projective limit of connections (Theorem 4.2.9) and the results of this section on the parallel displacements lead to interesting and some unexpected results, concerning the corresponding holonomy groups. Before exhibiting them, we recall a few facts from § 1.9, adapted to the present context.

Let $u = (u^i) \equiv (u^i)_{i \in \mathbb{N}} \in P$ and $b = \pi(u) = \left(\pi^i(u^i)\right)_{i \in \mathbb{N}} \in B$. If C_b is the *loop group at* b and $C_b^0 \subset C_b$ the group of 0-homotopic loops, we define the group homomorphism $k_u \colon \{\tau_\alpha \,|\, \alpha \in C_b\} \to G$ by

$$\tau_\alpha(u) = u \cdot k_u(\tau_\alpha), \qquad u \in \pi^{-1}(b).$$

Then $\boldsymbol{\Phi}_u := \{k_u(\tau_\alpha) \,|\, \alpha \in C_b\}$ is the *holonomy group of ω with reference point u*, while $\boldsymbol{\Phi}_u^0 = \{k_u(\tau_\alpha) \,|\, \alpha \in C_b^0\}$ is the respective *restricted holonomy group*.

Theorem 4.3.5 *The holonomy groups of ω are related with the corresponding ones of the component connections as follows:*

i) $\boldsymbol{\Phi}_u \subseteq \varprojlim \boldsymbol{\Phi}_{u^i}^i \subseteq \overline{\boldsymbol{\Phi}_u}$, $\boldsymbol{\Phi}_u^0 \subseteq \varprojlim (\boldsymbol{\Phi}_{u^i}^i)^0 \subseteq \overline{\boldsymbol{\Phi}_u^0}$, *where* $\overline{\boldsymbol{\Phi}_u}$ *(resp. $\overline{\boldsymbol{\Phi}_u^0}$) is the closure of $\boldsymbol{\Phi}_u$ (resp. $\boldsymbol{\Phi}_u^0$) in the topology of G.*

ii) $g^i(\boldsymbol{\Phi}_u) = \boldsymbol{\Phi}_{u^i}^i$, $g^i(\boldsymbol{\Phi}_u^0) = (\boldsymbol{\Phi}_{u^i}^i)^0$.

iii) If $\boldsymbol{\Phi}_u$ is open or closed in G, then $\boldsymbol{\Phi}_u = \varprojlim \boldsymbol{\Phi}_{u^i}^i$. As result, in both cases, $\boldsymbol{\Phi}_u$ is a Fréchet topological group.

iv) If the canonical projections p^{ji}, g^{ji} of the projective systems of bundles and groups involved are the natural embeddings, then the holonomy groups are Banach-Lie *groups.*

Proof i) Since $(p^{ji}, g^{ji}, \mathrm{id}_B)$ is a principal bundle morphism, we have in virtue of Proposition 4.3.2,

$$r \in \boldsymbol{\Phi}_{u^j}^j \;\Rightarrow\; \tau_\alpha^j(u^j) = u^j \cdot r \;\Rightarrow\; p^{ji}(\tau_\alpha^j(u^j)) = p^{ji}(u^j) \cdot g^{ji}(r)$$
$$\Rightarrow\; \tau_\alpha^i(u^i) = u^i \cdot g^{ji}(r) \;\Rightarrow\; g^{ji}(r) \in \boldsymbol{\Phi}_{u^i}^i;$$

in other words, $g^{ji}(\boldsymbol{\Phi}_{u^j}^j) \subseteq \boldsymbol{\Phi}_{u^i}^i$, for every $j \geq i$, thus we obtain the projective system $\{\boldsymbol{\Phi}_{u^i}^i; g^{ji}\}_{i,j \in \mathbb{N}}$ yielding the limit $\varprojlim \boldsymbol{\Phi}_{u^i}^i$. By similar arguments we obtain the limit of restricted holonomy groups $\varprojlim (\boldsymbol{\Phi}_{u^i}^i)^0$.

Now, if $r = (r^i)$ is an arbitrary element of $\boldsymbol{\Phi}_u$, there is a piecewise smooth curve α in B, such that $\alpha(0) = \alpha(1) = b$ and $\tau_\alpha(u) = u \cdot r$. Then $p^i(\tau_\alpha(u)) = p^i(u) \cdot g^i(r)$, equivalently $\tau_\alpha^i(u^i) = u^i \cdot r^i$; hence, $r^i \in \boldsymbol{\Phi}_{u^i}^i$ and $r = (r_i) \in \varprojlim \boldsymbol{\Phi}_{u^i}^i$. As a result,

(4.3.1) $$\boldsymbol{\Phi}_u \subseteq \varprojlim \boldsymbol{\Phi}_{u^i}^i.$$

To show that $\varprojlim \Phi^i_{u^i} \subseteq \overline{\Phi}_u$, we proceed as follows: For any $r = (r^i) \in \varprojlim \Phi^i_{u^i}$, there exist smooth curves $\beta_i \colon [0,1] \to B$ $(i \in \mathbb{N})$, such that

$$\beta_i(0) = \beta_i(1) = \pi^i(u^i) = b \quad \text{and} \quad \tau^i_{\beta_i}(u^i) = u^i \cdot r^i.$$

This is equivalent to saying that

(4.3.2) $$\widehat{\beta}^i_i(1) = u^i \cdot r^i,$$

where $\widehat{\beta}^i_i$ denotes the horizontal lift of β_i to the bundle P^i, with initial condition $\widehat{\beta}^i_i(0) = u^i$. Denoting, analogously, by $\widehat{\beta}^j_i$ the horizontal lift of β_i to P^j $(j > i)$, with $\widehat{\beta}^j_i(0) = u^j$, Proposition 4.3.1 implies that the limit

$$\gamma_i = \varprojlim_{j \in \mathbb{N}} \left(\widehat{\beta}^j_i \right) \colon [0,1] \longrightarrow P$$

exists, and coincides with the horizontal lift of β_i to P, with initial condition $\gamma_i(0) = \left(\widehat{\beta}^j_i(0) \right)_{j \in \mathbb{N}} = (u^j)_{j \in \mathbb{N}} = u$; that is,

$$\gamma_i = \varprojlim_{j \in \mathbb{N}} \left(\widehat{\beta}^j_i \right) = (\widehat{\beta}_i)_u.$$

Moreover, for every index $n \in \mathbb{N}$, there exists an element $z_n \in G$ such that $\gamma_n(1) = u \cdot z_n$, since $\gamma_n(1)$ and u belong to the same fibre $\pi^{-1}(b)$. Then, $\tau_{\gamma_n}(u) = u \cdot z_n$ and $(z_n)_{n \in \mathbb{N}}$ is a sequence of elements of Φ_u. We claim that the previous sequence converges to $r = (r^i) \in \varprojlim \Phi^i_{u^i}$. To this end, it suffices to prove that

$$\lim_{n \to +\infty} \left(g^i(z_n) \right) = r^i, \quad \forall\, i \in \mathbb{N}.$$

This is the case, since [see also (4.3.2)]

$$\gamma_n = \varprojlim_{j \in \mathbb{N}} \left(\widehat{\beta}^j_n \right) \;\Rightarrow\; p^i(\gamma_n(1)) = \widehat{\beta}^i_n(1) \;\Rightarrow\; p^i(u \cdot z_n) = u^i \cdot r^i$$

$$\Rightarrow\; u^i \cdot g^i(z_n) = u^i \cdot r^i \;\Rightarrow\; g^i(z_n) = r^i.$$

Therefore, $\varprojlim \Phi^i_{u^i} \subseteq \overline{\Phi}_u$, which, together with (4.3.1) proves the first assertion concerning Φ_u. The inclusions referring to the restricted holonomy group are proved in a similar way.

ii) From the proof of i) it is ensured that, for every $r^i \in \Phi^i_{u^i}$, there is at least one element $r \in \Phi_u$ such that $g^i(r) = r^i$; hence, $\Phi^i_{u^i} \subseteq g^i(\Phi_u)$. Conversely, in virtue of i), $g^i(\Phi_u) \subseteq g^i\left(\varprojlim \Phi^i_{u^i} \right) = \Phi^i_{u^i}$, thus we prove assertion ii).

iii) If Φ_u is closed in G, then the result is an immediate consequence of assertion i). On the other hand, if Φ_u is open, there is an open neighborhood U of the unit $e \in G$ such that $U \subset \Phi_u$. Then, for an arbitrarily chosen $r \in \varprojlim \Phi_{u^i}^i$, based on i), we check that there is a sequence $(z_n)_{n \in \mathbb{N}}$ in Φ_u with $\lim_{n \to +\infty} z_n = r$ and

$$\lim_{n \to +\infty} (z_n \cdot r^{-1}) = e$$

$$\Rightarrow z_n \cdot r^{-1} \in U \subset \Phi_u; \qquad n \geq n_0,$$

$$\Rightarrow r \in \Phi_u.$$

Therefore, $\Phi_u = \varprojlim \Phi_{u^i}^i$ in this case too. This completes the proof of assertion iii).

iv) Under the assumptions of the assertion iv), we have for every element $u \in P$:

$$P^1 \supseteq P^2 \supseteq \ldots, \quad \text{and} \quad \varprojlim P^i = \bigcap_{i \in \mathbb{N}} P^i,$$

$$G^1 \supseteq G^2 \supseteq \ldots, \quad \text{and} \quad \varprojlim G^i = \bigcap_{i \in \mathbb{N}} G^i,$$

$$\Phi_u^1 \supseteq \Phi_u^2 \supseteq \ldots, \quad \text{and} \quad \varprojlim \Phi_u^i = \bigcap \Phi_u^i.$$

Moreover, for every $s \in \Phi_u^1$, there exists a piecewise smooth curve $\alpha \colon [0, 1] \to B$ such that $\alpha(0) = \alpha(1) = b = \pi(u)$ and $\tau_\alpha^1(u) = u \cdot s$. However, $\tau_\alpha = \varprojlim \tau_\alpha^i$, which, in the case of a nested sequence as above, means that $\tau_\alpha^1|_{\pi^{-1}(b)} = \tau_\alpha$. In this way, for every $u \in \pi^{-1}(b)$, one has $\tau_\alpha(u) = u \cdot s$, which implies that $s \in \Phi_u$, thus $\Phi_u^1 \subseteq \Phi_u$. Since already $\Phi_u^1 \subseteq \Phi_u$, we conclude that $\Phi_u = \Phi_u^1$. This completes the proof. $\qquad \square$

Remark 4.3.6 Regarding the assertion i) of the previous theorem, we note that Theorem 4.5.5, in the end of § 4.5, provides a (counter)example showing that the equality $\Phi_u = \varprojlim \Phi_{u^i}^i$ is not necessarily true.

4.4 The curvature of a plb-connection

Let $\omega = \varprojlim \omega^i$ be a plb-connection (recall Theorem 4.2.9 and the terminology induced after its proof) on the plb-principal bundle $(P, G, B, \pi) = \left(\varprojlim P^i, \varprojlim G^i, B, \varprojlim \pi^i \right)$. The **curvature** of ω is the 2-form Ω, defined by $\Omega = d\omega \circ (h \times h$, where $h \colon TP \to HP$ is the projection to the horizontal subbundle. The curvature of ω^i is denoted by Ω^i, for all $i \in \mathbb{N}$.

To establish the point-wise convergence of (Ω^i) to Ω, we first need the following result, whose proof is quite lengthy and technical.

Lemma 4.4.1   *With the previous notations, if* $(d\omega^i)_{i\in\mathbb{N}}$ *are the exterior differentials of* $(\omega^i)_{i\in\mathbb{N}}$, *then, for every* $u^j \in P^j$ *and* $j \geq i$, *the following diagram is commutative:*

$$
\begin{array}{ccc}
T_{u^j}P^j \times T_{u^j}P^j & \xrightarrow{\ \ d\omega^j(u^j)\ \ } & \mathfrak{g}^j \\[2mm]
{\scriptstyle T_{u^j}p^{ji} \times T_{u^j}p^{ji}}\Big\downarrow & & \Big\downarrow{\scriptstyle T_{e^j}g^{ji}} \\[2mm]
T_{u^i}P^i \times T_{u^i}P^i & \xrightarrow[\ \ d\omega^i(u^i)\ \ ]{} & \mathfrak{g}^i
\end{array}
$$

where $u^i = p^{ji}(u^j)$.

Proof   Let $x := \pi^j(u^j) \in B$. We consider a chart (U, ϕ) of B containing x and a trivialization $(U, \Phi = \varprojlim \Phi^i)$ of P (we can take the same U for all of them). We choose an arbitrary $u = (u^i) \in \pi^{-1}(U)$ and a plb-chart $(V, \psi) = \left(\varprojlim V^i, \varprojlim \psi^i\right)$ of G such that $\mathrm{pr}_2(\Phi(u)) \in V$. Since u^j and $p^j(u)$ belong to the same fibre of P^j, there exists a (unique) $a^j \in G^j$ such that $u^j = p^j(u) \cdot a^j$, thus $\Phi_2^j(u^j) = \Phi_2^j(p^j(u)) \cdot a^j$, where $\Phi_2^j = \mathrm{pr}_2 \circ \Phi^j$. Then the pair $\left(V^j \cdot a^j, \psi^j \circ \rho_{(a^j)^{-1}}^j\right)$ is a chart of G^j containing $\Phi_2^j(u^j)$. Here $\rho_{(a^j)^{-1}}^j$ denotes the right translation of G^j by $(a^j)^{-1}$.

Also, the equality $u^j = p^j(u) \cdot a^j$ implies that $u^i = p^{ji}(u^j) = p^{ji}(p^j(u) \cdot a^j) = p^i(u) \cdot g^{ji}(a^j)$, thus the pair $\left(V \cdot g^{ji}(a^j), \psi^j \circ \rho_{g^{ji}(a^j)^{-1}}^j\right)$ is a chart of G^i containing $\Phi_2^i(u^i)$. Therefore, we may construct the charts (W^j, χ^j) and (W^i, χ^i) of P^j and P^i, respectively, with

$$W^j := (\Phi^j)^{-1}\left(U \times V^j \cdot a^j\right), \quad W^i := (\Phi^i)^{-1}\left(U \times V^i \cdot g^{ji}(a^j)\right),$$

$$\chi^j := \left(\phi \times \left(\psi^j \circ \rho_{(a^j)^{-1}}^j\right)\right) \circ \Phi^j, \quad \chi^i := \left(\phi \times \left(\psi^i \circ \rho_{g^{ji}(a^j)^{-1}}^i\right)\right) \circ \Phi^i,$$

as well as the associated charts $\left(\tau_{P^j}^{-1}(W^j), \widetilde{\chi}^j\right)$, $\left(\tau_{P^i}^{-1}(W^i), \widetilde{\chi}^i\right)$ of the tangent bundles (TP^j, P^j, τ_{P^j}) and (TP^i, P^i, τ_{P^i}).

Denoting by $\zeta^{ji} \colon \mathbb{G}^j \to \mathbb{G}^i$ the connecting morphisms of the model $\mathbb{G} = \varprojlim \mathbb{G}^i$ of G, we check that

$$(4.4.1) \qquad \left(\mathrm{id}_{\phi(U)} \times \zeta^{ji}\right) \circ \chi^j = \chi^i \circ p^{ji}, \qquad j \geq i.$$

Indeed,

$$
\begin{aligned}
\left(\mathrm{id}_{\phi(U)} \times \zeta^{ji}\right) \circ \chi^j &= \left(\mathrm{id}_{\phi(U)} \times \zeta^{ji}\right) \circ \left(\phi \times \left(\psi^j \circ \rho^j_{(a^j)-1}\right)\right) \circ \Phi^j \\
&= \left(\phi \times \left(\zeta^{ji} \circ \psi^j \circ \rho^j_{(a^j)-1}\right)\right) \circ \Phi^j \\
&= \left(\phi \times \left(\psi^i \circ g^{ji} \circ \rho^j_{(a^j)-1}\right)\right) \circ \Phi^j \\
&= \left(\phi \times \left(\psi^i \circ \rho^i_{g^{ji}(a^j)-1} \circ g^{ji}\right)\right) \circ \Phi^j \\
&= \left(\phi \times \left(\psi^i \circ \rho^i_{g^{ji}(a^j)-1}\right)\right) \circ (\mathrm{id}_U \times g^{ji}) \circ \Phi^j \\
&= \left(\phi \times \left(\psi^i \circ \rho^i_{g^{ji}(a^j)-1}\right)\right) \circ \Phi^i \circ p^{ji} \\
&= \chi^i \circ p^{ji}.
\end{aligned}
$$

Analogously, for the above charts of the tangent bundles,

$$(4.4.2) \qquad \left(\mathrm{id}_{\phi(U)} \times \zeta^{ji}\right) \times \left(\mathrm{id}_B \circ \zeta^{ji}\right) \circ \chi^j = \widetilde{\chi}^i \circ Tp^{ji}, \qquad j \geq i.$$

Taking now two arbitrary tangent vectors $X, Y \in T_{u^j} P^j$, we define the following vector fields:

$$
\begin{aligned}
\xi^j &: W^j \longrightarrow TP^j : z \mapsto \left(\widetilde{\chi}^j\right)^{-1}\left(\chi^j(z), \overline{\chi}^j_{u^j}(X)\right), \\
\xi^i &: W^i \longrightarrow TP^i : y \mapsto \left(\widetilde{\chi}^i\right)^{-1}\left(\chi^i(y), \overline{\chi}^i_{u^i}(T_{u^j} p^{ji}(X))\right), \\
\eta^j &: W^j \longrightarrow TP^j : z \mapsto \left(\widetilde{\chi}^j\right)^{-1}\left(\chi^j(z), \overline{\chi}^j_{u^j}(Y)\right), \\
\eta^i &: W^i \longrightarrow TP^i : y \mapsto \left(\widetilde{\chi}^i\right)^{-1}\left(\chi^i(y), \overline{\chi}^i_{u^i}(T_{u^j} p^{ji}(Y))\right),
\end{aligned}
$$

where, as in (1.1.4), $\overline{\chi}^r_{u^r} : T_{u^r} P^r \to \mathbb{B} \times \mathbb{G}^r$ is the linear isomorphism induced by the chart (W^r, χ^r). Then the definition of the tangent charts [see (1.1.7)] implies that

$$(4.4.3) \qquad \begin{aligned} \xi^j(u^j) &= X, \quad \xi^i(u^i) = T_{u^j} p^{ji}(X), \\ \eta^j(u^j) &= Y, \quad \eta^i(u^i) = T_{u^j} p^{ji}(Y). \end{aligned}$$

On the other hand, in virtue of (4.4.2), (4.4.1),

$$
\begin{aligned}
Tp^{ji} \circ \xi^j &= Tp^{ji} \circ \left(\chi^j\right)^{-1} \circ \left(\chi^j, \overline{\chi}^j_{u^j}(X)\right) \\
&= \left(\chi^i\right)^{-1} \circ \left(\mathrm{id}_{\phi(U)} \times \zeta^{ji}\right) \times \left(\mathrm{id}_\mathbb{B} \times \zeta^{ji}\right) \circ \left(\chi^j, \overline{\chi}^j_{u^j}(X)\right) \\
&= \left(\chi^i\right)^{-1} \circ \left(\left(\mathrm{id}_{\phi(U)} \times \zeta^{ji}\right) \circ \chi^j, \left(\mathrm{id}_B \times \zeta^{ji}\right)\left(\overline{\chi}^j_{u^j}(X)\right)\right) \\
&= \left(\chi^i\right)^{-1} \circ \left(\chi^i \circ p^{ji}, \overline{\chi}^j_{u^j}(T_{u^j} p^{ji}(X))\right) \\
&= \left(\chi^i\right)^{-1} \circ \left(\chi^i, \overline{\chi}^j_{u^j}(T_{u^j} p^{ji}(X))\right) \circ p^{ji} \\
&= \xi^i \circ p^{ji};
\end{aligned}
$$

that is, ξ^j and ξ^i $(j \geq i)$ are p^{ji}-*related*. Similarly, η^j and η^i are p^{ji}-related, and so are $[\xi^j, \eta^j]$ and $[\xi^i, \eta^i]$.

Now, going back to the desired commutativity of the statement, we observe that, for arbitrary $X, Y \in T_{u^j} P^j$ as before,

$$\left(T_{e^j} g^{ji} \circ d\omega^j(u^j)\right)(X, Y) = \left(d\omega^i(u^i) \circ (T_{u^j} p^{ji} \times T_{u^j} p^{ji})\right)(X, Y)$$

$$\Leftrightarrow \quad T_{e^j} g^{ji} \left(d\omega^j(u^j)(\xi^j(u^j), \eta^j(u^j))\right) = d\omega^i(u^i)\left(\xi^i(u^i), \eta^i(u^i)\right)$$

$$(*) \Leftrightarrow \quad T_{e^j} g^{ji} \left(d\omega^j(\xi^j, \eta^j)(u^j)\right) = d\omega^i(\xi^i, \eta^i)(u^i)$$

$$\Leftrightarrow \quad T_{e^j} g^{ji} \left(\left(\xi^j(\omega^j(\eta^j)) - \eta^j(\omega^j(\xi^j)) - \omega^j([\xi^j, \eta^j])\right)(u^j)\right) =$$
$$= \left(\xi^i(\omega^i(\eta^i)) - \eta^i(\omega^i(\xi^i)) - \omega^i([\xi^i, \eta^i])\right)(u^i)$$

$$\Leftrightarrow \quad T_{e^j} g^{ji}\left(T_{u^j}\omega^j(\eta^j)(\xi^j(u^j))\right) - T_{e^j} g^{ji}\left(T_{u^j}\omega^j(\xi^j)(\eta^j(u^j))\right) -$$
$$- T_{e^j} g^{ji}\left(\omega^j(\xi^j).[\xi^i, \eta^i](u^j)\right) =$$
$$= T_{u^i}\omega^i(\eta^i)\left(\xi^i(u^i)\right) - T_{u^i}(\omega^i(\xi^i)\left(\eta^i(u^i)\right) - \omega^i(u^i).[\xi^i, \eta^i](u^i).$$

Thus, applying (4.4.3), it suffices to verify the equalities

$$(4.4.4a) \qquad T_{e^j} g^{ji}\left(T_{u^j}\omega^j(\eta^j)(X)\right) = T_{u^i}\omega^i(\eta^i)\left(T_{u^j} p^{ji}(X)\right)$$

$$(4.4.4b) \qquad T_{e^j} g^{ji}\left(T_{u^j}\omega^j(\xi^j)(Y)\right) = T_{u^i}\omega^i(\xi^i)\left(T_{u^j} p^{ji}(Y)\right)$$

$$(4.4.4c) \qquad T_{e^j} g^{ji}\left(\omega^j(u^j).[\xi^j, \eta^j](u^j)\right) = \omega^i(u^i).[\xi^i, \eta^i](u^i)$$

The $(p^{ji}, g^{ji}, \mathrm{id}_B)$-relatedness of ω^j and ω^i implies

$$T_{e^j} g^{ji}\left(\omega^j(u^j).[\xi^j, \eta^j](u^j)\right) = \omega^i(u^i)\left(T_{u^j} p^{ji}[\xi^j, \eta^j](u^j)\right) =$$
$$= \omega^i(u^i).[\xi^j, \eta^j](p^{ji}(u^j)) = \omega^i(u^i).[\xi^i, \eta^i](u^i),$$

thus proving (4.4.4c). Equality (4.4.4a) will be a consequence of the commutative diagram

$$
\begin{array}{ccc}
T_{u^j} P^j & \xrightarrow{\;T_{u^j}\omega^j(\eta^j)\;} & T_{\omega^j(\eta^j)(Y)}\mathfrak{g}^j \\[2pt]
{\scriptstyle T_{u^j} p^{ji}}\Big\downarrow & & \Big\downarrow{\scriptstyle T_{\omega^j(\eta^j)(Y)}(T_{e^j} g^{ji})} \\[2pt]
T_{u^i} P^i & \xrightarrow[\;T_{u^i}\omega^i(\eta^i)\;]{} & T_{\omega^i(\eta^i)(Y^i)}\mathfrak{g}^i
\end{array}
$$

where $Y^i := T_{u^j} g^{ji}(Y)$, derived by differentiation of

$$(4.4.5) \qquad T_{e^j} g^{ji} \circ \omega^j(\eta^j) = \omega^i(\eta^i) \circ p^{ji}.$$

The preceding equality is satisfied, because, for every $z \in P^j$,

$$T_{e^j} g^{ji}.\omega^j(\eta^j)(z) = \omega^i(\eta^i).p^{ji}(z)$$

$$\Leftrightarrow \quad T_{e^j} g^{ji}\left(\omega_z^j(\eta^j(z))\right) = \omega_{p^{ji}(z)}^i\left(\eta^i(p^{ji}(z))\right)$$

$$\Leftrightarrow \quad \omega_{p^{ji}(z)}^j\left(T_z p^{ji}(\eta^j(z))\right) = \omega_{p^{ji}(z)}^i\left(\eta^i(p^{ji}(z))\right)$$

the last equality being true since, as already proved, η^j and η^i are p^{ji}-related vector fields. Therefore, after the identifications

$$T_{\omega^j(\eta^j)(Y)}\mathfrak{g}^j \equiv \mathfrak{g}^j, \; T_{\omega^i(\eta^i)(Y)}\mathfrak{g}^i \equiv \mathfrak{g}^i, \; T_{\omega^j(\eta^j)(Y)}\left(T_{e^j}g^{ji}\right) \equiv T^{e^j}g^{ji}$$

and differentiation of (4.4.5), we obtain (4.4.4a). The proof of (4.4.4b) is similar. This completes the proof of the lemma. $\qquad\square$

Lemma 4.4.2 *Let $d\omega$ be the exterior differential of $\omega = \varprojlim \omega^i$. Then, for every $u = (u^i) \in P = \varprojlim P^i$, the diagram*

$$
\begin{array}{ccc}
T_u P \times T_u P & \xrightarrow{\;d\omega(u)\;} & \mathfrak{g} \\
\downarrow{\scriptstyle T_u p^j \times T_u p^j} & & \downarrow{\scriptstyle T_e g^i} \\
T_{u^i} P^i \times T_{u^i} P^i & \xrightarrow{\;d\omega^i(u^i)\;} & \mathfrak{g}^i
\end{array}
$$

is commutative, where $p^i \colon P \to P^i$ and $g^i \colon G = \varprojlim G^i \to G^i$ are the canonical projections.

Proof Following the main lines of the previous proof, we consider a chart $(W, \chi) = \left(\varprojlim W^i, \varprojlim \chi^i\right)$ of P at u, as well as the induced charts $(\tau_P^{-1}(W), \widetilde{\chi})$ and $\left(((\tau_{P^i})^{-1}(W^i), \widetilde{\chi}^i)\right)$ of (TP, P, τ_P) and (TP^i, P^i, τ_{P^i}). For arbitrary $X, Y \in T_u P$, we define the vector fields

$$\xi \colon W \longrightarrow TP \colon z \mapsto \widetilde{\chi}^{-1}\left(\chi(z), \overline{\chi}_u(X)\right),$$

$$\xi^i \colon W^i \longrightarrow TP^i \colon y \mapsto \left(\widetilde{\chi}^i\right)^{-1}\left(\chi^i(y), \overline{\chi}_{u^i}^i(T_u p^i(X))\right),$$

$$\eta \colon W \longrightarrow TP \colon z \mapsto \widetilde{\chi}^{-1}\left(\chi(z), \overline{\chi}_u(Y)\right),$$

$$\eta^i \colon W^i \longrightarrow TP^i \colon y \mapsto \left(\widetilde{\chi}^y\right)^{-1}\left(\chi^i(y), \overline{\chi}_{u^i}^y(T_u p^i(Y))\right),$$

It is clear that

$$\xi(u) = X, \quad \xi^i(u^i) = T_u p^i(X), \quad \eta(u) = Y, \quad \eta^i(u^i) = T_u p^i(Y).$$

Moreover, ξ, ξ^i and η, η^i are p^i-related, and so are $[\xi, \eta]$ and $[\xi^i, \eta^i]$. Thus the desired commutativity of the diagram is equivalent to

$$T_e g^i \big(d\omega(u)(X,Y)\big) = d\omega^i(u^i) \big(T_u p^i(X), T_u p^i(Y)\big)$$
$$\Leftrightarrow \quad T_e g^i \big(d\omega(u)(\xi(u), \eta(u))\big) = d\omega^i(u^i)\big(\xi^i(u^i), \eta^i(u^i)\big)$$
$$\Leftrightarrow \quad T_e g^i \big(d\omega(\xi, \eta)(u)\big) = d\omega^i(\xi^i, \eta^i)(u^i)$$

The last equality is proved as its analog $(*)$ in the proof of Lemma 4.4.1

$\square$

Corollary 4.4.3   *With the notation of the preceding lemmata,*

$$d\omega(u) = \varprojlim d\omega^i(u^i); \qquad u = (u^i) \in P, \ i \in \mathbb{N}.$$

Proof   Lemma 4.4.1 implies the existence of $\varprojlim d\omega^i$. The latter coincides with $d\omega(u)$ in virtue of Lemma 4.4.2 and Proposition 2.3.5. $\square$

We prove now the following main result of this section.

Theorem 4.4.4   *Let $\omega = \varprojlim \omega^i$ be a connection on a plb-principal bundle $\ell = \varprojlim \ell^i$. If Ω is the curvature of ω and Ω^i the curvature of ω^i, for all $i \in \mathbb{N}$, then $\Omega(u) = \varprojlim \Omega^i(u^i)$, for every $u = (u^i) \in P$.*

Proof   First we check that, for every $u^j \in P^j$, the diagram

$$
\begin{array}{ccc}
T_{u^j} P^j \times T_{u^j} P^j & \xrightarrow{\ \Omega^j(u^j)\ } & \mathfrak{g}^j \\[2mm]
\Big\downarrow {\scriptstyle T_{u^j} p^{ji} \times T_{u^j} p^{ji}} & & \Big\downarrow {\scriptstyle T_{e^j} g^{ji}} \\[2mm]
T_{u^i} P^i \times T_{u^i} P^i & \xrightarrow[\ \Omega^i(u^i)\ ]{} & \mathfrak{g}^i
\end{array}
$$

is commutative, where $u^i = p^{ji}(u^j)$. Indeed, for any $X, Y \in T_{u^j} P^j$, we denote by X^h and Y^h their horizontal components. Then the definition of curvature implies that

$$
\begin{aligned}
\big(T_{e^j} g^{ji} \circ \Omega^j_{u^j}\big)(X, Y) &= T_{e^j} g^{ji} \big(d\omega^j(u^j)(X^h, Y^h)\big) \\
\text{[Lemma 4.4.1]} \qquad &= d\omega^i(u^i)\big(T_{u^j} p^{ji}(X^h), T_{u^j} p^{ji}(Y^h)\big) \\
\text{[see (1.7.19)]} \qquad &= d\omega^i(u^i)\big(T_{u^j} p^{ji}(X)^h, T_{u^j} p^{ji}(Y)^h\big) \\
&= \Omega^i(u^i)\big(T_{u^j} p^{ji}(X), T_{u^j} p^{ji}(Y)\big) \\
&= \big(\Omega^i(u^i) \circ (T_{u^j} p^{ji} \times T_{u^j} p^{ji})\big)(X, Y).
\end{aligned}
$$

For convenience, we have used the same superscript h to denote the horizontal projections of all the connections.

The previous conclusion implies the existence of $\varprojlim \Omega^i$. Thus, to prove the statement, it is sufficient to check also the commutativity of the diagram

$$
\begin{array}{ccc}
T_u P \times T_u P & \xrightarrow{\;\Omega(u)\;} & \mathfrak{g} \\[2pt]
\Big\downarrow{\scriptstyle T_u p^i \times T_u p^i} & & \Big\downarrow{\scriptstyle T_e g^i} \\[2pt]
T_{u^i} P^i \times T_{u^i} P^i & \xrightarrow[\;\Omega^i(u^i)\;]{} & \mathfrak{g}^i
\end{array}
$$

for every $i \in \mathbb{N}$. Indeed, for every $X, Y \in T_u P$,

$$
\begin{aligned}
(T_e g^i \circ \Omega(u))(X, Y) &= T_e g^i \big(d\omega(u)(X^h, Y^h) \big) \\
\text{[Lemma 4.4.2]} \qquad &= d\omega^i(u^i) \big(T_u p^i(X^h), T_u p^i(Y^h) \big) \\
&= d\omega^i(u^i) \big(T_u p^i(X)^h, T_u p^i(Y)^h \big) \\
&= \Omega^i(u^i) \big(T_{u^j} p^{ji}(X), T_{u^j} p^{ji}(Y) \big) \\
&= \big(\Omega^i(u^i) \circ (T_u p^i \times T_u p^i) \big)(X, Y). \qquad \square
\end{aligned}
$$

In view of the preceding result, under the reservations following Definition 4.2.1, we symbolically write $\Omega = \varprojlim \Omega^i$.

Corollary 4.4.5 Ω *is horizontal and* $R_g^* \Omega = \mathrm{Ad}(g^{-1}\Omega)$.

Proof By definition, Ω is horizontal. The second property is proved using the same arguments as in the proof of Lemma 4.2.4. $\square$

As expected, Ω satisfies the structural equation. But this requires the following:

Lemma 4.4.6 *Given a connection* $\omega = \varprojlim \omega^i$ *on* $P = \varprojlim P^i$, *it follows that* $[\omega, \omega](u) = \varprojlim [\omega^i, \omega^i]^i(u^i)$, *for every* $u = (u^i) \in P$, *where* $[\,,\,]$ *and* $[\,,\,]^i$ *are, respectively, the brackets of* $\mathfrak{g}$ *and* $\mathfrak{g}^i$, *for all* $i \in \mathbb{N}$.

Proof In analogy to previous lemmata, we need to prove the equalities

$$
(4.4.6) \qquad T_{e^j} g^{ji} \circ [\omega^j, \omega^j]^j = [\omega^i, \omega^i]^i \circ \big(T_{u^j} p^{ji} \times T_{u^j} p^{ji} \big),
$$

$$
(4.4.7) \qquad T_e g \circ [\omega, \omega] = [\omega^i, \omega^i]^i \circ \big(T_u p^i \times T_u p^i \big).
$$

Indeed, for every $X, Y \in T_{u^j} P^j$, the definition of the bracket of forms and the $(p^{ji}, g^{ji}, \mathrm{id}_B)$-relatedness of ω^j, ω^i $(j \geq i)$ yield:

$$
\begin{aligned}
\big(T_{e^j} g^{ji} \circ [\omega^j, \omega^j]^j \big)(X, Y) &= T_{e^j} g^{ji} \big([\omega^j(u^j)(X), \omega^j(u^j)(Y)]^j \big) \\
&= \big[\omega^i(u^i)\big(T_{u^j} p^{ji}(X)\big), \omega^i(u^i)\big(T_{u^j} p^{ji}(Y)\big) \big]^i \\
&= [\omega^i, \omega^i]^i \circ \big(T_{u^j} p^{ji} \times T_{u^j} p^{ji} \big)(X, Y),
\end{aligned}
$$

which leads to (4.4.6). The proof of (4.4.7) is similar, using the fact that ω and ω^i are $(p^i, g^i, \mathrm{id}_B)$-related. $\qquad\square$

Proposition 4.4.7 *With the previous assumptions, the curvature* $\Omega = \varprojlim \Omega^i$ *satisfies Cartan's (second) structure equation* $\Omega = d\omega + \frac{1}{2}[\omega, \omega]$.

Proof In virtue of the structure equation for Banach principal bundles (1.8.1), Corollary 4.4.3 and Lemma 4.4.6, we find for every $u = (u^i) \in P$:

$$\Omega(u) = \varprojlim \Omega^i(u^i) = \varprojlim \left(d\omega^i(u^i) + \frac{1}{2}[\omega^i, \omega^i]^i(u^i) \right) =$$

$$= \varprojlim \left(d\omega^i(u^i) \right) + \frac{1}{2} \varprojlim \left([\omega^i, \omega^i]^i(u^i) \right) = d\omega + \frac{1}{2}[\omega, \omega](u). \qquad\square$$

Corollary 4.4.8 *The curvature* $\Omega = \varprojlim \Omega^i$ *satisfies the Bianchi identity* $d\Omega = [\omega, \omega]$. *Equivalently,* $D\Omega = 0$.

Proof We proceed by applying the limit process to (1.8.4) and (1.8.4′). Note that the horizontal projection $h \colon T_u P \to H_u P$, $u = (u^i) \in P$, coincides with $\varprojlim h^i \colon T_{u^i} P^i \to H_{u^i} P^i$ by Proposition 4.2.7. $\qquad\square$

4.5 Flat plb-bundles

We shall discuss the relationship of flat connections with the holonomy groups and holonomy homomorphisms.

Throughout this section we fix a plb-bundle $\ell = (P, G, B, \pi)$ with *connected* base B. If $\omega = \varprojlim \omega^i$ is a connection on a P with curvature (form) Ω, then, according to the terminology of §§ 1.8.2 and 1.9, ω is called *flat* if $\Omega = 0$. In this case, the pair (P, ω) is called a *flat bundle*.

Proposition 4.5.1 *The following assertions are equivalent:*
 i) ω *is flat.*
 ii) Every ω^i, $i \in \mathbb{N}$, *is flat.*
 iii) The restricted holonomy group Φ_u^0 *of* ω *is trivial, for any* $u \in P$.

Proof i) $\Rightarrow$ ii): The assumption implies that $\varprojlim \Omega^i(u^i) = \Omega(u) = 0$, for every $u = (u^i) \in P$, thus

$$(4.5.1) \qquad\qquad \Omega_{u^i}^i = \Omega^i(u^i) = 0.$$

To prove ii), it suffices to show that $\Omega_{v^i}^i(X^i, Y^i) = 0$, for arbitrary $v^i \in P^i$ and $X^i, Y^i \in H_{v^i} P^i$. Indeed, if $\pi(v^i) = b \in B$, we consider also an arbitrary $u = (u^i) \in \pi^{-1}(b)$. Since $u^i, v^i \in (\pi^i)^{-1}(b)$, there is a (unique) $a \in G^i$ such that $v^i = u^i \cdot a$. Then, according to (1.7.4),

$T_{u^i} R_a^i(H_{u^i} P^i) = H_{v^i} P^i$; hence, there are $\bar{X}^i, \bar{Y}^i \in H_{u^i} P^i$ such that $T_{u^i} R_a^i(\bar{X}^i) = X^i$ and $T_{u^i} R_a^i(\bar{Y}^i) = Y^i$, where R_a^i is the right translation of P^i by $a \in G^i$. Moreover, $\pi \circ p^i = \pi$ yields $T_{u^i} \pi^i \circ T_u p^i = T_u \pi$, thus, restricting to the horizontal subspaces, we obtain the commutative diagram

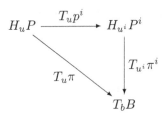

Since $T_u \pi$ and $T_{u^i} \pi^i$ are (top)linear isomorphisms, so is $T_u p^i$. Therefore, there are $X, Y \in T_u P$ such that $T_u p^i(X) = \bar{X}^i$ and $T_u p^i(Y) = \bar{Y}^i$. The previous considerations now yield:

$$
\begin{aligned}
\Omega_{v^i}^i(X^i, Y^i) &= \Omega_{u^i \cdot a}^i \left(T_{u^i} R_a^i \left(T_u p^i(X) \right), T_{u^i} R_a^i \left(T_u p^i(Y) \right) \right) \\
&= \left((R_a^i)^* \Omega^i \right)_{u^i} \left(T_u p^i(X), T_u p^i(Y) \right) \\
&= \mathrm{Ad}^i(a^{-1}) \left(\Omega_{u^i}^i \left(T_u p^i(X), T_u p^i(Y) \right) \right) \\
[\text{by (4.5.1)}] \quad &= \mathrm{Ad}^i(a^{-1})(0) = 0.
\end{aligned}
$$

ii) $\Rightarrow$ i): Obvious, in virtue of Theorem 4.4.4.

i) $\Rightarrow$ iii): By ii) and the properties of the holonomy group in the Banach context (see [Max72] and Theorem 1.9.1), we have that $(\boldsymbol{\Phi}_{u^i}^i)^0 = e^i$, for every $u^i \in P^i$. As a result, for any $u = (u^i) \in \mathbb{N}$, assertion i) of Theorem 4.3.5 implies that

$$\{e\} \subseteq \boldsymbol{\Phi}_u^0 \subseteq \varprojlim(\boldsymbol{\Phi}_{u^i}^i)^0 = \varprojlim(\{e^i\}) = \{e\},$$

thus $\boldsymbol{\Phi}_u^0 = e$.

iii) $\Rightarrow$ i): By the assumption and assertion ii) of Theorem 4.3.5,

$$(\boldsymbol{\Phi}_{u^i}^i)^0 = g^i \left(\boldsymbol{\Phi}_u^0 \right) = g^i(\{e\}) = \{e^i\};$$

hence ω^i is flat. This concludes the proof in virtue of i) $\Leftrightarrow$ ii). $\qquad \square$

We turn now to the holonomy homomorphism of a *flat* connection $\omega = \varprojlim \omega^i$ on a plb-bundle (P, G, B, π) with *connected* B. As we have sen in §1.10, $h_\omega \colon \pi_1(B) \to G$ and $h_{\omega^i} \colon \pi_1(B) \to G^i$ denote the corresponding holonomy homomorphisms of ω and ω^i ($i \in \mathbb{N}$).

Proposition 4.5.2 *With the previous notations, $h_\omega = \varprojlim h_{\omega^i}$.*

Proof In virtue of (1.10.1),

$$h_\omega([\alpha]) = k_u(\tau_\alpha), \quad h_{\omega^i}([\alpha]) = k^i_{u^i}(\tau^i_\alpha),$$

for any (fixed) $u = (u^i) \in P$ and every $[\alpha] \in \pi_1(P)$. Recall that k_u is defined by $\tau_\alpha(u) = u \cdot k_u(\tau_\alpha)$. The map $k^i_{u^i}$ is defined analogously.

To prove the statement, first we check that

(4.5.2) $$g^{ji} \circ h_{\omega^j} = h_{\omega^i}, \quad j \geq i.$$

To this end we observe that the existence of the limit $\tau_\alpha = \varprojlim \tau^i_\alpha$ (see Proposition 4.3.2) means that $p^{ji} \circ \tau^j_\alpha = \tau^i_\alpha \circ p^{ji}$, thus the bundle morphism $(p^{ji}, g^{ji}, \mathrm{id}_B)$ implies that

$$p^{ji}\big(\tau^j_\alpha(u^j)\big) = \big(u^j \cdot k^j_{u^j}(\tau^j_\alpha)\big) = u^i \cdot g^{ji}\big(k^j_{u^j}(\tau^j_\alpha)\big)$$
$$\Rightarrow \quad \tau^i_\alpha = u^i \cdot g^{ji}\big(k^j_{u^j}(\tau^j_\alpha)\big)$$
$$\Rightarrow \quad g^{ji}\big((k^j_{u^j}(\tau^j_\alpha)\big) = k^i_{u^i}(\tau^i_\alpha)$$
$$\Rightarrow \quad (g^{ji} \circ h_{\omega^j})([\alpha]) = h_{\omega^i}([\alpha]),$$

for every $[\alpha] \in \pi_1(B)$. This proves (4.5.2) ensuring, in turn, the existence of $\varprojlim h_{\omega^i}$. By similar arguments, we see that $g^i \circ h_{\omega^j} = h_{\omega^i}$, for every $i \in \mathbb{N}$. Therefore, Proposition 2.3.5 completes the proof. $\square$

Let G be a Fréchet-Lie group. Following the construction described in §1.10, a homomorphism $h : \pi_1(B) \to G$ determines a flat bundle (Q, θ) over B. In full terms, the bundle has the form $\ell = (Q, G, B, \pi_Q)$, with total space $Q = (\tilde{B} \times G)/\pi_1(B)$ and projection given by $\pi_Q([(\tilde{x}, s)]) = \tilde{p}(\tilde{x})$. More precisely, ℓ is associated, by the homomorphism h, to the principal bundle $(\tilde{B}, \pi_1(B), B, \tilde{p})$ determined by the universal covering $\tilde{B}$ of B. On the other hand, if $\kappa : \tilde{B} \times G \to (\tilde{B} \times G)/\pi_1(B)$ is the canonical map, and ω_o the canonical flat connection on the trivial bundle $\ell_o = (\tilde{B} \times G, \tilde{B}, \mathrm{pr}_1)$, then ω_o is $(\kappa, \mathrm{id}_G, \tilde{p})$-related with θ (actually this relationship determines θ). Recall that, if L_s is the left translation of G by $s \in G$, ω_o is given by

$$\omega_o(\tilde{x}, s).(X, Y) = T_s L_{s^{-1}}(Y),$$

for every $(\tilde{x}, s) \in \tilde{B} \times G$ and every $(X, Y) \in T_{\tilde{x}}\tilde{B} \times T_s G$.

In particular, if $G = \varprojlim G^i$, we know (see Remark 4.2.13) that ℓ is— up to an isomorphism—a plb-principal bundle and θ is related with a projective limit of connections. However, for such a G, we can define a concrete isomorphism relating (Q, θ) with a specific projective limit of Banach flat bundles. Before going into details, we define the group homomorphisms

$$(4.5.3) \qquad h^i := g^i \circ h : \pi_1(B) \longrightarrow G^i; \qquad i \in \mathbb{N},$$

where $g^i : G \to G^i$, are the canonical projections. Thus $h = \varprojlim h^i$.

Theorem 4.5.3 *Let B be a connected Banach manifold, $G = \varprojlim G^i$ a (Fréchet) plb-group, and $h : \pi_1(B) \equiv \pi_1(B, x_o) \to G$ a Lie group morphism. If (Q, θ) is the principal bundle induced by h, then there is a principal bundle isomorphism $(F, \mathrm{id}_G, \mathrm{id}_B)$ such that:*

i) (Q, G, B, π_Q) is $(F, \mathrm{id}_G, \mathrm{id}_B)$-isomorphic with $\left(\varprojlim P^i, G, B, \varprojlim \pi^i \right)$.

ii) θ is $(F, \mathrm{id}_G, \mathrm{id}_B)$-related with $\varprojlim \omega^i$, where (P^i, ω^i), with $P^i \equiv (P^i, G^i, B, \pi^i)$, is the flat bundle induced by h^i, for every $i \in \mathbb{N}$.

Proof We prove the properties of the statement in a series of steps.

Construction of $\varprojlim P^i$: Each h^i induces a Banach principal bundle $\ell^i = (P^i, G^i, B, \pi^i)$, equipped with a flat connection ω^i. In fact, we have that $P^i = (\tilde{B} \times G^i)/\pi_1(B)$ and $\pi^i([(\tilde{x}, s)]) = \tilde{p}(\tilde{x})$. The canonical flat connection ω_o^i of $\ell_o^i = (\tilde{B} \times G^i, \tilde{B}, \mathrm{pr}_1)$ is $(\kappa^i, \mathrm{id}_{G^i}, \tilde{p})$-related with ω^i, where $\kappa^i : \tilde{B} \times G^i \to (\tilde{B} \times G^i)/\pi_1(B)$ is the canonical map. Analogously, $\omega_o^i(\tilde{x}, a).(X, A) = T_a L_{a^{-1}}^i(X, A)$, for every $(\tilde{x}, a) \in \tilde{B} \times G^i$ and $(X, A) \in T_{\tilde{x}} \tilde{B} \times T_a G^i$, with L_a^i denoting the left translation of G^i by a.

For every $i, j \in \mathbb{N}$ with $j \geq i$, we define the map

$$(4.5.4) \qquad p^{ji} : P^j \longrightarrow P^i : [(\tilde{x}, s)]^j \mapsto [(\tilde{x}, g^{ji}(s))]^i.$$

We recall that the connecting morphisms of G, $g^{ji} : G^j \to G^i$, are group homomorphisms.

(a) p^{ji} *is well-defined*:

$$[(\tilde{x}, s)]^j = [(\tilde{y}, t)]^j$$
$$\Rightarrow \quad \exists \, [\gamma] \in \pi_1(B) : \tilde{x}, s) = (\tilde{y}, t) \cdot [\gamma]$$
$$\Rightarrow \quad (\tilde{x}, s) = \left(\tilde{y} \cdot [\gamma], h^j([\gamma])^{-1} \cdot t \right)$$
$$\Rightarrow \quad (\tilde{x}, g^{ji}(s)) = \left(\tilde{y} \cdot [\gamma], \left(g^{ji}(h^j([\gamma])) \right)^{-1} \cdot g^{ji}(t) \right)$$
$$\Rightarrow \quad (\tilde{x}, g^{ji}(s)) = \left(\tilde{y} \cdot [\gamma], h^i([\gamma])^{-1} \cdot g^{ji}(t) \right)$$
$$\Rightarrow \quad [(\tilde{x}, g^{ji}(s))]^i = [(\tilde{y}, g^{ji}(t))]^i.$$

(b) p^{ji} *commutes with the bundle projections, i.e. $\pi^i \circ p^{ji} = \pi^j$*:

$$(\pi^i \circ p^{ji})\left([(\tilde{x}, s)]^j\right) = \pi^i\left([(\tilde{x}, g^{ji}(s))]^j\right) = \tilde{p}(\tilde{x}) = \pi^j\left([(\tilde{x}, s)]^j\right),$$

for every $[(\tilde{x}, s)]^j \in P^j$.

(c) p^{ji} *is equivariant with respect to* G^j *and* G^i:

$$p^{ji}\big([(\tilde{x},s)]^j \cdot g\big) = p^{ji}\big([(\tilde{x},s\cdot g)]^j\big) = \big[(\tilde{x},g^{ji}(s\cdot g))\big]^i =$$

$$= \big[(\tilde{x},g^{ji}(s)\cdot g^{ji}(g))\big]^i = \big[(\tilde{x},g^{ji}(s))\big]^i \cdot g^{ji}(g)) = p^{ji}\big([(\tilde{x},s)]^j\big)\cdot g^{ji}(g)$$

for every $[(\tilde{x},s)]^j \in P^j$ and $g \in G^j$.

(d) p^{ji} *is smooth*: This is checked locally as follows. Adapting the constructions of § 1.6.6(a) to the present context, we see that a local trivialization (U,Φ) of $\tilde{B}$, with $\Phi\colon \tilde{B}|_U \to U\times\pi_1(B)$ and corresponding natural section $\sigma\colon U\to\tilde{B}$, induces on each P^i the local trivialization $(U,\overline{\Phi}^i)$, with $\overline{\Phi}^i\colon P^i|_U \to U\times G^i$ given by

$$\overline{\Phi}^i\big([(\tilde{x},s)]^i\big) = \big(\pi^i(\tilde{x}),h^i([\gamma])\cdot s\big) = \big(\tilde{p}(\tilde{x}),h^i([\gamma])\cdot s\big),$$

where $[\gamma] \in \pi_1(B)$ is determined by $\tilde{x} = \sigma(\tilde{p}(\tilde{x}))\cdot[\gamma]$. The inverse of the trivializing map is given by $\big(\overline{\Phi}^i\big)^{-1}(x,s) = [(\sigma(x),s)]^i$. Therefore, since $\overline{\Phi}^j(P^j|_U) \subseteq P^i|_U$, for every $j \geq i$, we obtain the following diagram actually yielding the local expression of p^{ji}, with respect to $(U,\overline{\Phi}^j)$ and $(U,\overline{\Phi}^i)$:

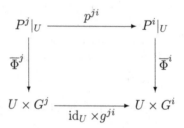

Indeed, for every $(x,s)\in U\times G^j$,

$$\big(\overline{\Phi}^i \circ p^{ji}\circ (\overline{\Phi}^j)^{-1}\big)(x,s) = \big(\overline{\Phi}^i\circ p^{ji}\big)\big([\sigma(x),s]^j\big) =$$

$$= \overline{\Phi}^i\big([\sigma(x),g^{ji}(s)]^i\big) = \big(\pi^i(\sigma(x)),h^i([\gamma])\cdot g^{ji}(s)\big) =$$

$$= \big(\tilde{p}(\sigma(x)),h^i([\gamma])\cdot g^{ji}(s)\big) = \big(x,h^i([\gamma])\cdot g^{ji}(s)\big)$$

where $[\gamma] \in G^i$ is now determined by $\sigma(x) = \sigma(x)\cdot[\gamma]$, thus $[\gamma] = [e_{x_o}]$ (: the unit of $\pi_1(B)$) and

$$\big(\overline{\Phi}^i \circ p^{ji}\circ (\overline{\Phi}^j)^{-1}\big)(x,s) = (x,g^{ji}(s)) = \big(\mathrm{id}_U \times g^{ji}\big)(x,s),$$

which proves the desired commutativity and, clearly, the smoothness of the connecting morphism p^{ji}.

As a consequence of (a)–(d), $F^{ji} := (p^{ji},g^{ji},\mathrm{id}_B)\colon \ell^j \to \ell^i$ is a principal bundle morphism, and $\big\{\ell^i;F^{ji}\big\}_{i,j\in\mathbb{N}}$ is a projective system of Banach principal bundles, in virtue of Definition 4.1.1. Accordingly,

$$\varprojlim \ell^i = \left(\varprojlim P^i, G = \varprojlim G^i, B, \varprojlim \pi^i \right)$$

is a well-defined plb-bundle. We note that (4.1.1) in the aforementioned definition, concerning the local trivializations of the limit bundle, is now satisfied by the local trivializations $\{(U, \overline{\Phi}^i)\}_{i \in \mathbb{N}}$ of P^i, derived each time by the same pair (U, Φ) as the latter is running the set of all trivializations of $\tilde{B}$. In fact, the existence of each $\varprojlim \overline{\Phi}^i$ is ensured by the preceding commutative diagram.

Construction of F: We define the maps

$$F^i \colon P \longrightarrow P^i \colon [(\tilde{x}, s)] \mapsto [(\tilde{x}, g^i(s))]^i,$$

for all $i \in \mathbb{N}$. Working as in the case of p^{ji}, we easily check that F^i is a well-defined, equivariant (with respect to G and G^i) map satisfying $\pi^i \circ F^i = \pi_Q$, for every $i \in \mathbb{N}$. It is also smooth because locally identifies with $\mathrm{id}_{\tilde{B}} \times g^i$. Note that the charts of Q have a form analogous to that of P^i mentioned earlier. Moreover, by the very definitions, $p^{ji} \circ F^j = F^i$. Consequently, we obtain the smooth limit map $F := \varprojlim F^i \colon P \to \varprojlim P^i$, commuting with the projections π_Q and $\varprojlim \pi^i$ of ℓ and $\varprojlim \ell^i$, respectively, since

$$\varprojlim \pi^i \circ F = \varprojlim \pi^i \circ \varprojlim F^i = \varprojlim (\pi^i \circ F^i) = \varprojlim \pi_Q = \pi_Q.$$

Hence, $(F, \mathrm{id}_G, \mathrm{id}_B)$ is a G-B-(iso)morphism of (Q, G, B, π_Q) onto $\varprojlim \ell^i$.

Construction of $\varprojlim \omega^i$: Let ω^i denote the flat connection on ℓ^i, induced by h^i. Since the ω_0^i is $(\kappa^i, \mathrm{id}_{G^i}, \tilde{p})$-related with ω^i, it follows that

$$(4.5.5) \qquad \omega_0^i(\tilde{x}, s) = \omega^i([(\tilde{x}, s)]^i) \circ T_{(\tilde{x}, s)} \kappa^i, \qquad (\tilde{x}, s) \in \tilde{B} \times G^i.$$

To show that ω^j and ω^i $(j \geq i)$ are $(p^{ji}, g^{ji}, \mathrm{id}_B)$-related, it suffices to check the commutativity of the diagram

$$
\begin{array}{ccc}
T_{u^j} P^j & \xrightarrow{\ \omega^j(u^j)\ } & \mathfrak{g}^j \\[4pt]
{\scriptstyle T_{u^j} p^{ji}} \Big\downarrow & & \Big\downarrow {\scriptstyle T_{e^j} g^{ji}} \\[4pt]
T_{u^i} P^i & \xrightarrow[\ \omega^i(u^i)\ ]{} & \mathfrak{g}^i
\end{array}
$$

for every $u^j = [(\tilde{x}, s)]^j \in P^j$, with $u^i = p^{ji}(u^j) = [(\tilde{x}, g^{ji}(s))]^i$. Indeed, for any tangent vector $W \in T_{u^j} P^j$, there are $X \in T_{\tilde{x}} \tilde{B}$ and $Y \in T_s G^j$

such that $T_{(\tilde{x},s)}\kappa^j(X,Y) = W$, because κ^j is a submersion. Hence, in virtue of the analog of (4.5.5) for ω_o^j and the definition of the canonical flat connection ω_o^j, we obtain:

$$
\begin{aligned}
(T_{e^j}g^{ji} \circ \omega^j(u^j))(W) &= T_{e^j}g^{ji}\left(\omega^j([(\tilde{x},s)]^j) \circ T_{(\tilde{x},s)}\kappa^j.(X,Y)\right) \\
&= T_{e^j}g^{ji}\left(\omega_o^j(\tilde{x},s).(X,Y)\right) \\
&= T_{e^j}g^{ji}\left(T_s L^j_{s-1}(Y)\right) \\
&= T_s(g^{ji} \circ L^j_{s-1})(Y).
\end{aligned}
$$

(4.5.6)

Similarly,

$$
\begin{aligned}
(\omega^i(u^i) \circ p^{ji})(W) &= \omega^i([(\tilde{x},g^{ji}(s))]^i)\left(T_{[(\tilde{x},s)]^j}p^{ji}(T_{(\tilde{x},s)}\kappa^j(X,Y))\right) \\
&= \omega^i([(\tilde{x},g^{ji}(s))]^i)\left(T_{(\tilde{x},s)}(p^{ji} \circ \kappa^j)(X,Y)\right)
\end{aligned}
$$

or, in virtue of $p^{ji} \circ \kappa^j = \kappa^i \circ (\mathrm{id}_{\tilde{B}} \times g^{ji})$,

$$
\begin{aligned}
(\omega^i(u^i) \circ p^{ji})(W) &= \left(T_{(\tilde{x},s)}(\kappa^i \circ (\mathrm{id}_{\tilde{B}} \times g^{ji}))(X,Y)\right) \\
&= \omega^i([(\tilde{x},g^{ji}(s))]^i)\left(T_{(\tilde{x},g^{ji}(s))}\kappa^i(X, T_s g^{ji}(Y))\right) \\
&= \omega_o^i(\tilde{x},g^{ji}(s))(X, T_s g^{ji}(Y)) \\
&= T_s(L^i_{g^{ji}(s)-1} \circ g^{ji})(Y).
\end{aligned}
$$

(4.5.7)

Since, by (3.4.3), $g^{ji} \circ L^j_{s-1} = L^i_{g^{ji}(s)-1} \circ g^{ji}$, equalities (4.5.6) and (4.5.7) prove the commutativity of the preceding diagram, implying in turn that $\{\omega^i\}_{i\in\mathbb{N}}$ is a projective system of connections as in Definition 4.2.1. Consequently, $\omega := \varprojlim \omega^i$ is a connection on $\varprojlim P^i$. In virtue of Proposition 4.5.1 ω is flat.

Relatedness of θ and $\varprojlim \omega^i$: For any $[(\tilde{x},s)] \in Q$ and $W \in T_{[(\tilde{x},s)]}Q$, there is a pair $(X,Y) \in T_{\tilde{x}}\tilde{B} \times T_s G$ such that $T_{(\tilde{x},s)}\kappa(X,Y) = W$. Then, the $(\kappa, \mathrm{id}_G, \tilde{p})$-relatedness of θ and ω_o implies

(4.5.8)
$$
\begin{aligned}
\theta_{[(\tilde{x},s)]}(W) &= \theta_{\kappa(\tilde{x},s)}\left(T_{(\tilde{x},s)}\kappa(X,Y)\right) = \\
&= \omega_o(\tilde{x},s).(X,Y) = T_s L_{s-1}(Y).
\end{aligned}
$$

Analogously,

(4.5.9) $$(F^*\omega)_{[(\tilde{x},s)]}(W) = \omega_{F([(\tilde{x},s)])}\left((T_{[(\tilde{x},s)]}F \circ T_{(\tilde{x},s)}\kappa)(X,Y)\right).$$

On the other hand, taking into account the identifications of Propositions 3.2.2 and 3.2.5, $F = \varprojlim F^i$ implies that

$$
\begin{aligned}
T_{[(\tilde{x},s)]}F(T_{(\tilde{x},s)}\kappa(X,Y)) &= \left(T_{[(\tilde{x},s)]}F^i\big(T_{(\tilde{x},s)}\kappa(X,Y)\big)\right)_{i\in\mathbb{N}} \\
&= \left(T_{(\tilde{x},s)}(F^i \circ \kappa)(X,Y)\right)_{i\in\mathbb{N}} \\
&= \left(T_{(\tilde{x},s)}\big(\kappa^i \circ (\mathrm{id}_{\tilde{B}} \times g^i)\big)(X,Y)\right)_{i\in\mathbb{N}} \\
&= \left(T_{(\tilde{x},g^i(s))}\big(X,T_s g^i(Y)\big)\right)_{i\in\mathbb{N}}.
\end{aligned}
$$

Substituting the latter in (4.5.9), we find that

$$
\begin{aligned}
(F^*\omega)_{[(\tilde{x},s)]}(W) &= (F^*\omega)_{[(\tilde{x},s)]}\left(\Big(T_{(\tilde{x},s)}\big(X,T_s g^i(Y)\big)\Big)_{i\in\mathbb{N}}\right) \\
&= \left(\omega^i_{\kappa^i(\tilde{x},g^i(s))}\big(T_{(\tilde{x},s)}\big(X,T_s g^i(Y)\big)\big)\right)_{i\in\mathbb{N}} \\
&= \left(\big((\kappa^i)^*\omega^i\big)_{(\tilde{x},g^i(s))}\big(X,T_s g^i(Y)\big)\right)_{i\in\mathbb{N}}
\end{aligned}
$$

or, by the relatedness of ω^i_{o} and ω^i,

$$
\begin{aligned}
(4.5.10) \qquad (F^*\omega)_{[(\tilde{x},s)]}(W) &= \left(\omega^i_{\mathrm{o}}(\tilde{x},g^i(s)).\big(X,T_s g^i(Y)\big)\right)_{i\in\mathbb{N}} \\
&= \left(T_{g^i(x)}L^i_{g^i(s)^{-1}}\big(T_s g^i(Y)\big)\right)_{i\in\mathbb{N}}.
\end{aligned}
$$

However, the aforementioned identifications, Corollary 3.2.6 and equality (3.4.1) imply that $Y = (T_s g^i(Y))_{i\in\mathbb{N}}$; therefore,

$$
\begin{aligned}
T_s L_{s^{-1}}(Y) &= T_s L_{s^{-1}}(Y)\big((T_s g^i(Y))_{i\in\mathbb{N}}\big) \\
[\text{see } (3.4.1)] \qquad &= \left(\varprojlim T_{g^i(s)}L^i_{g^i(s)^{-1}}\right)\big((T_s g^i(Y))_{i\in\mathbb{N}}\big) \\
&= \left(T_{g^i(s)}L^i_{g^i(s)^{-1}}\big(T_s g^i(Y)\big)\right)_{i\in\mathbb{N}}.
\end{aligned}
$$

Consequently, in virtue of the preceding, equalities (4.5.8) and (4.5.10) yield the desired relatedness and complete the proof. $\qquad\square$

Using some of the tools applied in the previous study of flat bundles, we are in a position to give the counterexample mentioned in Remark 4.3.6. Before this, we need a few technicalities:

Set $B := \mathbb{R}^2 - \{(\nu,0)\,|\,\nu \in \mathbb{N}\}$. Then B is a connected manifold whose fundamental group $\pi_1(B)$ is free with countable many generators $(x_1, x_2, \ldots)$. For every $i \in \mathbb{N}$, we define the map $a^i\colon \pi_1(B) \to \mathbb{R}$, where $a^i(\gamma)$ equals the sum of exponents of the x_i generator of γ (for simplicity we set $\gamma = [\gamma]$). Also, we define the maps

$$
h^i\colon \pi_1(B) \longrightarrow \mathbb{R}^i\colon \gamma \mapsto \big(a^1(\gamma), a^2(\gamma)), \ldots, a^i(\gamma)\big),
$$

for every $i \in \mathbb{N}$.

With the previous notations we prove:

Lemma 4.5.4   *The following assertions are true:*
i) Each h^i is a Lie group homomorphism.
ii) $h^i(\pi_1(B)) = \mathbb{Z}^i$.
iii) The homomorphism $h := \varprojlim h^i \colon \pi_1(B) \to \mathbb{R}^\infty$ is defined.
iv) $h(\pi_1(B)) \neq \varprojlim h^i(\pi_1(B))$.

Proof i) Considering, as usual, $\pi_1(B)$ with the discrete differential structure, it suffices to show that every a^i is a group homomorphism. Indeed, let any elements

$$\beta = x_{i_1}^{\varepsilon_i} \cdot x_{i_2}^{\varepsilon_2} \cdots x_{i_k}^{\varepsilon_k}, \quad \gamma = x_{j_1}^{\delta_1} \cdot x_{j_2}^{\delta_2} \cdots x_{j_\lambda}^{\delta_\lambda}$$

of $\pi_1(B)$. If x_1 appears in β at the positions $i_{p_1}, i_{p_2}, \ldots, i_{p_x}$ and in γ at the positions $j_{q_1}, j_{q_2}, \ldots, j_{q_y}$, then for

$$\beta \cdot \gamma = x_{i_1}^{\varepsilon_i} \cdot x_{i_2}^{\varepsilon_2} \cdots x_{i_k}^{\varepsilon_k} \cdot x_{j_1}^{\delta_1} \cdot x_{ij_2}^{\delta_2} \cdots x_{j_\lambda}^{\delta_\lambda}$$

the same x_1 appears at the $i_{p_1}, i_{p_2}, \ldots, i_{p_x}, j_{q_1}, j_{q_2}, \ldots, j_{q_y}$ positions. Therefore,

$$a^1(\beta \cdot \gamma) = \varepsilon_{p_1} + \varepsilon_{p_2} + \cdots + \varepsilon_{p_x} + \delta_{q_1} + \delta_{q_2} + \cdots \delta_{q_y} = a^1(\beta) + a^1(\gamma).$$

Similar arguments hold for every a^i, $i \geq 2$.

ii) Obviously, $h^i(\pi_1(B)) \subseteq \mathbb{Z}^i$ since $a^i(\gamma)$ is an integer, for every $\gamma \in (\pi_1(B))$. Conversely, for an arbitrary element $u = (k_1, k_2, \ldots, k_i) \in \mathbb{Z}^i$, it follows that $u = h^i\left(x_1^{k_1} \cdot x_2^{k_2} \cdots x_i^{k_i}\right) \in h^i(\pi_1(B))$.

iii) It suffices to show that the diagram

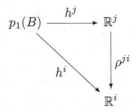

is commutative, for every $j \geq i$, where

$$\rho^{ji} \colon \mathbb{R}^j \longrightarrow \mathbb{R}^i \colon (t_1, t_2, \ldots, t_j) \mapsto (t_1, t_2, \ldots, t_i)$$

are the connecting morphisms of $\varprojlim \mathbb{R}^i = \mathbb{R}^\infty$. This is the case, because

$$\left(\rho^{ji} \circ h^j\right)(\gamma) = \rho^{ji}\left(a^1(\gamma), a^2(\gamma), \ldots, a^j(\gamma)\right)$$
$$= \left(a^1(\gamma), a^2(\gamma), \ldots, a^i(\gamma)\right)$$
$$= h^i(\gamma).$$

iv) We observe that $\varprojlim h^i\left(\pi_1(B)\right) = \varprojlim \mathbb{Z}^i = \mathbb{Z}^\infty$. Therefore, if we assume that $h\left(\pi_1(B)\right) = \varprojlim h^i\left(\pi_1(B)\right)$, we would have that $h\left(\pi_1(B)\right) = \mathbb{Z}^\infty$, thus there would be some $\gamma \in \pi_1(B)$ such that $h(\gamma) = (1, 1, \cdots)$ or, equivalently, $a^i(\gamma) = 1$, for every $i \in \mathbb{N}$. This means that all the generators x_i ($i \in \mathbb{N}$) would appear in the representation of γ, which is absurd. As a result, we do have that $h\left(\pi_1(B)\right) \neq \varprojlim h^i\left(\pi_1(B)\right)$. $\qquad \square$

Theorem 4.5.5 *There exists a plb-connection $\omega = \varprojlim \omega^i$ whose holonomy group does not coincide with the projective limit of the holonomy groups of the factor connections.*

Proof We consider the connected manifold $B := \mathbb{R}^2 - \{(\nu, 0) \mid \nu \in \mathbb{N}\}$, the plb-group $G := (\mathbb{R}^\infty, +) = \varprojlim G^i$, where $G^i = (\mathbb{R}^i, +)$, as well as the Lie group morphism $h := \varprojlim h^i \colon \pi_1(B) \to G$ defined in the previous lemma. Then, as in Theorem 4.5.3 and the discussion before it, h determines a plb-principal bundle (Q, θ) which is $(F, \mathrm{id}_G, \mathrm{id}_B)$-isomorphic with the plb-bundle $\left(\varprojlim P^i, \varprojlim \omega^i\right)$, induced by the family (h^i). Fixing the points $q \in Q$ and $F(q) =: u = (u^i)$, we obtain the corresponding holonomy groups of θ and ω, Φ_q and Φ_u. It is an immediate consequence of (1.9.5) (see also [KN68, §II.6]) that $\Phi_q = \Phi_u$. On the other hand, for each (P^i, h^i), induced by h^i, we obtain the holonomy group $\Phi^i_{u^i}$. We also have that $\Phi_q = h\left(\pi_1(B)\right)$ and $\Phi^i_{u^i} = h^i\left(\pi_1(B)\right)$. Therefore, if we assume that $\Phi_u = \varprojlim \Phi^i_{u^i}$, then we would have that

$$h\left(\pi_1(B)\right) = \Phi_q = \Phi_u = \varprojlim \Phi^i_{u^i} = \varprojlim h^i\left(\pi_1(B)\right),$$

which contradicts assertion iv) of Lemma 4.5.4. Hence, $\Phi_u \neq \varprojlim \Phi^i_{u^i}$, thus verifying the statement for the (flat) connection $\omega = \varprojlim \omega^i$ on the bundle $\left(\varprojlim P^i, G, B, \varprojlim \pi^i\right)$. $\qquad \square$

5

Projective systems
of vector bundles

Vector bundles of fibre type a Fréchet space $\mathbb{F}$ are difficult to handle because of the pathology of the structure group $GL(\mathbb{F})$. Therefore, the aim of the present chapter is to propose a method to address many relevant issues by using the language of projective limits.

The structure of vector bundles induced by projective limits of Banach vector bundles is a bit more complicated than that of the principal bundles (studied in Chapter 4), and it is completely determined by the group $\mathcal{H}_0(\mathbb{F})$ to be defined in Section 5.1 below, a group replacing the pathological $GL(\mathbb{F})$.

Important examples are the infinite jets of sections of a Banach vector bundle and spaces of linear and antisymmetric maps such as $\mathcal{L}(TB, \mathbb{F})$, $A_k(TB, \mathbb{F})$, fully described in Chapter 6.

The important geometric notion of a (linear) connection in the present context will be deferred to Chapter 7.

5.1 A particular Fréchet group

Before delving into the structure of Fréchet vector bundles in our framework, we introduce a particular type of a Fréchet topological group which will essentially play the role of the structural group of the bundles under study.

As usual, $\mathbb{F} = \varprojlim \mathbb{E}^i$ is a Fréchet space, projective limit of the Banach spaces $\mathbb{E}^i$, $i \in \mathbb{N}$. Referring to (2.3.3), we set

$$(5.1.1) \qquad \mathcal{H}^i(\mathbb{F}) := \mathcal{H}^i(\mathbb{F}, \mathbb{F}),$$

and

$$(5.1.2) \qquad \mathcal{H}_0^i(\mathbb{F}) := \mathcal{H}^i(\mathbb{F}) \cap \prod_{j=1}^{i} \mathcal{L}is(\mathbb{E}^j),$$

where $\mathcal{L}is(\mathbb{E})$ (also denoted by $\mathrm{GL}(\mathbb{E})$) is the group of invertible elements of $\mathcal{L}(\mathbb{E})$. Similarly, recalling (2.3.12), we set

$$(5.1.3) \qquad \mathcal{H}_0(\mathbb{F}) := \mathcal{H}(\mathbb{F}) \cap \prod_{j=1}^{\infty} \mathcal{L}is(\mathbb{E}^j),$$

Proposition 5.1.1 *The following assertions are true:*

i) Every $\mathcal{H}_0^i(\mathbb{F})$, $i \in \mathbb{N}$, is a Banach-Lie group modelled on $\mathcal{H}^i(\mathbb{F})$, while $\mathcal{H}_0(F)$ is a topological group with the relative topology of $\mathcal{H}(\mathbb{F})$.

ii) The projective limit $\varprojlim \mathcal{H}_0^i(\mathbb{F})$ exists and coincides, up to an isomorphism of topological groups, with $\mathcal{H}_0(\mathbb{F})$. Thus $\mathcal{H}_0(F)$ is a Fréchet topological group.

Proof Since each $\mathcal{L}is(\mathbb{E}^j)$ is open in $\mathcal{L}(E^j)$, (5.1.2) shows that $\mathcal{H}_0^i(\mathbb{F})$ is an open subset of $\mathcal{H}^i(\mathbb{F})$, $i \in \mathbb{N}$. Clearly, $\mathcal{H}_0^i(\mathbb{F})$ is a group with multiplication (the continuous bilinear) composition map:

$$\left(f^1, \ldots, f^i\right) \cdot \left(g^1, \ldots, g^i\right) := \left(f^1 \circ g^1, \ldots, f^i \circ g^i\right).$$

Therefore, $\mathcal{H}_0^i(\mathbb{F})$ is a Banach-Lie group modelled on $\mathcal{H}^i(\mathbb{F})$, for every $i \in \mathbb{N}$.

Assertion ii) is based on the proof of Theorem 2.3.10: Denoting again by $h^{ji} \colon \mathcal{H}^j(\mathbb{F}) \to \mathcal{H}^i(\mathbb{F})$ ($j \geq i$) the analogs of (2.3.6), we see that the maps

$$h_0^{ji} := h^{ji}\big|_{\mathcal{H}_0^j(\mathbb{F})} \colon \mathcal{H}_0^j(\mathbb{F}) \longrightarrow \mathcal{H}_0^i(\mathbb{F})$$

are morphisms of topological groups satisfying $h_0^{jk} = h_0^{ik} \circ h_0^{ji}$, for every $i, j, k \in \mathbb{N}$ with $j \geq i \geq k$; hence, we obtain the projective system $\left(\mathcal{H}_0^i(\mathbb{F}); h_0^{ji}\right)$ whose limit $\varprojlim \mathcal{H}_0^i(\mathbb{F})$ is a topological group.

On the other hand, $\mathcal{H}_0(\mathbb{F})$ is a topological group, with the obvious multiplication $(f^i)_{i \in \mathbb{N}} \cdot (g^i)_{i \in \mathbb{N}} := (f^i \circ g^i)_{i \in \mathbb{N}}$, and topology the relative topology as a subset of $\mathcal{H}(\mathbb{F})$. Also, we observe that the maps

$$h_0^k := h^k\big|_{\mathcal{H}_0(\mathbb{F})} \colon \mathcal{H}_0(\mathbb{F}) \longrightarrow \mathcal{H}_0^k(\mathbb{F}) \colon (f^i)_{i \in \mathbb{N}} \mapsto (f^1, \ldots, f^k); \qquad k \in \mathbb{N},$$

are morphisms of topological groups satisfying $h_0^{ji} \circ h_0^j = h_0^i$. Thus we obtain the morphism of topological groups

$$h_0 := \varprojlim h_0^i \colon \mathcal{H}_0(\mathbb{F}) \longrightarrow \varprojlim \mathcal{H}_0^i(\mathbb{F}).$$

Following the proof of Theorem 2.3.10, we check that h_0 is a bijection, and h_0^{-1} is continuous as the restriction of the continuous map $h^{-1}\colon \varprojlim \mathcal{H}^i(\mathbb{F}) \to \mathcal{H}(\mathbb{F})$ to $\varprojlim \mathcal{H}_0^i(\mathbb{F})$. Hence, $\mathcal{H}_0(\mathbb{F})$ and $\varprojlim \mathcal{H}_0^i(\mathbb{F})$ can be identified as topological groups. Since $\varprojlim \mathcal{H}_0^i(\mathbb{F})$ is a Fréchet topological group, we conclude the proof.   □

It should be noted that, although each $\mathcal{H}_0^i(\mathbb{F})$ is a Banach-Lie group, $\mathcal{H}_0(\mathbb{F}) \equiv \varprojlim \mathcal{H}_0^i(\mathbb{F})$ is not necessarily a Fréchet-Lie group in the sense of Proposition 3.4.1, since the projective system of the previous groups does not necessarily satisfy the conditions of a plb-manifold in the sense of Definition 3.1.2. More precisely, condition (2) of the aforementioned definition is in question because the projective limit of the *open* sets $\mathcal{H}_0^i(\mathbb{F})$ is not necessarily open.

5.2 Projective systems and Fréchet vector bundles

Let (E^i, B, π^i), $i \in \mathbb{N}$, be Banach vector bundles (over the same base B) of fibre type $\mathbb{E}^i$, respectively, and let $f^{ji}\colon E^j \to E^i$ ($j \geq i$), be vector bundle morphisms over the identity. For every open $U \subseteq B$, we set $E_U^i := (\pi^i)^{-1}(U)$.

Definition 5.2.1   By a *projective system of Banach vector bundles* $\{E^i; f^{ji}\}_{i,j \in \mathbb{N}}$ we mean a countable family $\{(E^i, B, \pi^i); f^{ji}\}_{i,j \in \mathbb{N}}$ satisfying the following additional conditions:

(PVB. 1) The Banach spaces $\mathbb{E}^i$ form a projective system with connecting morphisms ρ^{ji}.

(PVB. 2) For each $x \in B$, there exist local trivializations (U, τ^i) of E^i (with the same U), $i \in \mathbb{N}$, such that $x \in U$ and

$$\tau^i \circ f^{ji} = (\mathrm{id}_U \times \rho^{ji}) \circ \tau^j, \qquad j \geq i.$$

Condition (PVB. 1) implies that $\mathbb{F} := \varprojlim \mathbb{E}^i$ exists and has the structure of a Fréchet space. The equality figuring in (PVB.2) is pictured in the next commutative diagram:

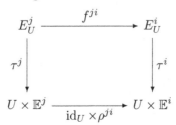

We note that Conditions (PVB. 1)–(PVB. 2), which enrich the set-theoretic definition of a projective system, are necessary to assure that the induced projective limits remain within the category of vector bundles.

Projective systems of vector bundles, as in Definition 5.2.1, are also called *strong* (see, e.g., [Gal98]). However, we adhere to the simpler terminology of our definition.

Proposition 5.2.2 *If $\{E^i; f^{ji}\}_{i,j\in\mathbb{N}}$ is a projective system of Banach vector bundles, then $E := \varprojlim E^i$ exists and is a plb-manifold; hence, by Proposition 3.1.3, E is a Fréchet manifold.*

Proof The existence of E follows from equality

$$(5.2.1) \qquad\qquad f^{jk} = f^{ik} \circ f^{ji}; \qquad j \geq i \geq k,$$

verified in the following way: Let $u \in E^j$, with $\pi^j(u) = x \in B$, and let (U, τ^i) be the trivializations satisfying condition (PVB. 2). Since $u \in E^j_U$, $f^{ji}(u) \in E^i_U$, and $f^{ik}(f^{ji}(u)) \in E^k_U$, we obtain

$$\begin{aligned}
\tau^k\big(f^{ik}(f^{ji}(u))\big) &= \big(\mathrm{id}_U \times \rho^{ik}\big)\big(\tau^i(f^{ji}(u))\big) \\
&= \big(\big(\mathrm{id}_U \times \rho^{ik}\big) \circ \big(\mathrm{id}_U \times \rho^{ji}\big)\big)(\tau^j(u)) \\
&= \big(\big(\mathrm{id}_U \times \rho^{jk}\big) \circ \tau^j\big)(u) = \tau^k\big(f^{jk}(u)\big)\big),
\end{aligned}$$

which leads to (5.2.1).

If $\mathbb{B}$ is the (Banach space) model of the base B, each bundle E^i has local charts with model $\mathbb{B} \times \mathbb{E}^i$. Obviously, $\{\mathbb{B} \times \mathbb{E}^i; \mathrm{id}_\mathbb{B} \times \rho^{ji}\}$ is a projective system of Banach spaces inducing the Fréchet space $\varprojlim(\mathbb{B} \times \mathbb{E}^i) = \mathbb{B} \times \mathbb{F}$.

The smooth structure of E is defined as follows: Let any $u = (u^i) \in E$. Since $\pi^j(u^j) = \pi^i(f^{ji}(u^j)) = \pi^i(u^i)$, we set $\pi^i(u^i) =: x$, for every $i \in \mathbb{N}$. If (U, τ^i), $i \in \mathbb{N}$, are the trivializations of (PV. 2) with $x \in U$, shrinking U (if necessary), we may also consider the chart (U, ϕ) of B at x. Then, $(E^i_U, \Phi^i) := \big(E^i_U, (\phi \times \mathrm{id}_{\mathbb{E}^i}) \circ \tau^i\big)$ is a chart of E^i. It is clear that $u^j \in E^j_U$, $f^{ji}(E^j_U) \subseteq E^i_U$, and the diagram

$$\begin{array}{ccc}
E^j_U & \xrightarrow{\quad f^{ji} \quad} & E^i_U \\
\Big\downarrow{\Phi^j} & & \Big\downarrow{\Phi^i} \\
\phi(U) \times \mathbb{E}^j & \xrightarrow[\mathrm{id}_{\phi(U)} \times \rho^{ji}]{} & \phi(U) \times \mathbb{E}^i
\end{array}$$

is commutative, as a consequence of equalities

$$
\begin{aligned}
\left(\mathrm{id}_{\mathbb{B}} \times \rho^{ji}\right) \circ \Phi^{j} &= \left(\mathrm{id}_{\mathbb{B}} \times \rho^{ji}\right) \circ (\phi \times \mathrm{id}_{\mathbb{E}^{j}}) \circ \tau^{j} \\
&= (\phi \times \rho^{ji}) \circ \tau^{j} \\
&= (\phi \times \mathrm{id}_{\mathbb{E}^{i}}) \circ (\mathrm{id}_{U} \times \rho^{ji}) \circ \tau^{j} \\
&= (\phi \times \mathrm{id}_{\mathbb{E}^{i}}) \circ \tau^{i} \circ f^{ji} = \Phi^{i} \circ f^{ji}.
\end{aligned}
$$

Then, $\left\{E_{U}^{i}, f^{ji}\right\}_{i \in \mathbb{N}}$ and $\{\Phi^{i}\}_{i \in \mathbb{N}}$ are projective systems and their respective limits $\varprojlim E_{U}^{i}, \varprojlim \Phi^{i}$ exist.

Similarly, the vb-morphisms f^{ji} imply that $\pi^{i} \circ f^{ji} = \pi^{j}$, for all indices i, j with $j \geq i$, thus we define the continuous map

$$
\pi := \varprojlim \pi^{i} \colon E = \varprojlim E^{i} \longrightarrow B.
$$

As a result,

$$
\varprojlim E_{U}^{i} = \varprojlim (\pi^{i})^{-1}(U) = \pi^{-1}(U),
$$

which means that $\varprojlim E_{U}^{i}$ is open in E. On the other hand,

$$
\varprojlim \left((\Phi(E_{U}^{i})\right) = \varprojlim \left(\phi(U) \times \mathbb{E}^{i}\right) = \phi(U) \times \mathbb{F};
$$

that is, $\varprojlim \left(\Phi(E_{U}^{i})\right)$ is an open subset of $\mathbb{B} \times \mathbb{F}$. The last arguments prove that the pairs $\left(\varprojlim E_{U}^{i}, \varprojlim \Phi\right)$ determine projective limit charts (see Definition 3.1.1) inducing the structure of a plb-manifold (Definition 3.1.2) on E. This concludes the proof. □

The previous proposition shows that $\{E^{i}; f^{ji}\}_{i,j \in \mathbb{N}}$ is a projective system in the ordinary sense. However, the conditions of Definition 5.2.1 ensure something more: the structure of a Fréchet manifold on $\varprojlim E^{i}$.

Corollary 5.2.3   *The map* $\pi := \varprojlim \pi^{i} \colon \varprojlim E^{i} \to B$ *is smooth as a pls-map.*

Proof   Immediate consequence of the equalities $\pi^{i} \circ f^{ji} = \pi^{j}$ $(j \geq i)$ and Proposition 3.1.8. □

Definition 5.2.4   Let $\{E^{i}; f^{ji}\}_{i,j \in \mathbb{N}}$ be a projective system of Banach vector bundles. The triplet $\left(E := \varprojlim E^{i}, B, \pi := \varprojlim \pi^{i}\right)$ is called a **plb-vector bundle**.

The term (plb-) vector bundle is justified by the following:

Theorem 5.2.5   *A plb-vector bundle* (E, B, π) *is a Fréchet vector bundle*

Proof We have already seen that E is a Fréchet manifold and π a smooth map. We need to show the Fréchet analogs of (VB. 1)–(VB. 3) in § 1.4.1.

First observe that, for each $x \in B$,

$$\left(E_x^i := (\pi^i)^{-1}(x); f^{ji}|_{E_x^j} \right)_{i,j \in \mathbb{N}}$$

is a projective system of Banach spaces, since $f^{ji}|_{E_x^j} : E_x^j \to E_x^i$ are continuous linear maps. Hence, $\varprojlim E_x^i$ is defined and

$$E_x := \pi^{-1}(x) = \left(\varprojlim (\pi^i)^{-1} \right)(x) = \varprojlim \left((\pi^i)^{-1}(x) \right) = \varprojlim E_x^i.$$

Indeed, if (U, τ^i) are the local trivializations of (PVB 2), with $x \in U$, then (see also the commutative diagram following Definition 5.2.1), for every $i, j \in \mathbb{N}$ with $j \geq i$, we obtain the diffeomorphism

$$\tau := \varprojlim \tau^i \colon \pi^{-1}(U) = \varprojlim \left((\pi^i)^{-1}(U) \right) \longrightarrow \varprojlim (U \times \mathbb{E}_i) = U \times \mathbb{F}.$$

Now, for every $u = (u^i) \in \pi^{-1}(U)$,

$$
\begin{aligned}
(\mathrm{pr}_1 \circ \tau)(u) &= \mathrm{pr}_1 \left((\tau^i(u^i))_{i \in \mathbb{N}} \right) \\
&= \mathrm{pr}_1 \left(\left(\pi^i(u^i), \mathrm{pr}_2(\tau^i(u^i)) \right)_{i \in \mathbb{N}} \right) \\
&= \left(\pi^i(u^i) \right)_{i \in \mathbb{N}} = \pi(u);
\end{aligned}
$$

that is, the following diagram is commutative:

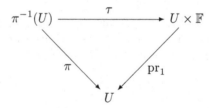

On the other hand, if $u = (u^i) \in E_x$ (thus $u^i \in E_x^i$, for every $i \in \mathbb{N}$), similar arguments show that

$$\tau(u) = \left(\tau_x^i(u^i) \right)_{i \in \mathbb{N}} = \left(\varprojlim \tau_x^i \right)(u);$$

hence,

$$\tau_x = \varprojlim \tau_x^i \colon E_x = \varprojlim E_x^i \longrightarrow \varprojlim \mathbb{E}^i = \mathbb{F}$$

is an isomorphism of Fréchet spaces. Therefore, for each $x \in B$, there is a trivialization (U, τ) of E satisfying the Fréchet analog of (VB. 1).

For the proof of (the analog of) (VB. 2), we consider two trivializations

(U, τ) and (V, σ), with $U \cap V \neq \emptyset$, $\tau = \varprojlim \tau^i$, $\sigma = \varprojlim \sigma^i$. Then, by the previous arguments, $\tau_x \circ \sigma_x^{-1} \colon \mathbb{F} \to \mathbb{F}$ is an isomorphism of Fréchet spaces, for every $x \in U$; hence, $\tau_x \circ \sigma_x^{-1} \in \mathcal{L}is(\mathbb{F}) \subset \mathcal{L}(\mathbb{F})$.

To complete the proof it remains to show that the transition functions

$$(5.2.2) \qquad T_{UV} \colon U \cap V \longrightarrow \mathcal{L}(\mathbb{F}) \colon x \mapsto T_{UV}(x) := \tau_x \circ \sigma_x^{-1},$$

are smooth, for all overlapping trivializations of E.

We note that $\mathcal{L}(\mathbb{F})$ is not necessarily a Fréchet space. Actually, it is a Hausdorff locally convex topological vector space, whose topology is determined by the uniform convergence on the bounded subsets of $\mathbb{F}$. Hence, for the desired smoothness, we first consider the map [see also (2.3.12) and (2.3.3)]

$$(5.2.3) \quad T_{UV}^* \colon U \cap V \colon \longrightarrow \mathcal{H}(\mathbb{F}) \colon x \mapsto T_{UV}^*(x) := \left(\tau_x^i \circ (\sigma_x^i)^{-1} \right)_{i \in \mathbb{N}},$$

and we claim it is smooth. To this end, we define the maps

$$(5.2.4) \qquad \begin{aligned} &(T_{UV}^*)^k \colon U \cap V \longrightarrow \mathcal{H}^k(\mathbb{F}), \quad \text{with} \\ &(T_{UV}^*)^k(x) := \left(\tau_x^1 \circ (\sigma_x^1)^{-1}, \ldots, \tau_x^i \circ (\sigma_x^k)^{-1} \right). \end{aligned}$$

Since $\tau_x^k \circ (\sigma_x^k)^{-1} = (T_{UV}^*)^k$ ($1 \leq k \leq i$), where $(T_{UV}^*)^k$ is the transition function of the Banach vector bundle (E^k, π^k, B) over $U \cap V$, then $(T_{UV}^*)^k \colon U \cap V \to \mathcal{L}(\mathbb{E}^k)$ is smooth, and so is $(T_{UV}^*)^i$ [recall from the proof of Theorem 2.3.10 that $\mathcal{H}^i(\mathbb{F})$ is a Banach space as a closed subspace of $\prod_{k=1}^i \mathcal{L}(\mathbb{E}^k)$]. Using now the notations of Section 5.1, we routinely verify that $h^{ji} \circ (T_{UV}^*)^j = (T_{UV}^*)^i$, thus $\varprojlim (T_{UV}^*)^i$ is defined. Since also $h^i \circ T_{UV}^* = (T_{UV}^*)^i$, Proposition 2.3.5 implies that $T_{UV}^* = \varprojlim (T_{UV}^*)^i$, from which the smoothness of T_{UV}^* follows.

Next we see that

$$(5.2.5) \qquad\qquad T_{UV} = \varepsilon \circ T_{UV}^*,$$

where $\varepsilon \colon \mathcal{H}(\mathbb{F}) \to \mathcal{L}(\mathbb{F})$ is the map $(f^i)_{i \in \mathbb{N}} \mapsto \varprojlim f^i$. Because the latter is continuous linear, thus smooth in the sense of our differentiability (see Section 2.2), we conclude that T_{UV} is indeed a smooth map. $\qquad \square$

Remarks 5.2.6 1) From equalities (5.2.3) and (5.1.3) we see that, in particular, the maps T_{UV}^* take values in $\mathcal{H}_0(\mathbb{F})$, i.e.,

$$T_{UV}^* \colon U \cap V \longrightarrow \mathcal{H}_0(\mathbb{F}) \subset \mathcal{H}(\mathbb{F}),$$

while, from (5.2.4) and (5.1.2),

$$(T_{UV}^*)^i \colon U \cap V \longrightarrow \mathcal{H}_0^i(\mathbb{F}) \subset \mathcal{H}^i(\mathbb{F}).$$

Since $\mathcal{H}_0(\mathbb{F})$ is only a topological group, T_{UV}^* is not smooth as an $\mathcal{H}_0(\mathbb{F})$-valued map. However, this does not affect the structure of (E, B, π), because only the smoothness of T_{UV}^* as an $\mathcal{H}(\mathbb{F})$-valued map matters. The significance of $\{T_{UV}^*\}$ will be further illustrated in subsequent results.

2) Using the arguments of the last part of the preceding proof, we see that, in fact,

$$T_{UV}^* = \varprojlim (T_{UV}^*)^i \colon U \cap V \longrightarrow \varprojlim \mathcal{H}_0^i(\mathbb{F}) = \mathcal{H}_0(\mathbb{F}).$$

3) To remember that $\{T_{UV}^*\}$ are smooth when they are considered as $\mathcal{H}(\mathbb{F})$-valued maps, but not smooth as $\mathcal{H}_0(\mathbb{F})$-valued ones, we say that $\{T_{UV}^*\}$ are **generalized smooth maps** in $\mathcal{H}_0(\mathbb{F})$.

Definition 5.2.7   The local trivializations of (E, B, π) of the form (U, τ), with $\tau = \lim \tau^i$, are called **plb-trivializations**. Analogously, the maps $T_{UV}^* \colon U \cap V \to \mathcal{H}_0(\mathbb{F})$, derived from such trivializations are called **plb-transition maps**. They clearly satisfy equality

$$T_{UV}^*(x) = T_{UW}^*(x) \circ T_{WV}^*(x); \qquad x \in U \cap V \cap W,$$

thus $\{T_{UV}^*\}$ is a cocycle with values in $\mathcal{H}_0(\mathbb{F})$.

More precisely:

Definition 5.2.8   Let B be a Banach manifold, $\mathcal{C} = \{U_\alpha\}_{\alpha \in I}$ an open cover of B, and let $\mathbb{F} = \varprojlim \mathbb{E}^i$ be a Fréchet space, where $\{\mathbb{E}^i; \rho^{ji}\}$ is a projective system of Banach spaces. An $\mathcal{H}_0(\mathbb{F})$-**valued cocycle** of B, with respect to the cover $\mathcal{C}$, is a family of smooth maps

$$T_{\alpha\beta}^* \colon U_{\alpha\beta} := U_\alpha \cap U_\beta \longrightarrow \mathcal{H}(\mathbb{F}),$$

such that

$$T_{\alpha\beta}^*(x) \in \mathcal{H}_0(\mathbb{F}); \qquad x \in U_{\alpha\beta},$$
$$T_{\alpha\gamma}^*(x) = T_{\alpha\beta}^*(x) \circ T_{\beta\gamma}^*(x); \qquad x \in U_{\alpha\beta\gamma}.$$

Theorem 5.2.9   *Let $\{T_{\alpha\beta}^*\}$ be an $\mathcal{H}_0(\mathbb{F})$-valued cocycle over an open cover $\mathcal{C} = \{U_\alpha\}_{\alpha \in I}$ of a Banach manifold B. Then there exists a unique —up to isomorphism—plb-vector bundle with plb-transition maps $\{T_{\alpha\beta}^*\}$.*

Proof   For every $k \in \mathbb{N}$ and $\alpha, \beta \in I$, we define the map

$$T_{\alpha\beta}^k := \mathrm{pr}_k \circ T_{\alpha\beta}^* \colon U_{\alpha\beta} \longrightarrow \mathcal{L}(\mathbb{E}^k),$$

where

$$\mathrm{pr}_k \colon \mathcal{H}(\mathbb{F}) \longrightarrow \mathcal{L}(\mathbb{E}^k) \colon (f^i)_{i \in \mathbb{N}} \mapsto f^k.$$

$T^k_{\alpha\beta}$ being the composite of smooth maps (in the sense of § 2.2), is also smooth in the sense of Banach manifolds. Thus $\{T^k_{\alpha\beta}\}_{\alpha,\beta \in I}$ is a cocycle of B over $\mathcal{C}$, with values in $\mathcal{L}is(\mathbb{E}^k) \subset \mathcal{L}(\mathbb{E}^k)$, and determines a Banach vector bundle (E^k, B, π^k), whose transition maps are precisely $\{T^k_{\alpha\beta}\}_{\alpha,\beta \in I}$.

According to the construction expounded in § 1.4.2,

$$E^k = \bigcup_{\alpha,\beta \in I} \left(\{\alpha\} \times U_\alpha \times \mathbb{E}^k \right) \Big/ \sim_k$$

where

$$(\alpha, x, u) \sim_k (\beta, y, v) \quad \Leftrightarrow \quad x = y, \ v = T^k_{\beta\alpha}(u).$$

Denoting by $[(\alpha, x, u)]_k$ the equivalence class of (α, x, u), we define the projection $\pi^k \colon E^k \to B$ by setting $\pi^k([(\alpha, x, u)]_k) := x$, and the trivializations $(U_\alpha, \tau^k_\alpha)$, with

$$\tau^k_\alpha \colon (\pi^k)^{-1}(U_\alpha) \longrightarrow U_\alpha \times \mathbb{E}^k \ : \ \tau^k_\alpha([(\beta, x, u)]_k) := (x, T^k_{\alpha\beta}(x)(u).$$

For every $i, j \in \mathbb{N}$ with $i \leq j$, we further define the map

$$f^{ji} \colon E^j \longrightarrow E^i \colon [(\alpha, x, u)]_j \mapsto [(\alpha, x, \rho^{ji}(u))]_i.$$

The reader will have no difficulty to verify that f^{ji} is well-defined and $\pi^i \circ f^{ji} = \pi^j$; thus, with respect to $(U_\alpha, \tau^j_\alpha)$, we have that $f^{ji}(E^j_{U_\alpha}) \subseteq E^i_{U_\alpha}$. It is also easy to check that

$$(\mathrm{id}_{U_\alpha} \times \rho^{ji}) \circ \tau^j_{U_\alpha} = \tau^i_{U_\alpha} \circ f^{ji}; \qquad j \geq i,$$

from which follows that f^{ji} is smooth, and its restriction to every fibre E^j_x is precisely the continuous linear map

$$f^{ji}\big|_{E^j_x} = \left(\tau^i_{\alpha,x}\right)^{-1} \circ \rho^{ji} \circ \tau^j_{\alpha,x}.$$

Moreover, by appropriate restrictions, we may consider the trivializations (over the same $U_\alpha \in \mathcal{C}$) $(U_\alpha, \tau^j_\alpha)$ and $(U_\alpha, \tau^i_\alpha)$ of E^j and E^i, respectively; hence, the map

$$\tau^{ji}_\alpha \colon U_\alpha \ni x \longmapsto \tau^i_{\alpha,x} \circ f^{ji}\big|_{E^j_x} \circ \left(\tau^j_{\alpha,x}\right)^{-1}$$

is constant; in fact, $\tau^{ji}_\alpha(x) = \rho^{ji}$, for every $x \in U_\alpha$, thus it is smooth. The preceding arguments prove that conditions (VBM. 1)–(VBM. 2) of § 1.4.3 are fulfilled and, in turn, (f^{ji}, id_B) is a morphism of Banach vector bundles.

The previous constructions result in the projective system of Banach vector bundles $\{E^i; f^{ji}\}$, consequently Proposition 5.2.2 determines the plb-vector bundle (E, B, π) with $E = \varprojlim E^i$ and $\pi = \varprojlim \pi^i$.

Assume now that $\{T'_{\alpha\beta}\}_{\alpha,\beta \in I}$ are the transition functions of E. By definition,

$$T'_{\alpha\beta}(x) = \tau_{\alpha,x} \circ (\tau_{\beta,x})^{-1} = \left(\tau^i_{\alpha,x} \circ (\tau^i_{\beta,x})^{-1} \right)_{i \in \mathbb{N}}.$$

But, for every $u \in \mathbb{E}^i$,

$$\left(\tau^i_{\alpha,x} \circ (\tau^i_{\beta,x})^{-1} \right)(u) = \tau^i_{\alpha,x}([(\beta, x, u)]_i) = T^i_{\alpha\beta}(x)(u),$$

therefore,

$$T'_{\alpha\beta}(x) = \left(T^i_{\alpha\beta}(x) \right)_{i \in \mathbb{N}} = T^*_{\alpha\beta}(x);$$

that is, the transition maps of E are exactly the given ones.

Finally, assume that there is another plb-vector bundle (E', B, π) with transition functions $\{T^*_{\alpha\beta}\}_{\alpha,\beta \in I}$ over $\mathcal{C}$. Then, by Theorem 5.2.5, E and E' are Fréchet vector bundles with the same transition maps $\{T_{\alpha\beta} = \varepsilon \circ T^*_{\alpha\beta}\}_{\alpha,\beta \in I}$. Thus, applying to the Fréchet framework the arguments of § 1.4.3, relating cocycles and vb-isomorphisms, we conclude that the bundles E and E' are isomorphic. $\quad\square$

Theorem 5.2.10   *Let (E, B, π) be a Fréchet vector bundle of fibre type $\mathbb{F}$ and base a Banach manifold B. Then E is a plb-vector bundle if and only if the transition maps $T_{\alpha\beta} \colon U_{\alpha\beta} \to \mathcal{L}(\mathbb{F})$ $(\alpha, \beta \in I)$ over an open cover $\mathcal{C}$ (determined by the local trivializations of the bundle) have a decomposition of the form $T_{\alpha\beta} = \varepsilon \circ T^*_{\alpha\beta}$, where $\{T^*_{\alpha\beta}\}_{\alpha,\beta \in I}$ is a plb-cocycle of B over the cover $\mathcal{C}$.*

Proof   If E is a plb-vector bundle, the desired decomposition is actually equality (5.2.5) obtained in the proof of Theorem 5.2.5.

Conversely, assume that we have the decomposition of the statement. By Proposition 5.2.9, the cocycle $\{T^*_{\alpha\beta}\}_{\alpha\beta}$ determines a plb-vector bundle, say, (E', B, π') whose plb-transition maps are $\{T^*_{\alpha\beta}\}_{\alpha\beta}$ (see also Definition 5.2.7). Then, as proved in Theorem 5.2.5, E' is a Fréchet bundle with (ordinary) transition maps $T_{\alpha\beta} = \varepsilon \circ T^*_{\alpha\beta}$. Hence, $E \cong E'$. The identification induces on E the structure of a plb-vector bundle. $\quad\square$

5.3   Morphisms of plb-vector bundles

Let $\mathcal{S} = \{E^i; f^{ji}\}_{i,j \in \mathbb{N}}$ and $\bar{\mathcal{S}} = \{\bar{E}^i; \bar{f}^{ji}\}_{i,j \in \mathbb{N}}$ be two projective systems of Banach vector bundles, over the Banach manifolds B and $\bar{B}$, respectively, and corresponding fibre types $\{E^i\}$, $\{\bar{E}^i\}$ (see Definition 5.2.1).

Definition 5.3.1 A *morphism* of $\mathcal{S}$ into $\bar{\mathcal{S}}$ is a family $\{(g^i, h)\}_{i \in \mathbb{N}}$ satisfying the following conditions:

(**PVBM. 1**) Each (g^i, h) is a vector bundle morphism from (E^i, π^i, B) to $(\bar{E}^i, \bar{\pi}^i, \bar{B})$.

(**PVBM. 2**) The limit map $\varprojlim g^i \colon \varprojlim E^i \to \varprojlim \bar{E}^i$ exists.

Of course, condition (PVBM. 2) is equivalent to the commutativity of the diagram ($j \geq i$):

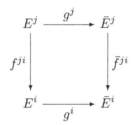

The next result, beside its interest per se, combines many technicalities met so far.

Proposition 5.3.2 *With the previous notations,* $\left(\varprojlim g^i, h\right)$ *is a morphism between the Fréchet vector bundles* (E, B, π) *and* $(\bar{E}, \bar{B}, \bar{\pi})$, *determined by the systems* $\mathcal{S}$ *and* $\bar{\mathcal{S}}$, *respectively, in virtue of Theorem 5.2.5.*

Proof By Proposition 3.1.8, g is smooth. Also, in the proof of Theorem 5.2.5, we found that $E_x = \varprojlim E_x^i$, and similarly $\bar{E}_{h(x)} = \varprojlim \bar{E}_{h(x)}^i$, for every $x \in B$. Condition (PVBM. 2) implies that

$$(5.3.1) \qquad \bar{f}^{ji} \circ g_x^j = g_x^i \circ f^{ji},$$

for every $x \in B$ (with f^{ji} and $\bar{f}^{ji}$ now restricted to the respective fibres), thus $\varprojlim g_x^i$ exists. But $\bar{f}^i \circ g_x = g_x^i \circ f^i$, where $f^i \colon E \to E^i$, $\bar{f}^i \colon \bar{E} \to \bar{E}^i$ are the natural projections (also restricted to the fibres); hence (see Proposition 2.3.5), $g_x = \varprojlim g_x^i$, which shows that the map $g_x \colon E_x \to \bar{E}_{h(x)}$ is continuous linear, for every $x \in B$. The previous arguments prove the Fréchet analog of (VB. 1) in § 1.4.3.

For the analog of (VB. 2), we proceed as follows: If $x \in B$ is an arbitrary point, we choose a plb-trivialization $(V, \tau = \varprojlim \tau^i)$ of E with $x \in V$, and a plb-trivialization $(\bar{U}, \bar{\tau} = \varprojlim \bar{\tau}^i)$ of $\bar{E}$ with $h(x) \in \bar{U}$. Taking $U = V \cap g^{-1}(\bar{U})$ and appropriately restricting τ, we obtain the plb-trivialization $(U, \tau = \varprojlim \tau^i)$ of E with $x \in U$ (for simplicity we use the same symbol for the trivializing maps), such that $h(U) \subseteq \bar{U}$. We define the map

$$G \colon U \longmapsto \mathcal{L}(\mathbb{F}, \bar{\mathbb{F}}) \colon x \mapsto \bar{\tau}_{h(x)} \circ g_x \circ \tau_x^{-1},$$

where $\mathbb{F} = \varprojlim \mathbb{E}^i$ and $\bar{\mathbb{F}} = \varprojlim \bar{\mathbb{E}}^i$ are the fibre types of the bundles E and $\bar{E}$, respectively. To achieve our goal, we need to show that G is smooth. To this end we consider the maps

$$R^i \colon U \longrightarrow \mathcal{L}(\mathbb{E}^i, \bar{\mathbb{E}}^i) \colon x \mapsto \bar{\tau}^i_{h(x)} \circ g^i_x \circ (\tau^i_x)^{-1}, \qquad i \in \mathbb{N}.$$

They are smooth because every (g^i, h) is a vb-morphism of E^i into $\bar{E}^i$. Therefore, the map

$$G^i \colon U \longrightarrow \mathcal{L}(\mathbb{E}^1, \bar{\mathbb{E}}^1) \times \cdots \times \mathcal{L}(\mathbb{E}^i, \bar{\mathbb{E}}^i),$$

with $G^i := (R^1, \ldots, R^i)$, is also smooth. But G^i takes values in $\mathcal{H}^i(\mathbb{F}, \bar{\mathbb{F}})$. Indeed, if $\rho^{ji} \colon \mathbb{E}^j \to \mathbb{E}^i$ and $\bar{\rho}^{ji} \colon \bar{\mathbb{E}}^j \to \bar{\mathbb{E}}^i$ are the connecting morphisms for $j \geq i$, we check that

$$\bar{\rho}^{jk} \circ R^j(x) = \bar{\rho}^{jk} \circ \bar{\tau}^j_{h(x)} \circ g^j_x \circ (\tau^j_x)^{-1}$$

[since $\bar{E} = \varprojlim \bar{\mathbb{E}}^i$] $\qquad = \bar{\tau}^k_{h(x)} \circ \bar{f}^{jk} \circ g^j_x \circ (\tau^j_x)^{-1}$

[by (5.3.1)] $\qquad = \bar{\tau}^k_{h(x)} \circ g^k_x \circ f^{jk} \circ (\tau^j_x)^{-1};$

hence, by the fibre-wise restriction of the commutative diagram following Definition 5.3.1,

$$\bar{\rho}^{jk} \circ R^j(x) = \bar{\tau}^k_{h(x)} \circ g^k_x \circ (\tau^k_x)^{-1} \circ \rho^{jk} = R^i(x) \circ \rho^{jk}.$$

The latter, in virtue of (2.3.5), proves the claim. Moreover, since $\mathcal{H}^i(\mathbb{F}, \bar{\mathbb{F}})$ is a closed subspace of $\prod_{k=1}^i \mathcal{L}(\mathbb{E}^k, \bar{\mathbb{E}}^k)$, every $G^i \colon U \to \mathcal{H}^i(\mathbb{F}, \bar{\mathbb{F}})$ is smooth.

Now, using the connecting morphisms $h^{ji} \colon \mathcal{H}^j(\mathbb{F}, \bar{\mathbb{F}}) \to \mathcal{H}^i(\mathbb{F}, \bar{\mathbb{F}})$ [see (2.3.6)], we immediately check that $h^{ji} \circ G^j = G^i$, for every $j \geq i$, thus we obtain the pls-map $G^* = \varprojlim G^i \colon U \to \mathcal{H}(\mathbb{F}, \bar{\mathbb{F}}) \equiv \varprojlim \mathcal{H}^i(\mathbb{F}, \bar{\mathbb{F}})$ [see also equality (2.3.8)].

We further verify that the diagram

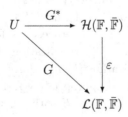

is commutative [recall that ε is given by (2.3.4)]. This is the case because, for every $x \in U$,

$$(\varepsilon \circ G^*)(x) = \varepsilon \left(\left(G^i(x) \right)_{i \in \mathbb{N}} \right) = \varepsilon \left(\left((R^1(x), \ldots, R^i(x)) \right)_{i \in \mathbb{N}} \right).$$

However, after the identification $\mathcal{H}(\mathbb{F}, \bar{\mathbb{F}}) \equiv \varprojlim \mathcal{H}^i(\mathbb{F}, \bar{\mathbb{F}})$, we have that

$$\left((R^1(x), \ldots, R^i(x)) \right)_{i \in \mathbb{N}} \equiv \left(R^i(x) \right)_{i \in \mathbb{N}}.$$

As a result,

$$(\varepsilon \circ G^*)(x) = \varepsilon \left(\left(R^i(x) \right)_{i \in \mathbb{N}} \right) = \varprojlim R^i(x)$$

$$= \varprojlim \left(\bar{\tau}^i_{h(x)} \circ g^i_x \circ (\tau^i_x)^{-1} \right)$$

$$= \bar{\tau}_{h(x)} \circ g_x \circ \tau_x^{-1} = G(x),$$

from which, along with the smoothness of G^* and ε, we conclude that G is smooth. This completes the proof. $\qquad\square$

Definition 5.3.3   A morphism $\left(\varprojlim g^i, h \right)$, as in Proposition 5.3.2, will be called a **plb-morphism**. If all (g^i, id_B) are vb-isomorphisms, then $(\varprojlim g^i, h)$ is called a **plb-isomorphism**.

With the cohomological classification of plb-vector bundles in mind, in the remainder of this section we consider bundles over the same base and of the same fibre type. More precisely, let $\left(E = \varprojlim E^i, B, \pi = \varprojlim \pi^i \right)$ and $\left(\bar{E} = \varprojlim \bar{E}^i, B, \bar{\pi} = \varprojlim \bar{\pi}^i \right)$ be plb-vector bundles of fibre type $\mathbb{F} = \varprojlim \mathbb{E}^i$. Then we prove:

Lemma 5.3.4   *Every plb-bundle isomorphism $(g = \varprojlim g^i, \mathrm{id}_B)$ of E onto $\bar{E}$ corresponds bijectively to a family $\{h_\alpha \colon U_\alpha \to \mathcal{H}_0(\mathbb{F})\}_{\alpha \in I}$ of generalized smooth maps over an open cover $\{U_\alpha\}_{\alpha \in I}$ of B such that*

$$(5.3.2) \qquad \bar{T}^*_{\alpha\beta}(x) = h_\alpha(x) \circ T^*_{\alpha\beta}(x) \circ h_\beta(x)^{-1}; \qquad x \in U_{\alpha\beta},$$

if $\{T^*_{\alpha\beta}\}$ and $\{\bar{T}^*_{\alpha\beta}\}$ are the $\mathcal{H}_0(\mathbb{F})$-valued cocycles of E and $\bar{E}$, respectively.

As usual, cocycles satisfying (5.3.2) are said to be **cohomologous**. For the definition of generalized smooth $\mathcal{H}_0(\mathbb{F})$-valued maps we refer to Remark 5.2.6(3).

Proof   Since each $g^i : E^i \to \bar{E}^i$ is an isomorphism between Banach bundles, there exists a family of smooth maps

$$h^i_\alpha \colon U_\alpha \longrightarrow \mathrm{GL}(\mathbb{E}^i); \qquad \alpha \in I,$$

given by

$$h_\alpha^i(x) = \bar{\tau}_{\alpha,x}^i \circ g_x^i \circ \left(\tau_{\alpha,x}^i\right)^{-1},$$

where $(U_\alpha, \tau_\alpha^i)_{\alpha \in I}$, $(U_\alpha, \bar{\tau}_\alpha^i)_{\alpha \in I}$ are trivializations of E^i and $\bar{E}^i$, respectively (recall the discussion on the cohomological classification of Banach vector bundles in §1.4.3). The trivializations can be suitably chosen so that condition (PVB. 2) of Definition 5.2.1 be satisfied for both families of trivializations. If $\rho^{ji} \colon \mathbb{E}^j \to \mathbb{E}^i$ are the connecting morphisms of the projective system $\{\mathbb{E}^i\}_{i \in \mathbb{N}}$, then applying (PVB. 2) and (PVBM. 2) of Definition 5.3.1 fibre-wise, we obtain

$$\begin{aligned}
\rho^{ji} \circ h_\alpha^j(x) &= \rho^{ji} \circ \bar{\tau}_{\alpha,x}^j \circ g_x^j \circ \left(\tau_{\alpha,x}^j\right)^{-1} \\
&= \bar{\tau}_{\alpha,x}^i \circ \bar{f}^{ji} \circ g_x^j \circ \left(\tau_{\alpha,x}^j\right)^{-1} \\
&= \bar{\tau}_{\alpha,x}^i \circ g_x^i \circ f^{ji} \circ \left(\tau_{\alpha,x}^j\right)^{-1} \\
&= \bar{\tau}_{\alpha,x}^i \circ g_x^i \circ \left(\tau_{\alpha,x}^i\right)^{-1} \circ \rho^{ji} \\
&= h_\alpha^i(x) \circ \rho^{ji},
\end{aligned}$$

for every $i, j \in \mathbb{N}$ with $j \geq i$. As a result, the linear isomorphism

$$h_\alpha(x) := \varprojlim h_\alpha^i(x) \colon \mathbb{F} = \varprojlim \mathbb{E}^i \longrightarrow \varprojlim \mathbb{E}^i = \mathbb{F}$$

exists for each $x \in U_\alpha$, thus we may define the generalized smooth map

$$(5.3.3) \qquad h_\alpha \colon U_\alpha \longrightarrow \mathcal{H}_0(\mathbb{F}) \colon x \mapsto \left(h_\alpha^1(x), \left(h_\alpha^1(x), h_\alpha^2(x)\right), \ldots\right).$$

On the other hand, for every $\alpha, \beta \in I$,

$$\begin{aligned}
\overline{T}_{\alpha\beta}^*(x) \circ h_\beta(x) \circ T_{\beta\alpha}^*(x) &= \left(\overline{T}_{\alpha\beta}^i(x) \circ h_\beta^i(x) \circ T_{\beta\alpha}^i(x)\right)_{i \in \mathbb{N}} \\
&= \left(\bar{\tau}_{\alpha,x}^i \circ \left(\bar{\tau}_{\beta,x}^i\right)^{-1} \circ h_\beta^i(x) \circ \tau_{\beta,x}^i \circ \left(\tau_{\alpha,x}^i\right)^{-1}\right)_{i \in \mathbb{N}} \\
&= \left(\bar{\tau}_{\alpha,x}^i \circ g_x^i \circ \left(\tau_{\alpha,x}^i\right)^{-1}\right)_{i \in \mathbb{N}} \\
&= \left(h_\alpha^i(x)\right)_{i \in \mathbb{N}} = h_\alpha(x),
\end{aligned}$$

which is precisely (5.3.2).

Conversely, every family $\{h_\alpha \colon U_\alpha \to \mathcal{H}_0(\mathbb{F})\}_{\alpha \in I}$ of generalized smooth maps, given by (5.3.3) and satisfying the compatibility condition (5.3.2), gives rise, for each $i \in \mathbb{N}$, to a corresponding family of smooth maps (relative to the bundle E^i)

$$h_\alpha^i \colon U_\alpha \longrightarrow \mathrm{GL}(\mathbb{E}^i); \qquad \alpha \in I,$$

so that the equality

$$\overline{T}^i_{\alpha\beta}(x) \circ h^i_\beta(x) \circ T^i_{\beta\alpha}(x) = h^i_\alpha(x); \qquad \alpha \in I,$$

holds for every $x \in U_\alpha$, where $\{T^i_{\alpha\beta}\}$ and $\{\overline{T}^i_{\alpha\beta}\}$ are the ordinary cocycles of E^i and $\bar{E}^i$, respectively. Therefore, following the discussion in § 1.4.3, we define the bundle isomorphism (over B) $g^i \colon E^i \to \bar{E}^i$ by setting

$$g^i_x = \left(\bar{\tau}^i_{\alpha,x}\right)^{-1} \circ h^i_\alpha(x) \circ \tau^i_{\alpha,x}, \qquad x \in B.$$

Then, working as in the first part of the proof, we have (for $j \geq i$):

$$\begin{aligned}
\bar{f}^{ji} \circ g^j_x &= \bar{f}^{ji} \circ \left(\bar{\tau}^j_{\alpha,x}\right)^{-1} \circ h^j_\alpha(x) \circ \tau^j_{\alpha,x} \\
&= \left(\bar{\tau}^i_{\alpha,x}\right)^{-1} \circ \rho^{ji} \circ h^j_\alpha(x) \circ \tau^j_{\alpha,x} \\
&= \left(\bar{\tau}^i_{\alpha,x}\right)^{-1} \circ h^i_\alpha(x) \circ \tau^i_{\alpha,x} \circ f^{ji} \\
&= g^i_x \circ f^{ji},
\end{aligned}$$

for every $x \in B$, thus $\bar{f}^{ji} \circ g^j = g^i \circ f^{ji}$. This equality ensures the existence of the map $g = \varprojlim g^i \colon E \to \bar{E}$. Now, since both the conditions of Definition 5.3.1 are fulfilled, Proposition 5.3.2 implies that (g, id_B) is a vb-isomorphism of E onto $\bar{E}$. $\qquad \square$

From the category of vector bundles (E, B, π) over the Banach manifold B, and of fibre type the Fréchet space $\mathbb{F}$, we single out those obtained by projective systems of Banach bundles in the sense of Theorem 5.2.5. We denote their set by $\mathcal{VB}_B(\mathbb{F})$. Considering the obvious equivalence relation induced by the plb-isomorphisms, we obtain the corresponding quotient space $\mathcal{VB}_B(\mathbb{F})/_\sim$. Then we obtain the following cohomological classification theorem.

Theorem 5.3.5 *If $\mathfrak{H}_0(\mathbb{F})$ denotes the sheaf of germs of $\mathcal{H}_0(\mathbb{F})$-valued generalized smooth maps on B, then*

$$\mathcal{VB}_B(\mathbb{F})/_\sim = H^1(B, \mathfrak{H}_0(\mathbb{F}))$$

within a bijection.

Here $H^1(B, \mathfrak{H}_0(\mathbb{F}))$ is the first cohomology set of B with coefficients in $\mathfrak{H}_0(\mathbb{F})$ (see also the last part of § 1.4.3).

Proof Let $[E]$ be the equivalence class of a bundle E of the prescribed type. If $\{T^*_{\alpha\beta}\}$ is the cocycle of E over a trivializing open cover $\mathfrak{U}$ of B, then we define the map

$$VB_B(\mathbb{F}) \ni [E] \longmapsto [\{T^*_{\alpha\beta}\}] \in H^1(B, \mathfrak{H}_0(\mathbb{F})).$$

It is a matter of routine to verify that, in virtue of Lemma 5.3.4, this is a well-defined bijection. $\square$

5.4 The sections of plb-vector bundles

Let $\{E^i; f^{ji}\}_{i,j\in\mathbb{N}}$ be a projective system of Banach vector bundles and (E, B, π) the induced plb-vector bundle (see Definitions 5.2.1, 5.2.4). We denote by $\Gamma(E^i)$ and $\Gamma(E)$ the $\mathcal{C}^\infty(B, \mathbb{R})$-modules of smooth sections of E^i and E, respectively.

Given a family of sections $\{\xi^i \in \Gamma(E^i) \mid i \in \mathbb{N}\}$, an obvious consequence of the definitions and the smoothness of the limits of smooth maps is that $\varprojlim \xi^i \in \Gamma(E)$. The converse is also true;, namely, we have:

Lemma 5.4.1   *Every section $\xi \in \Gamma(E)$ has the form $\xi = \varprojlim \xi^i$, where $\xi^i \in \Gamma(E^i)$, $i \in \mathbb{N}$.*

Proof   For every $i \in \mathbb{N}$, we set $\xi^i := f^i \circ \xi$ (recall that $f^i \colon E = \varprojlim E^i \to E^i$ is the canonical projection). Obviously, every ξ^i is smooth and satisfies

$$\pi^i \circ \xi^i = \pi^i \circ f^i \circ \xi = \pi \circ \xi = \mathrm{id}_B; \qquad i \in \mathbb{N},$$

thus $\xi^i \in \Gamma(E^i)$. On the other hand,

$$f^{ji} \circ \xi^j = f^{ji} \circ f^j \circ \xi = f^i \circ \xi = \xi^i; \qquad j \geq i,$$

thus $\varprojlim \xi^i$ exists. Since $\xi^i = f^i \circ \xi$, Proposition 2.3.5 (applied for B and $\{E^i\}$) implies that $\xi = \varprojlim \xi^i$. $\square$

Proposition 5.4.2   *The $\mathcal{C}^\infty(B, \mathbb{R})$-modules $\Gamma(E)$ and $\varprojlim \Gamma(E^i)$ coincide within an isomorphism.*

Proof   For every $j \geq i$, we define the maps

$$\gamma^{ji} \colon \Gamma(E^j) \longrightarrow \Gamma(E^i) \colon \xi \mapsto f^{ji} \circ \xi.$$

Since f^{ji} is continuous linear on the fibres, it follows that γ^{ji} is a morphism of $\mathcal{C}^\infty(B, \mathbb{R})$-modules. Moreover, $\gamma^{ik} \circ \gamma^{ji} = \gamma^{jk}$ ($k \leq i \leq j$). Therefore $\{\Gamma(E^j); \gamma^{ji}\}_{i,j\in\mathbb{N}}$ is a projective system of $\mathcal{C}^\infty(B, \mathbb{R})$-modules with limit $\varprojlim \Gamma(E^i)$.

On the other hand, the maps

$$\gamma^i \colon \Gamma(E) \longrightarrow \Gamma(E^i) \colon \xi \mapsto f^i \circ \xi; \qquad i \in \mathbb{N},$$

satisfy $\gamma^{ji} \circ \gamma^j = \gamma^i$ $(i \leq j)$, thus inducing

$$\gamma := \varprojlim \gamma^i \colon \Gamma(E) \longrightarrow \varprojlim \Gamma(E^i).$$

The preceding γ is 1–1, since, for every $\xi, \eta \in \Gamma(E)$,

$$\gamma(\xi) = \gamma(\eta) \quad \Leftrightarrow \quad \gamma^i(\xi) = \gamma^i(\eta); \qquad i \in \mathbb{N},$$
$$\Leftrightarrow \quad f^i \circ \xi = f^i \circ \eta; \qquad i \in \mathbb{N},$$

[Lemma 5.4.1] $\qquad\qquad \Leftrightarrow \quad \xi = \varprojlim \xi^i = \varprojlim \eta^i = \eta.$

Also, γ is onto, for if $(\xi^i)_{i \in \mathbb{N}}$ is an arbitrary element of $\Gamma(E^i)$, then

$$\gamma^{ji} \circ \xi^j = \xi^i \quad \Leftrightarrow \quad f^{ji} \circ \xi^j = \xi^i, \qquad j \geq i;$$

hence, we obtain the section $\varprojlim \xi^i$ such that

$$\gamma(\xi) = (\gamma^i(\xi))_{i \in \mathbb{N}} = (f^i \circ \xi)_{i \in \mathbb{N}} = (\xi^i)_{i \in \mathbb{N}}.$$

Finally, the linearity of every f^i on the fibres of E implies that every γ^i is a morphism of $\mathcal{C}^\infty(B, \mathbb{R})$-modules, and so is γ. $\qquad\qquad\qquad \square$

5.5 The pull-back of plb-vector bundles

The construction of the pull-back of a plb-vector bundle is a useful application, enlightening the methods expounded so far. Since some of the anticipated properties (analogous to the ones described in § 1.4.4) have quite lengthy proofs, we proceed by exhibiting them in separate statements.

Let $\{E^i; f^{ji}\}_{i,j \in \mathbb{N}}$ be a projective system of Banach vector bundles and (E, B, π) the induced plb-vector bundle (see Definitions 5.2.1 and 5.2.4). If $g \colon Y \to B$ is a smooth map of Banach manifolds, then, for every $i \in \mathbb{N}$, we obtain the pull-back of the bundle (E^i, π^i, B). This will be represented by the triplet $(g^*(E^i), \pi^i_*, B)$ (for the sake of conformity, we adorn the projections—as well as other relevant quantities below—with a star, put as a subscript to avoid double superscripts). Given a trivialization, say, (U, τ^i) of E^i, we construct the trivialization $(g^{-1}(U), \sigma^i)$ of $g^*(E^i)$, where the trivializing map

$$(5.5.1) \qquad \sigma^i \colon \left(\pi^i_*\right)^{-1}\left(g^{-1}(U)\right) = g^*(U) \times_U E^j_U \longrightarrow g^*(U) \times \mathbb{E}^i$$

is defined by

$$(5.5.2) \qquad \sigma^i(y, u) := \left(y, \tau^i_{g(y)}(u)\right), \qquad (y, u) \in g^*(U) \times_U E^j_U.$$

Proposition 5.5.1 *With the previous notations, for every $i, j \in \mathbb{N}$ with $i \leq j$, there is a morphism of Banach vector bundles $f_*^{ji} \colon g^*(E^j) \to g^*(E^i)$ such that $\left\{g^*(E^i); f_*^{ji}\right\}_{i,j\in}$ is a projective system.*

Proof We set

$$f_*^{ji}(y, u) := \left(y, f^{ji}(u)\right), \qquad (y, u) \in g^*(E^j) = Y \times_B E^j.$$

The map f_*^{ji} indeed takes values in $g^*(E^i)$ because, for every (y, u) as above, $g(y) = \pi^j(u)$. Since $f^{ji} \colon E^j \to E^i$ is a vb-morphism (over id_B), it follows that $g(y) = \pi^i\left(f^{ji}(u)\right)$, thus $\left(y, f^{ji}(u)\right) \in g^*(E^i)$.

We should first show that $\left(f_*^{ji}, \mathrm{id}_Y\right)$ is a vb-morphism. To this end, we check that f_*^{ji} is smooth: Let any $(y_0, u_0) \in g^*(E^j)$. Since $\{E^i; f^{ji}\}_{i,j\in\mathbb{N}}$ is a projective system of vector bundles, we can find trivializations (U, τ^j) and (U, τ^i) of E^j and E^i, respectively, satisfying condition (PVB. 2) of Definition 5.2.1, with $g(y_0) = \pi^j(u_0) \in U$. We consider the corresponding trivializations $(g^{-1}(U), \sigma^j)$ and $(g^{-1}(U), \sigma^i)$ of $g^*(E^j)$ and $g^*(E^i)$. As in the beginning of the proof,

$$f_*^{ji}\left(\left(\pi_*^j\right)^{-1}\left(g^{-1}(U)\right)\right) \subseteq \left(\pi_*^i\right)^{-1}\left(g^{-1}(U)\right),$$

from which, by direct application of the definitions and (PVB. 2), we verify that the next diagram is commutative.

$$
\begin{array}{ccc}
\left(\pi_*^j\right)^{-1}\left(g^{-1}(U)\right) & \xrightarrow{\;\;f_*^{ji}\;\;} & \left(\pi_*^i\right)^{-1}\left(g^{-1}(U)\right) \\
{\scriptstyle \sigma^j}\Big\downarrow & & \Big\downarrow{\scriptstyle \sigma^i} \\
g^{-1}(U) \times \mathbb{E}^j & \xrightarrow[\mathrm{id}_{g^{-1}(U)} \times \rho^{ji}]{} & g^{-1}(U) \times \mathbb{E}^i
\end{array}
$$

Hence, f_*^{ji} is smooth on the (open) neighborhood $\left(\pi_*^j\right)^{-1}\left(g^{-1}(U)\right)$ of (the arbitrary) $(y_0, u_0) \in g^*(E^j)$.

Moreover, for every $y \in g^{-1}(U)$, the restriction $f_{*,y}^{ji}$ of f_*^{ji} to the fibre $\left(\pi_*^j\right)^{-1}(y) = \{y\} \times \left(\pi^j\right)^{-1}(y) = \{y\} \times E_{g(y)}^j$ over y is the map

$$(y, u) \longmapsto \left(y, f_*^{ji}\big|_{E_{g(y)}^j}(u)\right)$$

(y fixed each time), thus $f_{*,y}^{ji}$ is continuous linear. Also, we immediately see that $\pi^i \circ f^{ji} = \pi^i$. So we have proved condition (VBM.1) of §1.4.3.

Next we have to prove condition (VBM. 2). First observe that for a trivialization $(g^{-1}(U), \sigma^i)$, as in (5.5.1) and (5.5.2), we define the map

$\sigma_y^i := \mathrm{pr}_2 \circ \sigma^i \colon \{y\} \times E_{g(y)}^i \to \mathbb{E}^i$. Thus $\sigma_y^i(y, u) = \tau_{g(y)}^i(u)$, which means that σ_y^i is continuous linear. The inverse of σ_y^i is given by

$$\left(\sigma_y^i\right)^{-1}(e) = \left(y, \left(\tau_{g(y)}^i\right)^{-1}(e)\right) = \left(y, (\tau^i)^{-1}(g(y), e)\right), \qquad e \in \mathbb{E}^i.$$

Now, for an arbitrary $y_0 \in g^*(E^j)$, we consider again the trivializations $(g^{-1}(U), \sigma^j)$ and $(g^{-1}(U), \sigma^i)$, with $y_0 \in g^{-1}(U)$. The desired condition now translates into showing the smoothness of

$$(5.5.3) \qquad g^{-1}(U) \ni y \longmapsto \sigma_y^i \circ f_{*,y}^{ji} \circ (\sigma_y^j)^{-1} \in \mathcal{L}(\mathbb{E}^j, \mathbb{E}^i).$$

However, for every $e \in \mathbb{E}^j$,

$$\left(\sigma_y^i \circ f_{*,y}^{ji} \circ (\sigma_y^j)^{-1}\right)(e) = \sigma_y^i\left(y, \left(f^{ji} \circ \left(\tau_{g(y)}^j\right)\right)^{-1}(e)\right)$$

$$= (\mathrm{pr}_2 \circ \sigma^i)\left(y, \left(f^{ji} \circ \left(\tau_{g(y)}^j\right)\right)^{-1}(e)\right)$$

$$= \left(\tau_{g(y)}^i \circ f^{ji} \circ \left(\tau_{g(y)}^j\right)^{-1}\right)(e).$$

Since (PVB. 2), applied to the fibres of the system $\{E^i; f^{ji}\}$, yields

$$(5.5.4) \qquad \tau_{g(y)}^i \circ f^{ji} = \rho^{ji} \circ \tau_{g(y)}^j,$$

it follows that $\left(\sigma_y^i \circ f_{*,y}^{ji} \circ (\sigma_y^j)^{-1}\right)(e) = \rho^{ji}(e)$. Therefore, (5.5.3) is smooth since it coincides with the constant map $y \mapsto \rho^{ji}$.

Having shown that each f_*^{ji} is a vb-morphism, to complete the proof of the statement we should verify the analogs of (PVB. 1) and (PVB. 2) of Definition 5.2.1 for the pull-back bundles. The first is satisfied because each $g^*(E^i)$ has fibre type $\mathbb{E}^i$ and $\{\mathbb{E}^i; \rho^{ji}\}$ is a projective system. The second is precisely the commutativity of the previous diagram, obtained from the analogous trivializations of the pull-back bundles, chosen with respect to every $y \in Y$. $\qquad \square$

Before proceeding, we see that

$$g^*\left(\varprojlim E^i\right) = Y \times_B E = Y \times_B \varprojlim E^i \cong \varprojlim(Y \times_B E^i) = \varprojlim g^*(E^i),$$

a fact implying that

the pull-back functor commutes with projective limits.

However, the next result gives something more than a simple set-theoretic bijection:

Proposition 5.5.2 *Let* $(g^*(E), \pi^*, Y)$ *be the pull-back of the plb-bundle* $\left(E = \varprojlim E^i, B, \pi = \varprojlim \pi^i\right)$ *by* $g \colon Y \to B$. *Then* $g^*(E)$ *and* $\varprojlim g^*(E^i)$ *are isomorphic Fréchet vector bundles.*

Proof For every $i \in \mathbb{N}$, we define the map

$$h^i \colon g^*(E) \longrightarrow g^*(E^i) \colon (y, u) \mapsto h^i(y, u) := (y, f^i(u)),$$

where $f^i \colon E = \varprojlim E^i \to E^i$, $i \in \mathbb{N}$, are the canonical projections. Since $f^{ji} \circ f^j = f^i$ ($j \geq i$), it follows that $f_*^{ji} \circ h^j = h^i$ (see the definition of f_*^{ji} in the preceding proof), consequently we obtain the limit map

$$h := \varprojlim h^i \colon g^*(E) \longrightarrow \varprojlim g^*(E^i).$$

We intend to show that (h, g) is the desired vb-isomorphism.

For this purpose we first prove that h is smooth: Take an arbitrary $(y_0, u_0) \in g^*(E)$ and choose any trivialization $(U, \tau) = (U, \varprojlim \tau^i)$ of E with $g(y_0) = \pi(u_0) \in U$. Then $(g^{-1}(U), \sigma)$, where

$$\sigma \colon (\pi^*)^{-1}(U) = g^{-1}(U) \times_U E_U \longrightarrow g^{-1}(U) \times \mathbb{F} \colon$$
$$(y, u) \longmapsto \sigma(y, u) := \big(y, \tau_{g(y)}(u)\big)$$

is a trivialization of $g^*(E)$, with $y_0 \in g^{-1}(U)$. On the other hand, as we have seen in the proof of Proposition 5.5.1, the trivializations $(g^{-1}(U), \sigma^i)$ of $g^*(E^i)$, for all $i \in \mathbb{N}$, satisfy the conditions of Definition 5.2.1, thus we obtain the plb-trivialization $\big(g^{-1}(U), \varprojlim \sigma^i\big)$ of the bundle $\big(\varprojlim g^*(E^i), \varprojlim \pi_*^i, Y\big)$.

We check that

$$h\big(g^{-1}(U) \times_U E_U\big) \subseteq \big(\varprojlim \pi_*^i\big)^{-1}\big(g^{-1}(U)\big) = \varprojlim \Big(\big(\pi_*^i\big)^{-1}\big(g^{-1}(U)\big)\Big).$$

This is so, because, for every $(y, u) \in g^{-1}(U) \times_U E_U$,

$$\begin{aligned}(5.5.5) \quad &\big(\varprojlim \pi_*^i\big)(h(y, u)) = \big(\varprojlim \pi_*^i\big)\Big(\big(h^i(y, u)\big)_{k \in \mathbb{N}}\Big) = \\ &= \big(\varprojlim \pi_*^i\big)\Big(\big(y, f^k(u)\big)_{k \in \mathbb{N}}\Big) = \big(\pi_*^i(y, f^i(u))\big)_{i \in \mathbb{N}} = y \in g^{-1}(U),\end{aligned}$$

from which we get $h(y, u) \in \big(\varprojlim \pi_*^i\big)^{-1}\big(g^{-1}(U)\big)$. On the other hand, equality $\big(\varprojlim \pi_*^i\big)^{-1}\big(g^{-1}(U)\big) = \varprojlim \big((\pi_*^i)^{-1}\big(g^{-1}(U)\big)\big)$, together with (5.5.5), yields the diagram

$$(\pi^*)^{-1}\big(g^{-1}(U)\big) \xrightarrow{\quad h \quad} \big(\varprojlim \pi_*^i\big)^{-1}\big(g^{-1}(U)\big)$$

$$\sigma \searrow \qquad \qquad \downarrow \varprojlim \sigma^i$$

$$g^{-1}(U) \times \mathbb{F}$$

which is commutative, because

$$
\begin{aligned}
\left(\left(\varprojlim \sigma^i\right) \circ h\right)(y, u) &= \left(\sigma^i\left(h^i(y, u)\right)\right)_{i \in \mathbb{N}} \\
&= \left(\sigma^i\left(y, f^i(u)\right)\right)_{i \in \mathbb{N}} \\
&= \left(y, \left(\tau^i_{g(y)}(f^i(u))\right)_{i \in \mathbb{N}}\right) \\
&= \left(y, \left(\varprojlim \tau^i_{g(y)}\right)\left((f^i(u))_{i \in \mathbb{N}}\right)\right) \\
&= \left(y, \tau_{g(y)}(u)\right) = \sigma(y, u).
\end{aligned}
$$

Therefore, h is smooth on the open neighbourhood $(\pi^*)^{-1}(g^{-1}(U))$ of the arbitrary $(y_0, u_0) \in g * (E)$.

In the course of the proof of (5.5.5) we found that $\left(\varprojlim \pi^i_*\right) \circ h = \pi^i_*$, while the previous diagram ensures that the restriction of h to the fibres is a continuous linear map.

Also, for an arbitrary $y_0 \in Y$ and the trivializations $(g^{-1}(U), \sigma)$ and $\left(g^{-1}(U), \varprojlim \sigma^i\right)$ as before, the map

(5.5.6) $\qquad g^{-1}(U) \longrightarrow \mathcal{L}(\mathbb{F}, \mathbb{F}) : y \mapsto \left(\varprojlim \sigma^i\right)_y \circ h_y \circ \sigma_y^{-1}$

is constantly equal to $\mathrm{id}_{\mathbb{F}}$, hence (5.5.6) is smooth. Note that we have set

$$
\left(\varprojlim \sigma^i\right)_y := \left(\varprojlim \sigma^i\right)\Big|_{\left(\varprojlim \pi^i_*\right)^{-1}(y)},
$$

$$
h_y := h\big|_{(\pi_*)^{-1}(y)}, \quad \sigma_y := \sigma\big|_{(\pi_*)^{-1}(y)}.
$$

The previous arguments altogether show that (h, id_Y) is a morphism between the Fréchet vector bundles $g^*(E)$ and $\varprojlim g^*(E^i)$.

To complete the proof, it remains to show that h is a bijection, so by similar arguments and using the same trivializations, we have that (h^{-1}, id_Y) is also a Fréchet vb-morphism. The injectivity of h is almost obvious:

$$
\begin{aligned}
h(y, u) = h(z, w) \quad &\Rightarrow \quad h^i(y, u) = h^i(z, w), \quad \forall\, i \in \mathbb{N} \\
&\Rightarrow \quad (y, f^i(u)) = (z, f^i(w)), \quad \forall\, i \in \mathbb{N} \\
[\text{since } u = (u^i), w = (w^i)] \quad &\Rightarrow \quad y = z, \ u = w.
\end{aligned}
$$

The surjectivity of h needs a bit of extra work: An arbitrary element of $\varprojlim g^*(E^i)$ has then form $\left(y^i, u^i\right)_{i \in \mathbb{N}}$, where each $(y^i, u^i) \in g^*(E^i)$ satisfies $f^{ji}_*(y^j, u^j) = (y^i, u^i)$, for every $j \geq i$; equivalently,

$$
\left(y^j, f^{ji}(u^j)\right) = (y^i, u^i); \qquad i, j \in \mathbb{N}, \ j \geq i.
$$

This implies that $y^1 = \cdots = y^i = \cdots =: y \in Y$ and $u = (u^i) \in \varprojlim E^i$. Moreover, $\pi(u) = \pi^i(u^i) = g(u^i) = g(y)$. Therefore, $(y, u) \in g^*(E)$. The latter element is mapped to $(y^i, u^i)_{i \in \mathbb{N}}$:

$$h(y, u) = \left(h^i(y, u) \right)_{i \in \mathbb{N}} = \left(y, f^i(u) \right)_{i \in \mathbb{N}} = \left(y^i, u^i \right)_{i \in \mathbb{N}},$$

which ends the proof. $\qquad\square$

Proposition 5.5.3 *The pair (g^*, g), with $g^* := \mathrm{pr}_2 |_{g^*(E)}$, is a vb-morphism of $g^*(E)$ into E.*

Proof Every pair (pr_2^i, g), $i \in \mathbb{N}$, with $\mathrm{pr}_2^i \colon g^*(E^i) \to E^i$, is a vb-morphism between the vector bundles $g^*(E^i)$ and E^i, such that the diagram

$$
\begin{array}{ccc}
g^*(E^i) & \xrightarrow{\;\;\mathrm{pr}_2^j\;\;} & E^j \\
{\scriptstyle f_*^{ji}}\downarrow & & \downarrow{\scriptstyle f^{ji}} \\
g^*(E^i) & \xrightarrow[\;\;\mathrm{pr}_2^i\;\;]{} & E^i
\end{array}
$$

is commutative; hence, $\varprojlim \mathrm{pr}_2^i$ is defined and $\left(\varprojlim \mathrm{pr}_2^i, g \right)$ is a vb-morphism of $g^*(E)$ and E. But the diagram

$$
\begin{array}{ccc}
g^*(E) & \xrightarrow{\;\;\mathrm{pr}_2\;\;} & E \\
{\scriptstyle h^i}\downarrow & & \downarrow{\scriptstyle f^i} \\
g^*(E^i) & \xrightarrow[\;\;\mathrm{pr}_2^i\;\;]{} & E^i
\end{array}
$$

is also commutative, since

$$\left(\mathrm{pr}_2^i \circ h^i \right)(y, u) = \mathrm{pr}_2^i \left(y, f^i(u) \right) = f^i(u) = \left(f^i \circ \mathrm{pr}_2 \right)(y, u),$$

for every $(y, u) \in g^*(E)$. Therefore, by Proposition 2.3.5, $\mathrm{pr}_2 = \varprojlim \mathrm{pr}_2^i$; that is,

$$(g^*, g) = (\mathrm{pr}_2, g) = \left(\varprojlim \mathrm{pr}_2^i, g \right)$$

is a vb-morphism. $\qquad\square$

Proposition 5.5.4 *The pull-back of a plb-bundle satisfies the following universal property: Let $\left(\bar{E} = \varprojlim \bar{E}^i, Y \right), \bar{\pi} = \varprojlim \bar{\pi}^i$ be a plb-bundle* *and* $\bar{g}\colon \bar{E} \to E$ *a smooth map such that* $(\bar{g}, g)$ *is a plb-vector bundle morphism, i.e.* $\bar{g} = \varprojlim \bar{g}^i$, *with* $\bar{g}^i\colon \bar{E}^i \to E^i$ ($i \in \mathbb{N}$). *Then there exists a unique smooth map* $\tilde{\pi}\colon \bar{E} \to g^*(E)$ *such that* $(\tilde{\pi}, \mathrm{id}_Y)$ *is a vb-morphism of* $\bar{E}$ *into* $g^*(E)$, *and* $g^* \circ \tilde{\pi} = \bar{g}$, *where* $g^* = pr_2|_{g^*(E)}$.

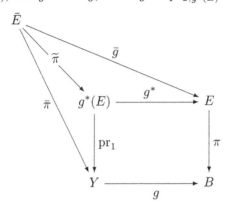

Proof By the universal property of $g^*(E^i)$, for every $i \in \mathbb{N}$, there is a unique vector bundle morphism $(\tilde{\pi}^i, \mathrm{id}_Y)\colon (\bar{E}^i, \bar{Y}, \pi^i) \to (g^*(E^i), Y \, \mathrm{pr}_1^i)$, where $\tilde{\pi}^i = (\bar{\pi}^i, \bar{g}^i)$. Since, by assumption, the limits $\varprojlim \bar{\pi}^i$ and $\varprojlim \bar{g}^i$ exist, we set $\tilde{\pi} := \varprojlim \tilde{\pi}^i \equiv \left(\varprojlim \bar{\pi}^i, \varprojlim \bar{g}^i \right)$. Then the plb-morphism

$$(\tilde{\pi}, \mathrm{id}_Y) = \left(\varprojlim \tilde{\pi}^i, \mathrm{id}_Y \right) \equiv \varprojlim \left(\tilde{\pi}^i, \mathrm{id}_Y \right)$$

exists since the limit $\varprojlim \tilde{\pi}^i$ is ensured by the commutativity of

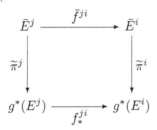

Indeed, for every $j \geq i$ and $u \in \bar{E}^j$, the existence of $\varprojlim \bar{\pi}^i$ and $\varprojlim \bar{g}^i$ implies

$$\left(f_*^{ji} \circ \tilde{\pi}^j \right)(u) = f_*^{ji}\left(\bar{\pi}^j(u), \bar{g}^j(u) \right)$$

[see Proposition 5.5.1]
$$= \left(\bar{\pi}^j(u), f^{ji}(\bar{g}^j(u)) \right)$$
$$= \left(\bar{\pi}^i\left(\bar{f}^{ji}(u) \right), g^i\left(\bar{f}^{ji}(u) \right) \right)$$
$$= \left(\bar{\pi}^i, \bar{g}^i \right)\left(\bar{f}^{ji}(a) \right) = \left(\tilde{\pi}^i \circ \bar{f}(ji) \right)(u).$$

The uniqueness of $\tilde{\pi}$ (as in the ordinary case) is immediate. $\qquad\square$

6
Examples of projective systems of bundles

We elaborate a number of examples of projective systems of vector and principal bundles. The most important among them are the infinite bundle of jets of sections of a Banach vector bundle, the generalized bundle of frames of a projective limit bundle, and a generalized bundle associated to an arbitrary Fréchet principal bundle (P, G, B, π) by means of an appropriate representation of G in a Fréchet space $\mathbb{F}$. The jet bundle is of particular interest because essentially it motivates the requirements of a limit vector bundle. On the other hand, the two aforementioned generalized bundles are non trivial examples of limit bundles with structure group $\mathcal{H}_0(\mathbb{F})$.

6.1 Trivial examples of plb-vector bundles

i) Every Banach vector bundle is obviously a plb-vector bundle.

ii) Every trivial bundle $(B \times \mathbb{F}, B, \mathrm{pr}_1)$ where B is a Banach manifold and $\mathbb{F}$ a Fréchet space, is a plb-vector bundle. This is the case, because, by Theorem 2.3.8, $\mathbb{F} = \varprojlim \mathbb{E}^i$, where $\{\mathbb{E}^i; \rho^{ji}\}$ is a projective system of Banach spaces. Considering the trivial Banach vector bundles $\ell^i = (B \times \mathbb{E}^i, \mathrm{pr}_1, B)$, $i \in \mathbb{N}$, and the vb-morphisms $(\mathrm{id}_B \times \rho^{ji}, \mathrm{id}_B) \colon \ell^j \to \ell^i$, $j \geq i$, we obtain the projective system of vector bundles $\{B \times \mathbb{E}^i; \mathrm{id}_B \times \rho^{ji}\}$ [it suffices now to use the trivializations $(B, \mathrm{id}_B \times \mathrm{id}_{\mathbb{E}^i})$]. This produces the plb-vector bundle

$$(B \times \mathbb{F}, B, \mathrm{pr}_1) = \left(B \times \varprojlim \mathbb{E}^i, B, \mathrm{pr}_1 \right).$$

iii) If (E, B, π) is a plb-vector bundle and $U_0 \subseteq B$ any open set of the base, then $\left(\pi^{-1}(U_0), \pi|_{\pi^{-1}(U_0)}, U_0 \right)$ is a plb-vector bundle as the limit of the projective system of vector bundles

$$\left(\left(\pi^i \right)^{-1} (U_0), \pi^i |_{(\pi^i)^{-1}(U_0)}, U_0 \right) ; \qquad i \in \mathbb{N},$$

with connecting morphisms

$$f_0^{ji} |_{(\pi^j)^{-1}(U_0)} : \left(\pi^j \right)^{-1} (U_0) \longrightarrow \left(\pi^i \right)^{-1} (U_0), j \geq i.$$

This is so because both E^i and $\left(\pi^i \right)^{-1} (U_0)$ have the same fibre type $\mathbb{F} = \varprojlim \mathbb{E}^i$, and (PVB. 2) is easily checked using the trivializations $\left(U, \varprojlim \tau^i \right)$ and $\left(U \cap U_0, \varprojlim \tau^i |_{(\pi^i)^{-1}(U \cap U_0)} \right)$.

6.2 Plb-vector bundles of maps

Let B be a *Banach manifold* with model $\mathbb{B}$, and let $\mathbb{E}$ be a Banach space. By what have we seen in Examples (c)–(e) of § 1.4.4, we can construct the Banach vector bundle $(L(TB, \mathbb{E}), B, \pi)$, where

$$L(TB, \mathbb{E}) = \bigcup_{x \in B} \mathcal{L}(T_x B, \mathbb{E}),$$

$$\pi(f) := x, \quad \text{if } f \in \mathcal{L}(T_x B, \mathbb{E}).$$

The fibre type of this bundle is the Banach space $\mathcal{L}(\mathbb{B}, \mathbb{E})$.

If we replace $\mathbb{E}$ by a Fréchet space $\mathbb{F}$, the usual construction of the vector bundle structure on $L(TB, \mathbb{E})$ cannot be applied to the case of $L(TB, \mathbb{F})$. One serious obstacle is the differentiability of the transition functions now taking values in $\mathcal{L}(\mathcal{L}(\mathbb{B}, \mathbb{F}), \mathcal{L}(\mathbb{B}, \mathbb{F}))$. The latter is a Hausdorff locally convex space, not necessarily a Fréchet one. However, by exploiting the representation of a Fréchet space as a projective limit of Banach spaces, we shall endow $L(TB, \mathbb{F})$ with the structure of a plb-vector bundle, hence with that of a Fréchet bundle.

To this end assume that $\mathbb{F} = \varprojlim \mathbb{E}^i$, where $\{ \mathbb{E}^i, \rho^{ji} \}_{i,j \in \mathbb{N}}$ is a projective system of Banach spaces. For each $i \in \mathbb{N}$, we consider the Banach vector bundle $L^i = (L(TB, \mathbb{E}^i), B, \pi^i)$, of fibre type $\mathcal{L}(\mathbb{B}, \mathbb{E}^i)$. For every $i, j \in \mathbb{N}$ with $j \geq i$, we define the map

$$l^{ji} : L(TB, \mathbb{E}^j) \longrightarrow L(TB, \mathbb{E}^i) : f \mapsto \rho^{ji} \circ f.$$

Each pair (l^{ji}, id_B) is a vb-morphism of L^j into L^i: First, the smoothness of l^{ji} is checked at an arbitrary $f_0 \in \mathcal{L}(B, \mathbb{E}^i)$ as follows. If, in particular, $f_0 \in \mathcal{L}(T_{x_0} B, \mathbb{E}^i)$, by appropriate restrictions, if necessary, we may choose a chart (U, ϕ) at x_0, and trivializations (U, σ^j), (U, σ^i) of L^j and L^i, respectively, with

$$\sigma^j(g) := \left(x, g \circ \overline{\phi}_x^{-1} \right); \qquad g \in \mathcal{L}(T_x B, \mathbb{E}^j),$$

$$\sigma^i(h) := \left(y, h \circ \overline{\phi}_y^{-1} \right); \qquad h \in \mathcal{L}(T_y B, \mathbb{E}^i),$$

where $\overline{\phi}_z \colon T_z B \to \mathbb{B}$ is the isomorphism defined by (1.1.4). Without difficulty, we check that $l^{ji} \left((\pi^j)^{-1}(U) \right) \subseteq (\pi^i)^{-1}(U)$, and the diagram

$$
\begin{array}{ccc}
(\pi^j)^{-1}(U) & \xrightarrow{\;\;l^{ji}\;\;} & (\pi^i)^{-1}(U) \\
\Big\downarrow{\sigma^j} & & \Big\downarrow{\sigma^i} \\
U \times \mathcal{L}(\mathbb{B}, \mathbb{E}^j) & \xrightarrow[\mathrm{id}_U \times r^{ji}]{} & U \times \mathcal{L}(\mathbb{B}, \mathbb{E}^i)
\end{array}
$$

is commutative, where $r^{ji} \colon \mathcal{L}(\mathbb{B}, \mathbb{E}^j) \to \mathcal{L}(\mathbb{B}, \mathbb{E}^i)$ is given by $r^{ji}(f) := \rho^{ji} \circ f$. The above diagram implies the smoothness of l^{ji} at f_0. Moreover, $l^{ji} \circ \pi^i = \pi^j$, and the restriction of l^{ji} to the fibre $(\pi^j)^{-1}(x)$, for every $x \in B$, is continuous linear. Therefore, condition (VBM. 1) of a vb-morphism is satisfied.

Also, with respect to the previous trivializations, we consider the map

$$(6.2.1) \qquad U \ni x \longmapsto \sigma_x^i \circ l_x^{ji} \circ \left(\sigma_x^j \right)^{-1} \in \mathcal{L} \left(\mathcal{L}(\mathbb{B}, \mathbb{E}^j), \mathcal{L}(\mathbb{B}, \mathbb{E}^i) \right),$$

where l_x^{ji} is the restriction of l^{ji} to the fibre over x. For every $f \in \mathcal{L}(\mathbb{B}, \mathbb{E}^j)$, we have that

$$\left(\sigma_x^i \circ l_x^{ji} \circ \left(\sigma_x^j \right)^{-1} \right)(f) = \left(\sigma_x^i \circ l_x^{ji} \right)\left(f \circ \overline{\phi}_x \right) =$$
$$= \sigma_x^i \left(\rho^{ji} \circ f \circ \overline{\phi}_x \right) = \rho^{ji} \circ f = r^{ji}(f);$$

that is, (6.2.1) is smooth as a constant. This proves (VBM. 2); hence, (l^{ji}, id_B) is indeed a vb-morphism.

We further show that $\left\{ L^i = L(TB, \mathbb{E}^i); l^{ji} \right\}_{i,j \in \mathbb{N}}$ is a projective system of vector bundles in the sense of Definition 5.2.1: Condition (PVB. 1) is true, since $\left\{ \mathcal{L}(\mathbb{B}, \mathbb{E}^i); r^{ji} \right\}_{i,j \in \mathbb{N}}$ is a projective system of Banach spaces, thus $\varprojlim \mathcal{L}(\mathbb{B}, \mathbb{E}^i)$ exists. We verify (PVB. 2) by taking, for every $x \in B$, a chart (U, ϕ) at x and the trivializations (U, σ^j), (U, σ^i) considered earlier, which yield the above commutative diagram. As a result, we obtain the plb-vector bundle

$$L := \varprojlim L^i = \left(\varprojlim L(TB, \mathbb{E}^i), B, \varprojlim \pi^i \right).$$

Now, for every $i \in \mathbb{N}$, we define the map

$$l^i \colon L(TB, \mathbb{F}) \longrightarrow L(TB, \mathbb{E}^i) \colon f \mapsto \rho^i \circ f,$$

where $\rho^i \colon \mathbb{F} = \varprojlim \mathbb{E}^i \to \mathbb{E}^i$ is the canonical projection. Then, for every $g \in L(TB, \mathbb{F})$,

$$\left(l^{ji} \circ l^j\right)(g) = l^{ji}\left(\rho^j \circ g\right) = \rho^{ji} \circ \rho^j \circ g = \rho^i \circ g = l^i(g);$$

hence, the following limit of maps exists

$$l := \varprojlim l^i \colon L(TB, \mathbb{F}) \longrightarrow \varprojlim L(TB, \mathbb{E}^i).$$

We shall prove that l is a bijection. Before this, let us remark that an arbitrary element $g \in \varprojlim L(TB, \mathbb{F})$ has the form $(g^i)_{i \in \mathbb{N}}$, with $g^i \in L(TB, \mathbb{E}^i)$ such that $l^{ji}(g^j) = g^i$. Since

$$g^i \in L(TB, \mathbb{E}^i) = \bigcup_{x \in B} \mathcal{L}(T_x B, \mathbb{E}^i),$$

there is an $x \in B$ such that $g^i \in \mathcal{L}(T_x B, \mathbb{E}^i)$. Similarly, $g^j \in \mathcal{L}(T_y B, \mathbb{E}^j)$, for some $y \in B$. But

$$x = \pi^i(g^i) = \left(\pi^i \circ l^{ji}\right)(g^j) = \pi^j(g) = y;$$

that is, the equality $l^{ji}(g^j) = g^i$, for all i, j with $j \geq i$, implies that all g^i's have the same domain, $T_x B$.

We can now proceed to the injectivity of l: For $f, f' \in L(TB, \mathbb{F})$,

$$
\begin{aligned}
l(f) = l(f') \quad &\Rightarrow \quad \left(\varprojlim l^i\right)(f) = \left(\varprojlim l^i\right)(f') \\
&\Rightarrow \quad \left(l^i(f)\right)_{i \in \mathbb{N}} = \left(l^i(f')\right)_{i \in \mathbb{N}} \\
&\Rightarrow \quad \rho^i \circ f = \rho^i \circ f', \quad \forall\, i \in \mathbb{N},
\end{aligned}
$$

or, since f and f' have the same domain, $T_x B$,

$$
\begin{aligned}
&\Rightarrow \quad \rho^i(f(u)) = \rho^i(f'(u)), \quad \forall\, i \in \mathbb{N}, u \in T_x B \\
&\Rightarrow \quad f(u) = f'(u), \quad \forall\, u \in T_x B \\
&\Rightarrow \quad f = f'.
\end{aligned}
$$

On the other hand, let any $a \in \varprojlim L(TB, \mathbb{E}^i)$. As before, $a = (g^i)_{i \in \mathbb{N}}$. Since, by the above remark, all g^i's have the same domain, we obtain the continuous linear map $g := \varprojlim g^i \in \varprojlim \mathcal{L}(T_x B, \mathbb{F}) \subset L(TB, \mathbb{F})$. Then

$$l(g) = \left(l^i(g)\right)_{i \in \mathbb{N}} = \left(\left(\rho^i \circ g\right)\right)_{i \in \mathbb{N}} = \left(g^i\right)_{i \in \mathbb{N}} = a;$$

that is, l is also surjective.

By means of the bijection l we transfer the plb-vector bundle structure of $\varprojlim L(B,\mathbb{E}^i)$ to $L(TB,\mathbb{F})$ so that (l,id_B) is an isomorphism of Fréchet vector bundles. We note that the diagram

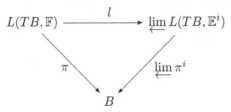

is commutative because, for every $f \in \mathcal{L}(T_xB,\mathbb{F})$,

$$\left(\varprojlim \pi^i \circ l\right)(f) = \left(\varprojlim \pi^i\right)\left(\left(l^j(f)\right)_{j\in\mathbb{N}}\right)$$
$$= \left(\varprojlim \pi^i\right)\left(\left(\rho^j \circ f\right)_{j\in\mathbb{N}}\right)$$
$$= \left(\pi^i(\rho^j \circ f)\right)_{j\in\mathbb{N}}$$
$$= x = \pi(f)$$

Without particular difficulty we extend the previous constructions to the case of k-*linear map bundle*

$$L_k(TB,\mathbb{F}) = \bigcup_{x\in B} \mathcal{L}_k(T_xB,\mathbb{F}),$$

and to the k-*alternating linear map bundle*

$$A_k(B,\mathbb{F}) = \bigcup_{x\in B} \mathcal{A}_k(T_xB,\mathbb{F}).$$

All of them become plb-bundles (in the sense of Definition 5.2.1); therefore, they are Fréchet vector bundles.

6.3 The infinite jet bundle

In §1.4.4 (f) we dealt with the structure of the bundle $J^k(\ell) := (J^kE, B, \pi^k)$ of k-jets of the local sections of a Banach vector bundle $\ell = (E,B,\pi)$. Here we want to show that

$$J^\infty(E) := \varprojlim J^kE$$

is a Fréchet vector bundle, as a plb-bundle.

For our purpose, we consider the maps

$$J^{lk}\colon J^l(E) \longrightarrow J^k(E)\colon j_x^l\xi \mapsto j_x^k\xi, \qquad l \geq k.$$

and obtain the projective system $\left\{ J^k(E); J^{lk} \right\}_{l,k \in \mathbb{N}}$, with limit is $J^\infty(E)$. We shall prove that this is in fact a projective system of Banach vector bundles.

Following the main lines of earlier proofs of the same nature, we first show that every J^{lk} is smooth. Indeed, if $j^l_{x_0}$ is an arbitrary jet of $J^l(E)$, we choose a vb-chart (U, ϕ, Φ) of E, with $x_0 \in U$, as well as the trivializations (U, σ^l) and (U, σ^k) of $J^l(E)$ and $J^k(E)$, respectively. We check that $J^{lk}\left((\pi^l)^{-1}(U) \right) \subseteq (\pi^k)^{-1}(U)$ $(l \geq k)$, thus we obtain the commutative diagram

$$
\begin{array}{ccc}
(\pi^l)^{-1}(U) & \xrightarrow{\;\;\;J^{lk}\;\;\;} & (\pi^k)^{-1}(U) \\
\Big\downarrow{\scriptstyle \sigma^l} & & \Big\downarrow{\scriptstyle \sigma^k} \\
U \times P^l(\mathbb{B}, \mathbb{E}) & \xrightarrow[\;\mathrm{id}_U \times P^{lk}\;]{} & U \times P^k(\mathbb{B}, \mathbb{E})
\end{array}
$$

where $P^{lk} \colon P^l(\mathbb{B}, \mathbb{E}) \to P^k(\mathbb{B}, \mathbb{E})$ is the continuous linear map given by $P^{lk}(f_1, \ldots, f_k, \ldots, f_l) := (f_1, \ldots, f_k)$. The commutativity of the diagram is checked as follows: For every $j^l_x \xi \in (\pi^l)^{-1}(U)$,

$$
\begin{aligned}
\left(\sigma^k \circ J^{lk} \right)\left(j^l_x \xi \right) &= \sigma^k \left(j^{lk}_x \xi \right) = \left(x; \xi_\phi(\phi(x)), D\xi_\phi(\phi(x)), \ldots, D^k\xi_\phi(\phi(x)) \right) \\
&= \left(\mathrm{id}_U \times P^{lk} \right)\left(x; \xi_\phi(\phi(x)), D\xi_\phi(\phi(x)), \ldots, D^l\xi_\phi(\phi(x)) \right) \\
&= \left(\left(\mathrm{id}_U \times P^{lk} \right) \circ \sigma^l \right)\left(j^l_x \xi \right).
\end{aligned}
$$

Consequently, J^{lk} is continuous on the neighborhood $(\pi^l)^{-1}(U)$ of $j^l_{x_0}$.

Obviously, $\pi^k \circ J^{lk} = \pi^l$, while the restriction of the previous diagram to each fibre $(\pi^l)^{-1}(x)$ yields

$$
J^{lk}_x := J^{lk} \big|_{(\pi^l)^{-1}(x)} = \left(\sigma^k_x \right)^{-1} \circ P^{lk} \circ \sigma^l_x.
$$

Therefore, for any $x \in B$, we choose the previous trivializations (U, σ^l) and (U, σ^k) and define the map

$$
J \colon U \longrightarrow L\left(P^l(\mathbb{B}, \mathbb{E}), P^k(\mathbb{B}, \mathbb{E}) \right) \colon y \mapsto \sigma^k_y \circ J^{lk}_y \circ \left(\sigma^l_y \right)^{-1}.
$$

We immediately see that $J(y) = P^{lk}$, for every $y \in U$, thus J is smooth, every (J^{lk}, id_B) is a vb-morphism, and $\left\{ J^k(E); J^{lk} \right\}_{l,k \in \mathbb{N}}$ is a countable family of vector bundles connected by vb-morphisms. It remains to show that it is a projective system of vector bundles in the sense of Definition 5.2.1.

Indeed, the maps $P^{lk}\colon P^l(\mathbb{B},\mathbb{E}) \to P^k(\mathbb{B},\mathbb{E})$ satisfy the equalities

$$P^{km} \circ P^{lk} = P^{lm}, \quad \forall\, k,l,m \in \mathbb{N}: l \geq k \geq m.$$

Thus $\left\{P^k(\mathbb{B},\mathbb{E}); P^{lk}\right\}$ is a projective system of Banach spaces, whose limit $\varprojlim P^k(\mathbb{B},\mathbb{E})$ is isomorphic to the Fréchet space

$$P^\infty(\mathbb{B},\mathbb{E}) = \mathbb{E} \times L(\mathbb{B},\mathbb{E}) \times L_s^2(\mathbb{B},\mathbb{E}) \times \cdots$$

[see Example 2.3.3(2)]. This implies condition (PVB. 1) of Definition 5.2.1. Moreover, for an arbitrary $x \in B$, condition (PVB. 2) is a consequence of the commutativity of the previous diagram with the same trivializations (U, σ^l), (U, σ^k). Therefore, $\left\{J^k(E); J^{lk}\right\}_{l,k\in\mathbb{N}}$ is a projective system of Banach vector bundles with corresponding limit the plb-bundle $\left(J^\infty(E), B, \varprojlim \pi^k\right)$. In particular, this is a Fréchet vector bundle.

Remark. It is worth noting that the smooth structure on $J^\infty(E) = \varprojlim J^k(E)$, derived from the previous approach (see also Proposition 5.2.2) is wider than the one defined by F. Takens in [Tak79]. The latter is obtained by declaring that a map $g\colon J^\infty(E) \to \mathbb{R}$ is smooth if, locally, there exist $k \in \mathbb{N}$, $U_k \subseteq J^k(E)$ open and $g^k\colon U_k \to \mathbb{R}$ smooth, so that $g|_{J_k^{-1}(U_k)} = g^k \circ J^k$, where $J^k\colon J^\infty(E) \to J^k(E)$ is the natural projection. This condition is satisfied if and only if g is the projective limit of the smooth maps $g^k \circ J^{lk}$ $(l \geq k)$; therefore, Taken's $\mathbb{R}$-valued smooth maps on $J^\infty(E)$ are necessarily projective limits of smooth maps. However, in our framework, smoothness is not restricted only to pls-maps (compare with Remark 3.1.9(1); see also Definition 3.1.7).

6.4 The tangent bundle of a plb-bundle

Let $\{E^i; f^{ji}\}_{i,j\in\mathbb{N}}$ be a projective system of vector bundles $\ell^i = (E^i, B, \pi)$, with connecting morphisms the vb-morphisms $(f^{ji}, \mathrm{id}_B)\colon \ell^j \to \ell^i$ $(j \geq i)$, with limit the plb -bundle

$$\ell = \varprojlim \ell^i = \left(E := \varprojlim E^i, B, \pi := \varprojlim \pi^i\right).$$

Applying the tangent operator, for every $i \in \mathbb{N}$, we obtain the vector bundle $T\ell^i := (TE^i, TB, T\pi^i)$. Accordingly, we have the smooth maps $Tf^{ji}\colon TE^j \to TE^i$, for all $j \geq i$.

In view of later applications, we want to show that $\{TE^i, Tf^{ji}\}_{i,j\in\mathbb{N}}$ is a plb-bundle. It is clear that, by Tf^{ji} in the system, we mean the vb-morphism $(Tf^{ji}, \mathrm{id}_{TB})$ of $T\ell^j \to T\ell^i$.

From $\pi^i \circ f^{ji} = \pi^j$, it follows that $T\pi^i \circ Tf^{ji} = T\pi^j$. On the other hand, for an arbitrary $v \in T_x B$, where TB is the base of all the bundles TE^i, we choose a trivialization $\left(U, \tau = \varprojlim \tau^i \right)$ of E, with $b \in U$, and consider the trivializations $\left(\tau_B^{-1}(U), \sigma^i \right)$ of TE^i, respectively (for all $i \in \mathbb{N}$), where $\tau_B \colon TB \to B$ is the projection of the tangent bundle TB (see § 1.1.5), and

$$\sigma^i \colon \left(T\pi^i \right)^{-1} \left(\tau_B^{-1}(U) \right) \longrightarrow \tau_B^{-1}(U) \times \mathbb{E}^i \times \mathbb{E}^i \colon$$

$$[(\alpha, u)] \longmapsto \sigma^i([(\alpha, u)]) := \left(T\pi^i([(\alpha, u)]), \tau_{\pi^i(u)}^i(u), (\mathrm{pr}_2 \circ \tau^i \circ \alpha)'(0) \right).$$

Recall that $[(\alpha, u)]$ is the equivalence class of $\alpha \colon (-\epsilon, \epsilon) \to TE^i$, a smooth curve with $\alpha(0) = u$. Also, $(\mathrm{pr}_2 \circ \tau^i \circ \alpha)'(0) = D(\mathrm{pr}_2 \circ \tau^i \circ \alpha)(0).(1)$.

Let $(\tau_B^{-1}(U), \sigma^j)$, $(\tau_B^{-1}(U), \sigma^i)$ $(j \geq i)$ be two trivializations, as above, and σ_v^j, σ_v^i their restrictions to the fibres $(T\pi^j)^{-1}(v)$ and $(T\pi^j)^{-1}(v)$, respectively. Note that $\sigma_v^j \colon (T\pi^j)^{-1}(v) \to \mathbb{E}^j \times \mathbb{E}^j$ is given by $\sigma_v^j = (\mathrm{pr}_2, \mathrm{pr}_3) \circ \sigma^j$. We show that the diagram

$$
\begin{array}{ccc}
(T\pi^j)^{-1}(v) & \xrightarrow{\ \ Tf^{ji}\ \ } & (T\pi^i)^{-1}(v) \\
\Big\downarrow{\sigma_v^j} & & \Big\downarrow{\sigma_v^i} \\
\mathbb{E}^j \times \mathbb{E}^j & \xrightarrow[\ \rho^{ji} \times \rho^{ji}\ ]{} & \mathbb{E}^i \times \mathbb{E}^i
\end{array}
$$

is commutative (of course, Tf^{ji} is also restricted to the fibres). Indeed, for every $[(\alpha, u)] \in (T\pi^j)^{-1}(v)$,

$$\left(\sigma_v^i \circ Tf^{ji} \right)([(\alpha, u)]) = \sigma_v^i \left([(f^{ji} \circ \alpha, f^{ji}(u))] \right)$$

$$= \left(\tau_{\pi^i(f^{ji}(u))}^i \left(f^{ji}(u) \right), \left(\mathrm{pr}_2 \circ \tau^i \circ f^{ji} \circ \alpha \right)'(0) \right)$$

or, by condition (PVB. 2) and the equality $\pi^i(u) = \pi^j(u)$,

$$= \left(\rho^{ji} \left(\tau_{\pi^j(u)}^j(u) \right), (\rho^{ji} \circ \mathrm{pr}_2 \circ \tau^j \circ \alpha)'(0) \right),$$

and, by the linearity of ρ^{ji},

$$= \left(\rho^{ji} \left(\tau_{\pi^j(u)}^j(u) \right), \rho^{ji} \circ (\mathrm{pr}_2 \circ \tau^j \circ \alpha)'(0) \right)$$

$$= (\rho^{ji} \times \rho^{ji}) \left(\tau_{\pi^j(u)}^j(u), (\mathrm{pr}_2 \circ \tau^j \circ \alpha)'(0) \right)$$

$$= \left((\rho^{ji} \times \rho^{ji}) \circ \sigma_v^j \right) ([(\alpha, u)]).$$

Therefore, the restriction of Tf^{ji} to the fibres of TE^j is a continuous linear map. With the same trivializations, we see that the map

$$G\colon \tau_{_B}^{-1}(U) \longrightarrow \mathcal{L}(\mathbb{E}^j \times \mathbb{E}^j, \mathbb{E}^i \times \mathbb{E}^i)\colon$$
$$w \longmapsto \sigma_w^i \circ Tf^{ji}\big|_{(T\pi^j)^{-1}(w)} \circ (\sigma_w^j)^{-1}$$

yields $G(w) = \rho^{ji} \times \rho^{ji}$, for all w, i.e. G is constant, thus smooth. Hence, we conclude that $(Tf^{ji}, \mathrm{id}_{TB})$ is a vb-morphism of TE^j into TE^i ($j \geq i$).

Each TE^i is of fibre type $\mathbb{E}^i \times \mathbb{E}^i$ and, obviously, the plb-space $\varprojlim(\mathbb{E}^i \times \mathbb{E}^i)$ is defined, thus (PVB. 1) of Definition 5.2.1 is satisfied. Finally, we verify (PVB. 2) by using the above trivializations $\{(\tau_{_B}^{-1}(U), \sigma^i)\}_{i \in \mathbb{N}}$ and following the procedure applied to the proof of the commutativity of the previous diagram. As a result, we obtain the plb-vector bundle

$$\varprojlim T\ell^i = \left(\varprojlim TE^i, TB, \varprojlim T\pi^i\right)$$

Wishing to show that the latter bundle is isomorphic to $(TE, TB, T\pi)$, we induce the maps $Tf^i\colon TE \to TE^i$, where $f^i\colon E = \varprojlim E^i \to E^i$ are the canonical projections, for all $i \in \mathbb{N}$. We observe that $f^{ji} \circ f^j = f^i$ implies $Tf^{ji} \circ Tf^j = Tf^i$, thus we obtain the limit map $h := \varprojlim Tf^i\colon TE \to \varprojlim TE^i$. We check that h is the diffeomorphism R of Theorem 3.2.8 (holding for arbitrary plb-manifolds). Indeed, for every $x = (x^i) \in E$,

$$h\big|_{T_xE} = \left(\varprojlim Tf^i\right)\big|_{T_xE} = \varprojlim \left(Tf^i\right)\big|_{T_x^iE^i} = \varprojlim T_x^iE^i = R_x,$$

since the morphism R_x is given by Corollary 3.2.6 with $\mu^i = f^i$, $i \in I$.

Next we immediately see that $\left(\varprojlim T\pi^i\right) \circ h = T\pi$. On the other hand, if $v \in TB$ is an arbitrary element of the base space, we consider the trivialization $(U, \tau = \varprojlim \tau^i)$ of E, with $\tau_B(v) \in U$, the corresponding trivialization $\left(\tau_{_B}^{-1}(U), \varprojlim \sigma^i\right)$ of $\varprojlim TE^i$, and the trivialization $\left(\tau_{_B}^{-1}(U), \sigma\right)$ of TE derived from (U, τ). We check that the diagram

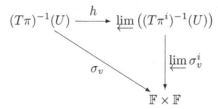

is commutative, because

$$\left(\varprojlim \sigma_v^i \circ h\right)\left([(\alpha, u)]\right) =$$

$$= \left(\sigma_v^i\left(Tf^i\left([(\alpha, u)]\right)\right)\right)_{i \in \mathbb{N}}$$

$$= \left(\sigma_v^i\left(\left[\left(f^i \circ \alpha, f^i(u)\right)\right]\right)\right)_{i \in \mathbb{N}}$$

$$= \left(\tau_{\pi^i(f^i(u))}^i\left(f^i(u)\right), \left(\mathrm{pr}_2 \circ \tau^i \circ f^i \circ \alpha\right)'(0)\right)_{i \in \mathbb{N}}$$

$$= \left(\left(\tau_{\pi^i(f^i(u))}^i \circ f^i\right)(u), \left(\mathrm{pr}_2 \circ (\mathrm{id}_U \times \rho^i) \circ \tau \circ \alpha\right)'(0)\right)_{i \in \mathbb{N}}$$

$$= \left(\rho^i\left(\tau_{\pi(u)}(u)\right), \left(\rho^i \circ \mathrm{pr}_2 \circ \tau \circ \alpha\right)'(0)\right)_{i \in \mathbb{N}}$$

$$= \left((\rho^i \times \rho^i)\left(\tau_{\pi(u)}(u), (\mathrm{pr}_2 \circ \tau \circ \alpha)'(0)\right)\right)_{i \in \mathbb{N}}$$

$$= \sigma_v\left([(\alpha, u)]\right).$$

This means that the restriction of h to the fibres of TE is a continuous linear map. Using once more the trivializations $\left(\tau_B^{-1}(U), \sigma\right)$ of TE and $\left(\tau_B^{-1}(U), \varprojlim \sigma^i\right)$ of $\varprojlim TE^i$, we see that the map

$$F \colon \tau_B^{-1}(U) \longrightarrow \mathcal{L}(\mathbb{F} \times \mathbb{F}, \mathbb{F} \times \mathbb{F}) \colon v \mapsto \varprojlim \sigma_v^i \circ h\big|_{(T\pi)^{-1}(v)} \circ \sigma_v^{-1}$$

is constantly $\mathrm{id}_{\mathbb{F} \times \mathbb{F}}$, thus F is smooth. As a result, (h, id_{TB}) is a morphism between the Fréchet vector bundles TE and $\varprojlim TE^i$. With the same trivializations, we prove, in a reverse way, that $(h^{-1}, \mathrm{id}_{TB})$ is also a vb-morphism; hence, $TE \equiv \varprojlim TE^i$ by means of the vb-isomorphism (h, id_{TB}).

6.5 The generalized frame bundle

Let $\{E^i, f^{ji}\}_{i,j \in \mathbb{N}}$ be a projective system of Banach vector bundles with limit the (Fréchet) plb-vector bundle (E, B, π). We want to define the frame bundle of (E, B, π). The pathology of the $\mathrm{GL}(\mathbb{F})$ compels us to a radical revision of the frame bundle by considering $\mathcal{H}_0(\mathbb{F})$ as the appropriate structure group within our framework.

Before proceeding, we introduce the following notation, combining (2.3.5) and (5.1.2): For two Fréchet spaces $\mathbb{F}_1$ and $\mathbb{F}_2$, we set

$$(6.5.1) \qquad \mathcal{H}_0^i(\mathbb{F}_1, \mathbb{F}_2) := \mathcal{H}^i(\mathbb{F}_1, \mathbb{F}_2) \cap \prod_{j=1}^{i} \mathcal{L}is(\mathbb{E}_1^j, \mathbb{E}_2^j)$$

Accordingly, we define the space

$$P(E^i) := \bigcup_{x \in B} \mathcal{H}_0^i(\mathbb{F}, E_x),$$

which is meaningful because $E_x = \varprojlim E_x^i$. Thus an element of $P(E^i)$ has the form $(q^1, \ldots, q^i)$, where the isomorphisms $q^j : \mathbb{E}^j \to E_x^j$ $(j = 1, \ldots, i)$ satisfy the equalities $f^{jk} \circ q^j = q^k \circ \rho^{jk}$, for every $1 \leq j, k \leq i$ with $k \leq j$. Recall that $f^{jk} : E^j \to E^k$ and $\rho^{jk} : \mathbb{E}^j \to \mathbb{E}^k$.

The bold typeface is used to distinguish $P(E^i)$ from the ordinary bundle of frames mentioned in § 1.6.5, which in the present context would be $P(E^i) := \bigcup_{x \in B} \mathcal{L}is(\mathbb{F}, E_x)$.

Proposition 6.5.1   *Each $P(E^i)$ $(i \in \mathbb{N})$ is a Banach principal bundle over B, with structure group $\mathcal{H}_0^i(\mathbb{F})$, and projection $p^i : P(E^i) \to B$, given by*

$$p^i\left(q^1, \ldots, q^i\right) := x, \;\; \text{if} \;\; \left(q^1, \ldots, q^i\right) \in \mathcal{H}_0^i(\mathbb{F}, E_x).$$

Proof   The smooth structure of $P(E^i)$ is defined as follows: For an arbitrary $(g^1, \ldots, g^i) \in P(E^i)$ with $p^i\left(g^1, \ldots, g^i\right) = x$, we choose a local trivialization $\left(U, \varprojlim \tau^i\right)$ of E, $x \in U$, and define the bijection

$$\Phi^i : (p^i)^{-1}(U) \longrightarrow U \times \mathcal{H}_0^i(\mathbb{F}):$$
$$\left(q^1, \ldots, q^i\right) \longmapsto \left(x; \tau_x^1 \circ q^1, \ldots \tau_x^i \circ q^i\right),$$

where $\tau_x^j : E_x^j \to \mathbb{E}^j$ $(j = 1, \ldots, i)$ is the isomorphism induced by the corresponding trivializations.

For another trivialization $\left(V, \varprojlim \sigma^i\right)$, with $x \in U \cap V$, and the corresponding bijection $\Psi^i : (p^i)^{-1}(V) \to V \times \mathcal{H}_0^i(\mathbb{F})$, we see that, on the overlapping,

$$
\begin{aligned}
(6.5.2) \quad & \left(\Psi^i \circ (\Phi^i)^{-1}\right)\left(x; h^1, \ldots, h^i\right) = \\
& = \left(x; \sigma_x^1 \circ (\tau_x^1)^{-1} \circ h^1, \ldots, \sigma_x^i \circ (\tau_x^i)^{-1} \circ h^i\right),
\end{aligned}
$$

which is a diffeomorphism. Then, in virtue of the gluing process (see, e.g., [Bou67, n° 5.2.4]) $P(E^i)$ is indeed a Banach manifold turning the quadruple $(P(E^i), \mathcal{H}_0^i(\mathbb{F}), B, p^i)$ into a Banach principal bundle, where $\mathcal{H}_0^i(\mathbb{F})$ acts on (the right of) $P(E^i)$ in the obvious way, i.e.

$$\left(q^1, \ldots, q^i\right) \cdot \left(g^1, \ldots, g^i\right) = \left(q^1 \circ g^1, \ldots, q^i \circ g^i\right),$$

for every $\left(q^1, \ldots, q^i\right) \in P(E^i)$ and $\left(g^1, \ldots, g^i\right) \in \mathcal{H}_0^i(\mathbb{F})$. $\square$

For later use we prove:

Corollary 6.5.2   *The transition functions* $\left\{g^i_{\alpha\beta}\colon U_{\alpha\beta} \to \mathcal{H}^i_0(\mathbb{F})\right\}_{\alpha,\beta \in I}$
of $\boldsymbol{P}(E^i)$, *over a trivializing cover* $\{U_\alpha\}_{\alpha \in I}$ *of* B, *are given by*

$$(6.5.3) \qquad g^i_{\alpha\beta}(x) = \left(g^1_{\alpha\beta}(x), \ldots, g^i_{\alpha\beta}(x)\right); \qquad x \in U_{\alpha\beta} = U_\alpha \cap U_\beta,$$

where $\left\{g^j_{\alpha\beta}\colon U_{\alpha\beta} \to \mathrm{GL}(\mathbb{E}^j)\right\}_{\alpha,\beta \in I}$ $(j = 1, \ldots, i)$ *are the transition functions of both* E^j *and the ordinary frame bundle* $P(E^j)$.

Proof   Direct consequence of (6.5.2) and the fact that $\left\{g^j_{\alpha\beta}\right\}_{\alpha\beta \in I}$ are the transition functions of both E^j and $P(E^j)$ [see (1.6.15)]. $\square$

For every $j \geq i$, we define the following connecting morphisms:

$$r^{ji}\colon P(E^j) \longrightarrow P(E^i)\colon \left(q^1, \ldots, q^j\right) \mapsto \left(q^1, \ldots, q^i\right),$$
$$h^{ji}_0\colon \mathcal{H}^j_0(\mathbb{F}) \longrightarrow \mathcal{H}^i_0(\mathbb{F})\colon \left(g^1, \ldots, g^j\right) \mapsto \left(g^1, \ldots, g^i\right),$$

[see also (2.3.6) and the notations of Proposition 5.1.1].

Lemma 6.5.3   *For every* $j \geq i$, *the triplet* $\left(r^{ji}, h^{ji}_0, \mathrm{id}_B\right)$ *is a principal bundle morphism of* $\left(\boldsymbol{P}(E^j), \mathcal{H}^j_0(\mathbb{F}), B, \boldsymbol{p}^j\right)$ *into* $\left(\boldsymbol{P}(E^i), \mathcal{H}^i_0(\mathbb{F}), B, \boldsymbol{p}^i\right)$

Proof   Immediate consequence of the preceding definitions. $\square$

Proposition 6.5.4   *The following assertions hold true:*
i) The collection

$$\left\{\left(\boldsymbol{P}(E^i), \mathcal{H}^i_0(\mathbb{F}), B, \boldsymbol{p}^i\right); \left(r^{ji}, h^{ji}_0, \mathrm{id}_B\right)\right\}_{i,j \in \mathbb{N}}$$

is a projective system of Banach principal bundles.

ii) The set $\boldsymbol{P}(E) := \varprojlim \boldsymbol{P}(E^i)$ *is the total space of a locally trivial principal bundle over* B, *with structure group* $\mathcal{H}_0(\mathbb{F})$, *called the* **generalized frame bundle** *of* E.

Proof   In virtue of Definition 4.1.1 and the preceding lemma, using the trivializations $\{U, \boldsymbol{\Phi}^i\}_{i \in \mathbb{N}}$ defined in the proof of Proposition 6.5.1, we obtain the first conclusion, thus $\boldsymbol{P}(E)$ exists.

For ii) take any $x \in B$ and consider the trivializations $\{U, \boldsymbol{\Phi}^i\}_{i \in \mathbb{N}}$, $x \in U$, as before. It is easily checked that the diagram

$$
\begin{array}{ccc}
(\boldsymbol{p}^j)^{-1}(U) & \xrightarrow{\ \boldsymbol{\Phi}^j\ } & U \times \mathcal{H}^j_0(\mathbb{F}) \\
{\scriptstyle r^{ji}}\Big\downarrow & & \Big\downarrow{\scriptstyle \mathrm{id}_U \times h^{ji}_0} \\
(\boldsymbol{p}^i)^{-1}(U) & \xrightarrow[\ \boldsymbol{\Phi}^i\ ]{} & U \times \mathcal{H}^i_0(\mathbb{F})
\end{array}
$$

is commutative. As a result, the morphism

$$(6.5.4) \qquad \Phi := \varprojlim \Phi^i \colon \varprojlim \left((p^i)^{-1}(U) \right) \longrightarrow U \times \mathcal{H}_0(\mathbb{F})$$

exists and determines a topological trivialization of $P(E)$ over U. The projection of $P(E)$ is $p = \varprojlim p^i$, while the action of $\mathcal{H}_0(\mathbb{F})$ on (the right) of $P(E)$ is the projective limit of the actions on the factors. $\qquad \square$

Remarks 6.5.5 1) The elements of $P(E)$ are of the form $(g^i)_{i \in \mathbb{N}}$, with $g^i \in P(E^i)$, since $\varprojlim g^i$ exists. In this respect see also the identifications (2.3.9), (2.3.9').

2) The homomorphism Φ defined by (6.5.4) is not smooth in the ordinary sense, since $\mathcal{H}_0(\mathbb{F})$ is not a Fréchet-Lie group. However, if Φ is considered as a $(U \times \mathcal{H}(\mathbb{F}))$-valued map ($\mathcal{H}(\mathbb{F})$ is a Fréchet space), then it is smooth in the sense of the differentiability defined in §2.2. Therefore, Φ is *generalized smooth* in the sense of Remark 5.2.6 (3). By the same token, the action of $\mathcal{H}_0(\mathbb{F})$ on $P(E)$ can be thought of as smooth.

3) In view of the preceding remark, $P(E)$ is a smooth Fréchet principal bundle, justifying the term *generalized frame bundle of E*.

4) In the next section we shall show that the original vector bundle E is associated with the generalized bundle of frames $P(E)$ (see Corollary 6.6.5 below).

6.6 Generalized associated bundles

Motivated by the construction of §1.6.6, we want to answer the following question: Given a Fréchet principal bundle (P, G, B, π) and a Fréchet space $\mathbb{F}$, is it possible to construct an associated vector bundle (of fibre type $\mathbb{F}$), from an arbitrary representation of G into $\mathbb{F}$? The answer is negative, if we try to imitate the classical pattern, since such a representation amounts to a homomorphism of the form $\varphi \colon G \to \mathrm{GL}(\mathbb{F})$, while, as we have explained on many occasions, $\mathrm{GL}(\mathbb{F})$ is too problematic. In fact, although the associated bundle $P \times_\varphi \mathbb{F}$ exists set-theoretically, it has in general no differential structure. As in the previous section, the replacement of $\mathrm{GL}(\mathbb{F})$ by $\mathcal{H}_0(\mathbb{F})$ is the key to a (partial) affirmative answer.

So, starting with a (not necessarily a projective limit) Fréchet principal bundle (P, G, B, π) over a *Banach* base, we consider a *representation* ϱ of G into $\mathbb{F}$; that is, a topological group homomorphism

$$(6.6.1) \qquad\qquad \varrho \colon G \longrightarrow \mathcal{H}_0(\mathbb{F}),$$

which is also smooth if considered as taking values in the Fréchet space $\mathcal{H}(\mathbb{F}) \supset \mathcal{H}_0(\mathbb{F})$ [see (2.3.12) and (2.3.3)].

As usual, $\mathbb{F} = \varprojlim \mathbb{E}^i$, where $\{\mathbb{E}^i, \rho^{ji}\}$ is a projective system of Banach spaces. Then, by the definition of $\mathcal{H}_0(\mathbb{F})$ [see (5.1.3)], we obtain the ordinary representations

$$(6.6.2) \qquad \varrho^i \colon G \longrightarrow \mathrm{GL}(\mathbb{E}^i) \colon g \mapsto \mathrm{pr}_i(\varrho(g)), \qquad i \in \mathbb{N}.$$

Here $\mathrm{pr}_i \colon \mathcal{H}_0(\mathbb{F}) \to \mathcal{Lis}(\mathbb{E}^i)$ denotes the projection to the i-th factor and $\mathrm{GL}(\mathbb{E}^i)$ is identified with $\mathcal{Lis}(\mathbb{E}^i)$. Therefore, in the notations of §1.6.6, we induce the Banach vector bundles (E^i, B, π^i), where

$$E^i := P \times_{\varrho^i} \mathbb{E}^i = \big\{ [(p, u^i)]_i \mid (p, u^i) \in P \times \mathbb{E}^i \big\}; \qquad i \in \mathbb{N},$$

and $\pi^i([(p, u^i)]) = \pi(p)$. Obviously, $[(p, u^i)]_i$ is the orbit of (p, u^i) with respect to the action $(p, u^i) \cdot g = (p \cdot g, \varrho^i(g^{-1})(u^i))$. Notice the use of a matching index in the corresponding equivalence class.

The Banach vector bundle structure of each E^i will be apparent in the proof of the next result.

Proposition 6.6.1 *The limit* $E := \varprojlim E^i$ *exists and admits the structure of a Fréchet vector bundle over* B.

Proof For every $i, j \in \mathbb{N}$, with $j \geq i$, we can define the map

$$f^{ji} \colon E^j \longrightarrow E^i \colon [(p, u^j)]_j \mapsto [(p, \rho^{ji}(u^j))]_i,$$

since $\varprojlim(\varrho^i(g)))$ exists for every $g \in G$. Also, $f^{ik} \circ f^{ji} = f^{jk}$, for every $j \geq i \geq k$, as a result of the analogous equalities for $\{\rho^{ji}\}$. Therefore, $\{E^i; f^{ji}\}_{i,j \in \mathbb{N}}$ is a projective system inducing $E := \varprojlim E^i$, though not yet satisfying the conditions of Definition 5.2.1.

The local structure of E is determined as follows: Assume that $\mathcal{C}$ is the open cover of B over the sets of which P is trivial. Let $x_0 \in B$ be an arbitrary point, and let (U, Φ) be a local trivialization of P, with $x_0 \in U$. Then, as in §1.6.6, for each E^i we have the corresponding local trivialization

$$\Psi^i \colon U \times \mathbb{E}^i \longrightarrow (\pi^i)^{-1}(U) \colon (x, u^i) \mapsto [(s(x), u^i)]_i,$$

where s is the natural section of P over U (with respect to Φ; see the beginning of §1.6.3). Immediate computations imply that Ψ^i is a bijection, and $f^{ji}\left((\pi^j)^{-1}(U)\right) \subseteq (\pi^i)^{-1}(U)$; thus the diagram

$$
\begin{array}{ccc}
U \times \mathbb{E}^j & \xrightarrow{\ \Psi^j\ } & (\pi^j)^{-1}(U) \\
\downarrow{\scriptstyle \mathrm{id}_U \times \rho^{ji}} & & \downarrow{\scriptstyle f^{ji}} \\
U \times \mathbb{E}^i & \xrightarrow[\ \Psi^i\ ]{} & (\pi^i)^{-1}(U)
\end{array}
$$

is commutative. Indeed,

$$(f^{ji} \circ \Psi^j)(x, u^j) = \big[\big(s(x), \rho^{ji}(u^j)\big)\big]_i = \big(\Psi^i \circ (\mathrm{id}_U \times \rho^{ji})\big)(x, u^j); \quad j \geq i,$$

for every $(x, u^j) \in U \times \mathbb{E}^j$. Hence, taking the inverse maps $\big\{\Phi^i = (\Psi^i)^{-1}\big\}_{i \in \mathbb{N}}$, we see that $\{E^i\}_{i \in \mathbb{N}}$ is a projective system of Banach vector bundles (in the sense of Definition 5.2.1), thus Theorem 5.2.5 concludes the proof. $\qquad\square$

To proceed further, we define the homomorphism

$$(6.6.3) \qquad\qquad \varphi := \varepsilon \circ \varrho,$$

where now

$$\varepsilon \colon \mathcal{H}_0(\mathbb{F}) \longrightarrow \mathrm{GL}(\mathbb{F}) \colon (f^i)_{i \in \mathbb{N}} \mapsto \varprojlim f^i,$$

[compare with the general case of (2.3.4)], and the action of G on (the right) of $P \times \mathbb{F}$, determined by $(p, u) \cdot g := \big(p \cdot g, \varphi(g^{-1})(u)\big)$.

Proposition 6.6.2 *The quotient (with respect to φ) $F := P \times_\varphi \mathbb{F}$, being in bijective correspondence with $E = \varprojlim E^i$, inherits the structure of a Fréchet vector bundle over B.*

Proof We define the mapping

$$f \colon P \times \mathbb{F} \longrightarrow E \colon (p, u) \mapsto \big([(p, \rho^i(u))]_i\big)_{i \in \mathbb{N}},$$

where $\rho^i \colon \mathbb{F} \to \mathbb{E}^i$ ($i \in \mathbb{N}$) are the canonical projections of $\mathbb{F}$. Since, for every $(p, u) \in P \times \mathbb{F}$ and $g \in G$,

$$
\begin{aligned}
f((p, u) \cdot g) &= f\big(p \cdot g, \varphi(g^{-1})(u)\big) = \big(\big[\big(p \cdot g, \rho^i(\varphi(g^{-1})(u))\big)\big]_i\big)_{i \in \mathbb{N}} \\
&= \big(\big[\big(p \cdot g, \varrho^i\,(g^{-1})\,(\rho^i(u))\big]_i\big)_{i \in \mathbb{N}} = \big([(p, \rho^i(u))]_i\big)_{i \in \mathbb{N}} \\
&= f(p, u),
\end{aligned}
$$

there exists a well-defined mapping $\widetilde{f}$ induced on the quotient F; that is, $\widetilde{f}([(p, u)]) = \big([(p, \rho^i(u))]_i\big)_{i \in \mathbb{N}}$, for every $[(p, u)] \in E$. We check that:

i) $\tilde{f}$ is 1–1: If $\tilde{f}([(p, u)]) = \tilde{f}([(q, v)])$, then $[(p, \rho^i(u))]_i = [(q, \rho^i(v))]_i$, for every $i \in \mathbb{N}$. Since $\pi(p) = \pi(q)$, there exists a unique $g \in G$ such that $q = p \cdot g$ and $\rho^i(v) = \varrho^i(g^{-1})\left(\rho^i(u)\right)$, for every $i \in \mathbb{N}$. Therefore,

$$\begin{aligned}
v = \left(\rho^i(u)\right)_{i \in \mathbb{N}} &= \left(\varrho_i(g^{-1})(\rho^i(u))\right)_{i \in \mathbb{N}} \\
&= \left(\varprojlim \varrho^i(g^{-1})\right)\left((\rho^i(u))_{i \in \mathbb{N}}\right) \\
&= \varepsilon\left(\varrho(g^{-1})\right)(u) = \phi(g^{-1})(u),
\end{aligned}$$

implying that $[(p, u)] = [(q, v)]$.

ii) $\tilde{f}$ is onto: Let an arbitrary $a \in E = \varprojlim E^i$. Then $a = \left([(p^i, u^i)]_i\right)_{i \in \mathbb{N}}$ such that, by the property of the elements of the limit with respect to the connecting morphisms,

$$f^{ji}\left([(p^j, u^j)]_j\right) = [(p^i, u^i)]_i; \qquad j \geq i,$$

while, by the definition of f^{ji},

$$f^{ji}\left([(p^j, u^j)]_j\right) = [(p^j, \rho^{ji}(u^j))]_i, \qquad j \geq i.$$

Thus, $[(p^i, u^i)]_i = [(p^j, \rho^{ji}(u^j))]_i$ implies the existence of a $g_{ji} \in G$ such that

$$(6.6.4) \qquad p^j = p^i \cdot g_{ji}, \quad \text{and} \quad \rho^{ji}(u^j) = \varrho^i\left(g_{ji}^{-1}\right)(u^i).$$

We set $p := p^1$ and $g_k := g_{k1}^{-1}$ ($k \in \mathbb{N}$). Then, the first of (6.6.4) yields (for $j = k, i = 1$)

$$(6.6.5) \qquad p = p^k \cdot g_{k1}^{-1} = p^k \cdot g_k, \qquad \forall\, k \in \mathbb{N}.$$

Applying (6.6.5) again to the first of (6.6.4), we see that

$$p^j = p^i \cdot g_{ji} \quad \Rightarrow \quad p \cdot g_j^{-1} = p \cdot g_i^{-1} \cdot g_{ji},$$

and, because G acts freely on P,

$$(6.6.6) \qquad g_j = g_{ji}^{-1} \cdot g_i, \qquad j \geq i.$$

Furthermore, by the compatibility of the actions with the connecting morphisms of $\{\mathbb{E}^i\}_{i \in \mathbb{N}}$, i.e. $\rho^{ji} \circ \varrho^j = \varrho^i$ (since $\{\varrho^i\}_{i \in \mathbb{N}}$ is a projective system), we have that

$$\rho^{ji}\left(\varrho^j(g_j^{-1})(u_j)\right) = \varrho^i(g_j^{-1})\left(\rho^{ji}(u^j)\right)$$

or, by the second equality of (6.6.4), and equality (6.6.6),

$$\rho^{ji}\left(\varrho^j(g_j^{-1})(u_j)\right) = \left(\varrho^i(g_i^{-1}) \circ \varrho^i(g_{ji}^{-1})\right)(u^i) = \varrho^i(g_i^{-1})(u_i).$$

Hence, the element $v := \left(\varrho_i \bigl(g_i^{-1} \bigr)(u_i) \right)_{i \in \mathbb{N}}$ belongs to $\mathbb{F}$, and

$$\widetilde{f}([(p, v)]) = \left(\bigl[(p_i \cdot g_i, \varrho_i \bigl(g_i^{-1} \bigr)(u_i)) \bigr]_i \right)_{i \in \mathbb{N}} = \left([(p_i, u_i)]_i \right)_{i \in \mathbb{N}} = a.$$

Consequently, $\widetilde{f}$ is the desired bijection which proves the statement. $\square$

Remark 6.6.3   The preceding proposition implies that

$$P \times_\varphi \left(\varprojlim \mathbb{E}^i \right) \cong \varprojlim \left(P \times_{\rho_i} \mathbb{E}^i \right)$$

as vector bundles. This formula generalizes the set-theoretical commutativity between inverse limits and cartesian products.

Propositions 6.6.1 and 6.6.2, combined together, are summarized in the following main result.

Theorem 6.6.4   *Let (P, G, B, π) be a Fréchet principal bundle over a Banach base, $\mathbb{F} \cong \varprojlim \mathbb{E}^i$ a Fréchet space and $\varphi : G \to GL(\mathbb{F})$ a representation of G in $\mathbb{F}$. If φ can be factored as in (6.6.3), then $F := P \times_\varphi \mathbb{F}$ admits the structure of a Fréchet vector bundle associated with P. In particular, F is identified with the projective limit of a system of Banach vector bundles $\{(E^i, B, \pi^i)\}_{i \in \mathbb{N}}$ of fibre type $\mathbb{E}^i$, respectively.*

We conclude with the following result, mentioned in Remark 6.5.5 (4).

Corollary 6.6.5   *Applying the technique of this section to the case of the generalized principal bundle of frames $\boldsymbol{P}(E)$ of a plb-vector bundle E, discussed in § 6.5, we readily verify that the associated vector bundle $\boldsymbol{P}(E) \times_\varphi \mathbb{F}$, where $\varphi = \varepsilon \circ \mathrm{id}_{\mathcal{H}_0(\mathbb{F})}$, coincides with E.*

7

Connections on plb-vector bundles

The objective of this chapter is to study projective systems of (linear) connections on plb-vector bundles. It will be shown that the derived limits are connections in the classical sense, characterized, however, by a generalized type of Christoffel symbols. The present category of connections entails important relevant geometric notions, like the parallel displacement along curves in the base space. The former cannot be approached, in general, because of the inherent difficulties in the study of differential equations in Fréchet spaces. The corresponding holonomy groups are also studied. These groups seem to live in the borders of the categories of plb-manifolds and algebraic groups as we explain at the end of § 7.2.

7.1 Projective limits of linear connections

For the convenience of the reader, we recall from § 1.5.1 that a (not necessarily linear) connection on a Banach vector bundle $\ell = (E, B, \pi)$, of fibre type $\mathbb{E}$ and base space model $\mathbb{B}$, is a bundle morphism $K \colon TE \longrightarrow E$. Fixing a vb-chart $(U_\alpha, \phi_\alpha, \Phi_\alpha)$ of E and the induced vb-chart of the tangent bundle TE, the local representation of K,

$$(7.1.1) \qquad K_\alpha \colon \phi_\alpha(U_\alpha) \times \mathbb{E} \times \mathbb{B} \times \mathbb{E} \longrightarrow \phi_\alpha(U_\alpha) \times \mathbb{E},$$

is given by

$$(7.1.2) \qquad K_\alpha(x, \lambda, y, \mu) = (x, \mu + \kappa_\alpha(x, \lambda).y),$$

where $\kappa_\alpha \colon \phi_\alpha(U_\alpha) \times \mathbb{E} \to \mathcal{L}(\mathbb{B}, \mathbb{E})$ is the (smooth) *local component* of K. If K is linear, then κ_α is linear with respect to the second variable and induces the Christoffel symbols $\left\{\Gamma_\alpha\colon \phi_\alpha(U_\alpha) \to \mathcal{L}(\mathbb{E}, \mathcal{L}(\mathbb{B}, \mathbb{E}))\right\}_{\alpha \in I}$ by setting

$$(7.1.3) \qquad \Gamma_\alpha(x).\lambda = \kappa_\alpha(x, \lambda), \qquad (x, \lambda) \in \phi_\alpha(U_\alpha) \times \mathbb{E}.$$

Focusing now on the category of projective limits of vector bundles, we consider a plb-vector bundle $\ell \equiv \varprojlim \ell^i = \varprojlim(E^i, B, \pi^i)$ with connecting morphisms $f^{ji}\colon E^j \to E^i$ ($j \ge i$) and fibre type the Fréchet space $\mathbb{F} = \varprojlim \mathbb{E}^i$. As in the case of plb-principal bundles, we assume that B is a Hausdorff space admitting smooth partitions of unity.

A *projective system of connections* is a sequence of connections K^i on ℓ^i ($i \in \mathbb{N}$) commuting with the connecting morphisms of the plb-bundles $E = \varprojlim E^i$ and $TE = \varprojlim TE^i$, i.e.

$$(7.1.4) \qquad f^{ji} \circ K^j = K^i \circ Tf^{ji}, \qquad j \ge i.$$

In virtue of (1.5.19), the preceding equality means that K^j and K^i are (f^{ji}, id_B)-related connections.

To show that such projective systems of connections lead to connections on the limit bundle, we need the equivalent of (7.1.4) in terms of the local components $\kappa_\alpha^i\colon \phi_\alpha(U_\alpha) \times \mathbb{E}^i \to \mathcal{L}(\mathbb{B}, \mathbb{E}^i)$, $i \in \mathbb{N}$, of the connections $\{K^i\}_{i \in \mathbb{N}}$, respectively, in analogy to the general formula (1.5.27). To this end, we first see that the connecting morphisms

$$(7.1.5) \qquad\qquad \rho^{ji}\colon \mathbb{E}^j \longrightarrow \mathbb{E}^i$$

of $\{\mathbb{E}^i\}_{i \in \mathbb{N}}$ induce the connecting morphisms

$$(7.1.6) \qquad r^{ji}\colon \mathcal{L}(\mathbb{B}, \mathbb{E}^j) \longrightarrow \mathcal{L}(\mathbb{B}, \mathbb{E}^i)\colon f \mapsto \rho^{ji} \circ f, \qquad j \ge i$$

of the system $\{\mathcal{L}(\mathbb{B}, \mathbb{E}^i)\}_{i \in \mathbb{N}}$ yielding $\mathcal{L}(\mathbb{B}, \mathbb{F}) \equiv \varprojlim \mathcal{L}(\mathbb{B}, \mathbb{E}^i)$.

Furthermore, from the definition of a projective system of vector bundles (see, in particular, condition (PVB. 2) of Definition 5.2.1), and the proof of Proposition 5.2.2, we obtain the trivializations $(U_\alpha, \tau_\alpha^i)$ of E^i, $i \in I$, and their corresponding vb-charts $\left(U_\alpha, \phi_\alpha, \Phi_\alpha^i := (\varphi_\alpha \times \mathrm{id}_{\mathbb{E}^i}) \circ \tau_\alpha^i\right)$, over the local charts (U_α, ϕ_α) of B. Regarding these vb-charts and the commutative diagram in the proof of Proposition 5.2.2, we verify that the local principal part of each connecting morphism f^{ji},

$$\left(f_\alpha^{ji}\right)^{\#}\colon \phi_\alpha(U_\alpha) \longrightarrow \mathcal{L}(\mathbb{E}^j, \mathbb{E}^i),$$

[see (1.5.22) and (1.5.23) together with (1.5.31)] is a constant map; namely,

$$\left(f_\alpha^{ji}\right)^{\#}(x) = \rho^{ji}, \qquad x \in \phi_\alpha(U_\alpha).$$

Therefore, (1.5.27) now becomes

(7.1.7) $\qquad r^{ji} \circ \kappa_\alpha^j = \kappa_\alpha^i \circ \left(\mathrm{id}_{\phi_\alpha(U_\alpha)} \times \rho^{ji} \right), \qquad j \geq i.$

In the case of a linear connection, taking the Christoffel symbols of the form $\left\{ \Gamma_\alpha^k \colon \phi_\alpha(U_\alpha) \to \mathcal{L}(\mathbb{E}^k, \mathcal{L}(\mathbb{B}, \mathbb{E}^k)) \right\}_{\alpha \in I}$ $(k = j, i)$, (7.1.7) is equivalent to

(7.1.8) $\qquad r^{ji} \circ \Gamma_\alpha^j(x) = \Gamma_\alpha^i(x) \circ \rho^{ji}; \qquad x \in \phi_\alpha(U_\alpha), j \geq i.$

In summary,

$$(7.1.4) \Leftrightarrow (7.1.7) \Leftrightarrow (7.1.8).$$

Proposition 7.1.1 *If $\{K^i\}_{i \in \mathbb{N}}$ is a projective system of connections, then the limit $K := \varprojlim K^i$ is a connection on the plb-vector bundle $\ell \equiv \varprojlim \ell^i$. K is called a* **plb-connection.**

Proof As we have proved in Proposition 5.2.2, the vector bundle charts $(U_\alpha, \phi_\alpha, \Phi_\alpha^i)$ converge projectively to the plb-vector chart $(U_\alpha, \phi_\alpha, \Phi_\alpha \equiv \varprojlim \Phi_\alpha^i)$ of ℓ. On the other hand, the induced chart of the limit tangent bundle TE is

$$\left(\tau_E^{-1}\left(\pi^{-1}(U) \right) = \varprojlim \tau_{E^i}^{-1}\left((\pi^i)^{-1}(U) \right), \Phi, \widetilde{\Phi} = \varprojlim \widetilde{\Phi}^i \right),$$

where each $\widetilde{\Phi}^i$ is defined as in (1.5.5).

An immediate consequence of (7.1.7) is that the local components

$$\left\{ \kappa_\alpha^i \colon \phi_\alpha(U_\alpha) \times \mathbb{E}^i \to \mathcal{L}(\mathbb{B}, \mathbb{E}^i) \right\}_{i \in \mathbb{N}},$$

of $\{K^i\}$ form (for each $\alpha \in I$) a projective system of smooth maps; hence,

(7.1.9) $\qquad \kappa_\alpha := \varprojlim \kappa_\alpha^i \colon \phi_\alpha(U_\alpha) \times \mathbb{F} \longrightarrow \mathcal{L}(\mathbb{B}, \mathbb{F})$

is a well-defined pls-map (and therefore smooth), such that

$$\kappa_\alpha(x, \lambda) = \left(\kappa_\alpha^i\left(x, \rho^i(\lambda) \right) \right)_{i \in \mathbb{N}}$$

$$\Leftrightarrow \rho^i \circ \kappa_\alpha(x, \lambda) = \kappa_\alpha^i\left(x, \rho^i(\lambda) \right); \quad i \in \mathbb{N},$$

where

(7.1.10) $\qquad \rho^i \colon \mathbb{F} = \varprojlim \mathbb{E}^i \longrightarrow \mathbb{E}^i; \qquad i \in \mathbb{N},$

are the canonical projections of the fibre type. Based on the latter equivalence, and using the limit charts of E mentioned in the beginning of the proof, we check that

$$
\begin{aligned}
K_\alpha(x,\lambda,y,\mu) &= \left(\Phi \circ K|_{\tau_E^{-1}(\pi^{-1}(U))} \circ \widetilde{\Phi}^{-1}\right)(x,\lambda,y,\mu)\\
&= \left(\left(\Phi^i \circ K^i|_{(\tau_{E^i}^i)^{-1}((\pi^i)^{-1}(U))} \circ (\widetilde{\Phi}^i)^{-1}\right)(x,\lambda^i,y,\mu^i)\right)_{i\in\mathbb{N}}\\
&= \left(K_\alpha^i(x,\lambda^i,y,\mu^i)\right)_{i\in\mathbb{N}}\\
&= \left(x,\mu^i + \kappa_\alpha^i(x,\lambda^i).y\right)_{i\in\mathbb{N}}\\
&= \left(x,(\mu^i)_{i\in\mathbb{N}} + \left(\rho^i(\kappa_\alpha(x,\lambda^i.y)_{i\in\mathbb{N}}\right)\right)\\
&= \left(x,\mu + \kappa_\alpha(x,\lambda).y\right),
\end{aligned}
$$

for every $x \in \phi_\alpha(U_\alpha)$, $y \in \mathbb{B}$, $\lambda = (\lambda^i)_{i\in\mathbb{N}}$, and $\mu = (\mu^i)_{i\in\mathbb{N}} \in \mathbb{F}$. As a result, the local characterization of a connection (via local components) implies that K is indeed a connection on the Fréchet vector bundle $E = \varprojlim E^i$. $\qquad\square$

In particular, we have:

Corollary 7.1.2 *Let $K = \varprojlim K^i$ be a projective limit of linear connections on a plb-vector bundle $\ell = \varprojlim \ell^i$. Then K is also a linear connection.*

Proof The linearity of the factor connections $\{K^i\}_{i\in\mathbb{N}}$ is equivalent to the fact that the local components

$$
\kappa_\alpha^i : \phi_\alpha(U_\alpha) \times \mathbb{E}^i \longrightarrow \mathcal{L}(\mathbb{B}, \mathbb{E}^i); \qquad i \in \mathbb{N},
$$

are linear with respect to the second variable, for every $\alpha \in I$. Since the local components of K are projective limits, i.e. $\kappa_\alpha = \varprojlim \kappa_\alpha^i$, the latter become also linear with respect to their second variable, thus K turns to be a linear connection. $\qquad\square$

Next we look at the Christoffel symbols of plb-connections. It will be shown that these connections are characterized by Christoffel symbols whose values are restricted to continuous linear maps represented by projective limits. More precisely, in accordance with the previous formalism [see also § 1.5.3 and (7.1.3)], the Christoffel symbols of a plb-connection $K = \varprojlim K^i$, over the vb-charts $(U_\alpha, \phi_\alpha, \Phi_\alpha \equiv \varprojlim \Phi_\alpha^i)$ of ℓ, are the maps

$$
\Gamma_\alpha : \phi_\alpha(U_\alpha) \longrightarrow \mathcal{L}(\mathbb{F}, \mathcal{L}(\mathbb{B}, \mathbb{F})),
$$

determined, as usual, by

$$
(7.1.11) \qquad \Gamma_\alpha(x).\lambda = \kappa_\alpha(x,\lambda); \qquad (x,\lambda) \in \phi_\alpha(U_\alpha) \times \mathbb{F}.
$$

Then we obtain the following preliminary result:

Proposition 7.1.3 $\Gamma_\alpha(x) = \varprojlim \Gamma_\alpha^i(x)$, *for every* $x \in \phi_\alpha(U_\alpha)$ *and* $\alpha \in I$.

Proof Because $\{\Gamma_\alpha^i(x)\}_{i \in \mathbb{N}}$ is a projective system, as an immediate consequence of (7.1.8), we need only to show that $\{\Gamma_\alpha^i(x)\}_{i \in \mathbb{N}}$ converges to $\Gamma_\alpha(x)$. Indeed, if we denote by

$$(7.1.12) \qquad r^i \colon \mathcal{L}(\mathbb{B}, \mathbb{F}) \longrightarrow \mathcal{L}(\mathbb{B}, \mathbb{E}^i) \colon f \mapsto \rho^i \circ f$$

the canonical projections of $\mathcal{L}(\mathbb{B}, \mathbb{F}) \equiv \varprojlim \mathcal{L}(\mathbb{B}, \mathbb{E}^i)$, then, by (7.1.9),

$$\left(r^i \circ \Gamma_\alpha(x)\right)(\lambda) = r^i\big(\kappa_\alpha(x, \lambda)\big) = \kappa_\alpha^i\big(x, \rho^i(\lambda)\big) = \big(\Gamma_\alpha^i(x) \circ \rho^i)\big)(\lambda),$$

for every $\lambda \in \mathbb{F}$ and every $i \in \mathbb{N}$. Therefore, Proposition 2.3.5 implies the assertion. $\qquad\qquad \square$

It is worth noticing here that, despite the previous "point-wise" convergence, the Christoffel symbols of a linear plb-connection themselves are not necessarily projective limits, i.e. equalities $\Gamma_\alpha = \varprojlim \Gamma_\alpha^i$ are not in general true. However, each Γ_α is *associated* to a limit of Christoffel-like maps in the following way: For every limit vb-chart $(U_\alpha, \phi_\alpha, \Phi_\alpha) \equiv \varprojlim \big(U_\alpha, \phi_\alpha, \Phi_\alpha^i\big)$ of $\ell \equiv \varprojlim \ell^i$ as before, we define the maps

$$(7.1.13) \qquad \begin{aligned} \Gamma_\alpha^{*i} \colon \phi_\alpha(U_\alpha) &\longrightarrow \mathcal{H}^i(\mathbb{E}^i, \mathcal{L}(\mathbb{B}, \mathbb{E}^i)) \colon \\ x &\mapsto \big(\Gamma_\alpha^1(x), \Gamma_\alpha^2(x), \ldots, \Gamma_\alpha^i(x)\big), \end{aligned}$$

for all $i \in \mathbb{N}$. Referring to Theorem 2.3.10, in particular to equalities (2.3.3) and (2.3.4), we obtain the limit space

$$\mathcal{H}(\mathbb{F}, \mathcal{L}(\mathbb{B}, \mathbb{F})) = \varprojlim \mathcal{H}^i(\mathbb{F}, \mathcal{L}(\mathbb{B}, \mathbb{E}^i))$$

and the continuous linear embedding

$$(7.1.14) \qquad \varepsilon \colon \mathcal{H}(\mathbb{F}, \mathcal{L}(\mathbb{B}, \mathbb{F})) \hookrightarrow \mathcal{L}(\mathbb{F}, \mathcal{L}(\mathbb{B}, \mathbb{F})) \colon (g^i)_{i \in \mathbb{N}} \mapsto \varprojlim g^i.$$

Note that the previous considerations are meaningful because $\mathcal{L}(\mathbb{B}, \mathbb{F}) \equiv \varprojlim \mathcal{L}(\mathbb{B}, \mathbb{E}^i)$ while in (7.1.13) we have also applied (2.3.9), (2.3.9').

A direct consequence of Theorem 2.3.10 is now the next result.

Proposition 7.1.4 *With the previous notations we have:*
i) The pls-map

$$\Gamma_\alpha^* := \varprojlim \Gamma_\alpha^{*i} \colon \phi_\alpha(U_\alpha) \longrightarrow \mathcal{H}(\mathbb{F}, \mathcal{L}(\mathbb{B}, \mathbb{F}))$$

can be defined.

ii) The usual Christoffel symbols $\{\Gamma_\alpha\colon \phi_\alpha(U_\alpha) \to \mathcal{L}(\mathbb{F}, \mathcal{L}(\mathbb{B}, \mathbb{F}))\}_{\alpha\in I}$ of the plb-connection $K = \varprojlim K^i$ factorize in the form

$$\Gamma_\alpha = \varepsilon \circ \Gamma_\alpha^*.$$

The maps $\{\Gamma_\alpha^*\}$ will be called **generalized** or **plb-Christoffel symbols**. They are important because they characterize plb-connections, as shown in the next result.

Proposition 7.1.5 *Let K be an arbitrary linear connection on a plb-vector bundle $\ell \equiv \varprojlim \ell^i$. If the ordinary Christoffel symbols of K,*

$$\{\Gamma_\alpha\colon \phi_\alpha(U_\alpha) \longrightarrow \mathcal{L}(\mathbb{F}, \mathcal{L}(\mathbb{B}, \mathbb{F}))\}_{\alpha\in I},$$

factor into $\Gamma_\alpha = \varepsilon \circ \Gamma_\alpha^*$, where

$$\Gamma_\alpha^* := \varprojlim \Gamma_\alpha^{*i}\colon \phi_\alpha(U_\alpha) \longrightarrow \mathcal{H}(\mathbb{F}, \mathcal{L}(\mathbb{B}, \mathbb{F}))$$

are pls-maps, for all vb-charts $(U_\alpha, \phi_\alpha, \Phi_\alpha) \equiv \varprojlim (U_\alpha, \phi_\alpha, \Phi_\alpha^i)$, then K coincides with a linear plb-connection, i.e. $K = \varprojlim K^i$.

Proof The assumption that $\Gamma_\alpha^* = \varprojlim \Gamma_\alpha^{*i}$ is a pls-map means that

$$\Gamma_\alpha^{*i}\colon \phi_\alpha(U_\alpha) \longrightarrow \mathcal{H}^i(\mathbb{E}^i, \mathcal{L}(\mathbb{B}, \mathbb{E}^i));$$

thus, for every $x \in \phi_\alpha(U_\alpha)$,

$$\Gamma_\alpha^{*i}(x) = \big(q^1(x), \ldots, q^i(x)\big) \in \prod_{k=1}^{i} \mathcal{L}(\mathbb{E}^k, \mathcal{L}(\mathbb{B}, \mathbb{E}^k)),$$

so that $r^{jk} \circ q^j(x) = q^k(x) \circ \rho^{jk}$ holds true for every $j, k = 1, \ldots, i$, with $j \geq k$. If

$$\mathcal{P}r_k\colon \prod_{j=1}^{i} \mathcal{L}\big(\mathbb{E}^j, \mathcal{L}(\mathbb{B}, \mathbb{E}^j)\big) \longrightarrow \mathcal{L}\big(\mathbb{E}^k, \mathcal{L}(\mathbb{B}, \mathbb{E}^k)\big)$$

is the k-th projection, we define the maps

$$\Gamma_\alpha^i := \mathcal{P}r_i \circ \Gamma_\alpha^{*i}; \qquad i \in \mathbb{N},$$

which, by their construction, satisfy

$$(7.1.15) \qquad r^{ji} \circ \Gamma_\alpha^j(x) = \Gamma_\alpha^i(x) \circ \rho^{ji}; \quad x \in \phi_\alpha(U_\alpha), j \geq i.$$

Since $\{\Gamma_\alpha\}_{\alpha\in I}$ are the Christoffel symbols of K, thus they satisfy the analog of the compatibility condition (1.5.12), it is readily verified that a similar condition (within the Banach framework) holds for the maps $\{\Gamma_\alpha^i\}_{\alpha\in I}$. Therefore, the factor Banach vector bundles ℓ^i admit respective linear connections K^i $(i \in \mathbb{N})$, with Christoffel symbols the given families of maps. Observing that (7.1.15) is precisely (7.1.8), we conclude that $\{K^i\}$ is a projective system with limit the plb-connection $\varprojlim K^i$.

We shall show that $\varprojlim K^i$ coincides with the initial connection K. By Proposition 2.3.5 and the local characterization of connections via the local components, along with (7.1.10) and (7.1.12), it suffices to verify the equality

$$r^i \circ \kappa_\alpha = \kappa_\alpha^i \circ \left(\mathrm{id}_{\phi_\alpha(U_\alpha)} \times \rho^i \right).$$

Equivalently, it suffices to show that, for every pair $\left(x, \lambda = (\lambda^i)\right) \in \phi_\alpha(U_\alpha) \times \mathbb{F}$,

$$r^i(\kappa_\alpha(x, \lambda)) = \kappa_\alpha^i(x, \lambda^i)$$
$$\Leftrightarrow \quad r^i(\Gamma_\alpha(x).\lambda) = \Gamma_\alpha^i(x).\lambda^i$$
$$\Leftrightarrow \quad r^i \circ \Gamma_\alpha(x) = \Gamma_\alpha^i(x) \circ \rho^i.$$

The last equality is a consequence of the factorization assumption of Γ_α. As a matter of fact,

$$\Gamma_\alpha(x) = \varepsilon\left(\Gamma_\alpha^*(x)\right)$$
$$\Leftrightarrow \quad \Gamma_\alpha(x) = \varprojlim \Gamma_\alpha^i(x)$$
$$\Leftrightarrow \quad r^i \circ \Gamma_\alpha(x) = \varprojlim \Gamma_\alpha^i(x) \circ \rho^i. \qquad \square$$

Thinking of $\mathcal{H}(\mathbb{F}, \mathcal{L}(\mathbb{B}, \mathbb{F}))$ as a subspace of $\mathcal{L}(\mathbb{F}, \mathcal{L}(\mathbb{B}, \mathbb{F}))$ by the embedding (7.1.14), we summarize the preceding two propositions in the following main result.

Theorem 7.1.6 *A linear connection K on a plb-vector bundle $\ell \equiv \varprojlim \ell^i$ is a linear plb-connection if and only if its Christoffel symbols take values in the subspace $\mathcal{H}(\mathbb{F}, \mathcal{L}(\mathbb{B}, \mathbb{F}))$ of $\mathcal{L}(\mathbb{F}, \mathcal{L}(\mathbb{B}, \mathbb{F}))$.*

7.2 Parallel displacement and holonomy groups

As we mentioned in the introduction to the present chapter, the parallel displacement along curves of the base space cannot be ensured. Consequently, holonomy groups in the classical sense cannot even be defined. The projective limit approach gives a way out in this case too. If we are restricted to plb-vector bundles and connections, the above important groups can be recovered and yield results very close to those obtained in the Banach case, bypassing thus the problems concerning differential equations in the (Fréchet) models.

Keeping up the formalism of plb-vector bundles and connections applied in the previous section, we prove the following first result on parallel sections of a limit vector bundle. In this respect we also refer to § 1.5.5 for the usual definitions.

Lemma 7.2.1 *Let $\beta\colon [0,1] \to B$ be a smooth curve in the base of a plb-vector bundle $\ell = (E, B, \pi) \equiv \varprojlim(E^i, B, \pi^i) = \varprojlim \ell^i$, and let $K = \varprojlim K^i$ be a linear plb-connection on ℓ. Then*

i) Every section $\xi\colon [0,1] \to E$ along β (: $\xi \in \Gamma_\beta(E)$) is realized as a projective limit of corresponding sections on the factor bundles ℓ^i, i.e.

$$\xi = \varprojlim \xi^i, \qquad \xi \in \Gamma_\beta(E^i).$$

ii) A section $\xi \in \Gamma_\beta(E)$ is parallel (with respect to K), if and only if $\xi^i \in \Gamma_\beta(E^i)$ is parallel (with respect to K^i), for every $i \in \mathbb{N}$.

Proof i) The factor sections are obtained by projecting ξ to the factor bundles; that is,

$$\xi^i := f^i \circ \xi\colon [0,1] \to E^i; \qquad i \in \mathbb{N},$$

where $f^i\colon E = \varprojlim E^i \to E^i$ are the canonical projections. They are smooth as composites of smooth maps and are projected to β since

$$\pi^i \circ \xi^i = \pi^i \circ f^i \circ \xi = \pi \circ \xi = \beta.$$

Besides, their relation with the connecting morphisms $f^{ji}\colon E^j \to E^i$, namely

$$f^{ji} \circ \xi^j = f^{ji} \circ f^j \circ \xi = f^i \circ \xi = \xi^i; \qquad j \geq i,$$

ensures that $\varprojlim \xi^i$ is defined and coincides with ξ by the very definition of $(\xi^i)_{i\in\mathbb{N}}$.

ii) Assume first that each $\xi^i \in \Gamma_\beta(E^i)$ is parallel, thus by (1.5.16),

$$K^i \circ T\xi^i \circ \partial = 0.$$

Hence, in virtue of (3.2.8),

$$K \circ T\xi \circ \partial = \varprojlim K^i \circ \varprojlim T\xi^i \circ \partial = \varprojlim(K^i \circ T\xi^i \circ \partial) = 0,$$

which means that K is parallel.

Conversely, if $\xi \in \Gamma_\beta(E)$ is parallel, then

$$K(T\xi(\partial_t)) = 0 \in E_{\pi(\xi(t))} = E_{\beta(t)} = \pi^{-1}(\beta(t)); \qquad t \in [0,1],$$

and, by Theorem 3.2.8,

$$0 = f^i\big(K(T\xi(\partial_t))\big) = K^i\big(Tf^i(T\xi(\partial_t))\big) = K^i\big(T\xi^i(\partial_t)\big) \in E^i_{\beta(t)},$$

for all $t \in [0,1]$. This shows that every ξ^i is parallel. $\square$

Using the preceding lemma we obtain now the following main result of this section.

Theorem 7.2.2 *Let $\ell = \varprojlim \ell^i$ be a plb-vector bundle endowed with a linear plb-connection $K = \varprojlim K^i$. If $\beta \colon [0,1] \to B$ is a smooth curve and $u \in E_{\beta(0)}$ an arbitrarily chosen point, then there exists a unique parallel section of ℓ along β, satisfying the initial condition $(0,u)$.*

Proof By the first part of the proof of Theorem 5.2.5 (referring the fibres of a plb-bundle), $E_{\beta(0)} = \varprojlim E^i_{\beta(0)}$. Therefore, any $u \in E_{\beta(0)}$ takes the form $u = (u^i)_{i \in \mathbb{N}}$, where $u^i \in E^i_{\beta(0)}$ and $f^{ji}(u^j) = u^i$ $(j \geq i)$. Since each E^i is a Banach vector bundle, there is a unique parallel $\xi^i \in \Gamma_\beta E^i$ such that $\xi^i(0) = u^i$. Then, for every $j \geq i$, the map $f^{ji} \circ \xi^j \colon [0,1] \to E^i$ is a parallel section along β, since

$$\pi^i \circ (f^{ji} \circ \xi^j) = \pi^j \circ \xi^j = \beta,$$

$$K^i \circ T(f^{ji} \circ \xi^j) \circ \partial = K^i \circ Tf^{ji} \circ T\xi^j \circ \partial =$$
$$= f^{ji} \circ K^j \circ T\xi^j \circ \partial = f^{ji} \circ 0 = 0,$$

the last equality being a consequence of the fibre-wise linearity of the connecting morphisms. Moreover,

$$(f^{ji} \circ \xi^j)(0) = f^{ji}(u^j) = u^i = \xi^i(0).$$

Hence, by the uniqueness of the parallel section with initial condition $(0, u^i)$,

$$f^{ji} \circ \xi^j = \xi^i; \qquad j \geq i,$$

from which we deduce that the smooth map $\xi := \varprojlim \xi^i \colon [0,1] \to E$ is defined. The latter is a section of E along β, because

$$(\pi \circ \xi)(t) = \pi\big((\xi^i(t))_{i \in \mathbb{N}}\big) = \big(\pi^i(\xi^i(t))\big)_{i \in \mathbb{N}} \equiv \beta(t).$$

It is also parallel according to Lemma 7.2.1 and satisfies the desired initial condition

$$\xi(0) = \big((\xi^i)(0)\big)_{i \in \mathbb{N}} = (u^i)_{i \in \mathbb{N}} = u.$$

Finally, assume that there is another parallel section η of E along β such that $\eta(0) = u$. Then η determines a family $(\eta^i)_{i \in \mathbb{N}}$ of analogous parallel sections on the factor bundles E^i, with $\eta^i(0) = u^i$ ($i \in \mathbb{N}$). Once again, the uniqueness of parallel sections in Banach vector bundles, yields $\eta^i = \xi^i$, for all $i \in \mathbb{N}$. Thus $\xi = \varprojlim \xi^i = \varprojlim \eta^i = \eta$, which completes the proof.   □

Definition 7.2.3   Let $\beta \colon [0,1] \to B$ be a smooth curve in the base of a plb-vector bundle $E = \varprojlim E^i$ endowed with a linear connection K (thus $K \equiv \varprojlim K^i$). Then, analogously to (1.5.18), the **parallel displacement** or **translation along** β is the map

$$\tau_\beta \colon E_{\beta(0)} \longrightarrow E_{\beta(1)} \colon u \mapsto \xi_u(1),$$

where ξ_u is the unique parallel section of E along B, with $\xi_u(0) = u$.

The parallel displacement remains also within the category of projective limits because of the following result.

Proposition 7.2.4   *Let $E = \varprojlim E^i$ be plb-vector bundle endowed with a linear connection $K = \varprojlim K^i$. For every smooth curve $\beta \colon [0,1] \to B$ in the base of E, the parallel displacement along β in E coincides with the projective limit of the corresponding parallel displacements in the Banach factor bundles, i.e. $\tau_\beta = \varprojlim \tau_\beta^i$*

Proof   In conjunction with Lemma 7.2.1 and Theorem 7.2.2, it suffices to check the compatibility of $\left(\tau_\beta^i\right)$ with the connecting morphisms $f^{ji} \colon E^j \to E^i$ and the canonical projections $f^i \colon E = \varprojlim E^i \to E^i$, restricted to the fibres $E_{\beta(0)} = \varprojlim E_{\beta(0)}^i$ and $E_{\beta(1)} = \varprojlim E_{\beta(1)}^i$; in other words, we should verify the commutativity of the diagrams

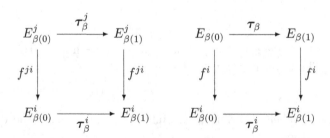

Let any $v \in E_{\beta(0)}^j$. If $\xi_v^j \colon [0,1] \to E^j$ is the (unique) parallel section of the bundle E^j along β with $\xi_v^j(0) = v$, then, as we pointed out in the proof of Theorem 7.2.2, the corresponding section of E^i along β with

initial condition $(0, f^{ji}(v))$ is precisely $f^{ji} \circ \xi_v^j$. Therefore,

(7.2.1)
$$\left(f^{ji} \circ \tau_\beta^j\right)(v) = f^{ji}\left(\xi_v^j(1)\right) = \xi_{f^{ji}(v)}^i(1)$$
$$= \tau_\beta^i\left(f^{ji}(v)\right) = \left(\tau_\beta^i \circ f^{ji}\right)(v),$$

thus proving the commutativity of the first diagram and the existence of the limit $\varprojlim \tau_\beta^i$.

Similarly, for any $u = (u^i) \in E$, $\xi_u = \varprojlim \xi_{u^i}^i$. In particular, for every $u = (u^i) \in E_{\beta(0)}$,

(7.2.2) $\quad \left(f^i \circ \tau_\beta\right)(u) = f^i(\xi_u(1)) = \xi_{u^i}^i(1) = \tau_\beta^i(u^i) = \tau_\beta^i\left(f^i(u)\right),$

implying the commutativity of the second diagram and the equality of the statement. $\qquad \square$

We define the **holonomy group** Φ_b of a plb-connection $K = \varprojlim K^i$, with reference point $b \in B$, by setting

$$\Phi_b := \left\{\tau_\beta \colon E_b \overset{\simeq}{\longrightarrow} E_b \ (\text{toplinear isomorphism})\right\}$$

for all smooth curves $\beta \colon [0,1] \to B$ with $\beta(0) = \beta(1) = b$. Since we are dealing with a fixed linear connection K, and there is no danger of confusion with connections on principal bundles here, we simply write Φ_b instead of $^K\Phi_b$, the latter being defined in § 1.5.5.

The **restricted holonomy group** Φ_b^0 is defined analogously, by considering closed curves at b, homotopic to zero.

To obtain substantial properties of the holonomy groups we need to assume the following strong condition, which is not in general true:

(7.2.3) The connecting morphisms $f^{ji} \colon E^j \to E^i$ and the canonical projections $f^i \colon E = \varprojlim E^i \to E^i$ of the projective system of vector bundles $\{E^i\}_{i \in \mathbb{N}}$ are *surjective* maps.

Then we are in a position to prove:

Proposition 7.2.5 *If $\{\Phi_b^i\}_{i \in \mathbb{N}}$ is the family of corresponding holonomy groups of the factor connections, the following assertions are true:*

i) The projective limit group $\varprojlim \Phi_b^i$ exists.

ii) Φ_b is a subgroup of $\varprojlim \Phi_b^i$ by means of an isomorphism.

Proof For the first assertion we define the maps

$$\sigma^{ji} \colon \Phi_b^j \longrightarrow \Phi_b^i \colon \tau_\beta^j \mapsto \tau_\beta^i, \qquad j \geq i.$$

They are well-defined, for if

(7.2.4) $$\tau_\beta^j = \tau_\gamma^j,$$

we have to show that

$$\tau_\beta^i = \sigma^{ji}(\tau_\beta^j) = \sigma^{ji}(\tau_\gamma^j) = \tau_\gamma^i,$$

for any smooth curves β and γ closed at b. Indeed, because of the existence of the limits $\tau_\beta = \varprojlim \tau_\beta^i$ and $\tau_\gamma = \varprojlim \tau_\gamma^i$, equality (7.2.4) yields

$$f^{ji} \circ \tau_\beta^j = f^{ji} \circ \tau_\gamma^j,$$

which, in virtue of (7.2.1), turns into

$$\tau_\beta^i \circ f^{ji} = \tau_\gamma^i \circ f^{ji}.$$

This proves the desired equality $\tau_\beta^i = \tau_\gamma^i$ as a result of the surjectivity of the connecting morphisms [recall condition (7.2.3)].

The maps σ^{ji} are also group morphisms, since the properties of the ordinary (in Banach bundles) parallel displacement (1.5.18) imply that

$$\sigma^{ji}(\tau_\beta^j \circ \tau_\gamma^j) = \sigma^{ji}(\tau_{\beta*\gamma}^j) = \tau_{\beta*\gamma}^i = \tau_\beta^i \circ \tau_\gamma^i = \sigma^{ji}(\tau_\beta^j) \circ \sigma^{ji}(\tau_\gamma^j), \quad j \geq i.$$

On the other hand, for every triplet of indices $j \geq i \geq k$,

$$\left(\sigma^{ik} \circ \sigma^{ji}\right)\left(\tau_\beta^j\right) = \sigma^{ik}\left(\sigma^{ji}(\tau_\beta^j)\right) = \sigma^{ik}\left(\tau_\beta^i\right) = \tau_\beta^k = \sigma^{jk}\left(\tau_\beta^j\right).$$

The previous arguments prove that the projective limit of groups $\varprojlim \mathbf{\Phi}_b^i$ can be defined.

For the second assertion we project the holonomy group $\mathbf{\Phi}_b$ onto the corresponding factor groups by means of the maps

$$h^i \colon \mathbf{\Phi}_b \longrightarrow \mathbf{\Phi}_b^i \colon \tau_\beta \mapsto \tau_\beta^i, \qquad i \in \mathbb{N}.$$

Working as in the proof of the first assertion, using (7.2.2) and the surjectivity of the canonical projections restricted to the fibres, we prove that $(h^i)_{i\in\mathbb{N}}$ are well-defined group morphisms yielding the projective limit

$$(7.2.5) \qquad\qquad h := \varprojlim h^i \colon \mathbf{\Phi}_b \longrightarrow \varprojlim \mathbf{\Phi}_b^i.$$

Concerning the kernel of h we observe that

$$\tau_\beta \in \operatorname{Ker} h \;\Leftrightarrow\; h(\tau_\beta) = \left(\operatorname{id}_{E_b^i}\right)_{i\in\mathbb{N}} \;\Leftrightarrow\; h^i(\tau_\beta) = \operatorname{id}_{E_b^i} \;\Leftrightarrow\; \tau_\beta^i = \operatorname{id}_{E_b^i},$$

for all $i \in \mathbb{N}$. However, according to Proposition 7.2.4, $\tau_\beta = \varprojlim \tau_\beta^i$; therefore, $\tau_\beta = \operatorname{id}_{E_b}$, i.e. the kernel of h is the trivial group and h an injective homomorphism by which $\mathbf{\Phi}_b$ can be identified with a subgroup of $\varprojlim \mathbf{\Phi}_b^i$. $\qquad\square$

Remarks 7.2.6 1) From the proof of assertion ii) it is clear that Φ_b coincides, up to a group isomorphism, with a subgroup of a projective limit of Banach-Lie groups. However, Φ_b fails to be a Fréchet-Lie group itself because the existence of appropriate limit charts cannot be assured. The same remark applies to the holonomy groups of connections on plb-principal bundles whose structure has been described in Theorem 4.3.5.

Both cases clarify now the claim that the holonomy groups of projective limit connections live between the categories of plb-manifolds and topological groups, as commented in the introduction to the present chapter.

2) Under the identification induced by h [see (7.2.5)], Φ_b becomes a topological group, in contrast to $\mathrm{GL}(\mathbb{F})$ (in which Φ_b embeds) that does not admit any reasonable topological group structure.

3) The assumption (7.2.3) is always fulfilled in the case of topological spaces via the construction of an appropriate system of topological spaces whose limit is homeomorphic to the initial limit space (for details we refer to [Dug75, Appendix Two, § 2.8]. This is not necessarily true for more complicated structures such as (Banach) vector bundles.

7.3 Connections on plb-vector and frame bundles

Linear connections on finite-dimensional or Banach vector bundles can be approached by means of the general theory of (infinitesimal) connections on principal bundles by associating vector bundles with their bundles of frames. The same is true in our setting if we consider the *generalized frame bundles* defined in § 6.5, where we have also explained the reasons necessitating the introduction of these bundles.

More precisely, we establish here a correspondence between linear connections on a limit vector bundle E and connections (in the sense of § 4.2) on the generalized bundle of frames $\boldsymbol{P}(E)$, defined in Proposition 6.5.4 [see also Remark 6.5.5(3)]. To this end we fix a plb-vector bundle

$$\ell = (E, B, \pi) \equiv \varprojlim \ell^i = \varprojlim(E^i, B, \pi^i)$$

of fibre type the Fréchet space $\mathbb{F} = \varprojlim \mathbb{E}^i$, with connecting morphisms $f^{ji} \colon E^j \to E^i$, $j \geq i$, and canonical projections $f^i \colon E \to E^i$, $i \in \mathbb{N}$. As we have seen in Proposition 6.5.4,

$$\left\{ \left(\boldsymbol{P}(E^i), \mathcal{H}_0^i(\mathbb{F}), B, \boldsymbol{p}^i\right) ; (\boldsymbol{r}^{ji}, h_0^{ji}, \mathrm{id}_B) \right\}_{i,j \in \mathbb{N}}$$

is a projective system of Banach principal bundles inducing the plb-principal bundle

$$\big(\boldsymbol{P}(E) = \varprojlim \boldsymbol{P}(E^i), \mathcal{H}_0(\mathbb{F}), B, \boldsymbol{p} = \varprojlim \boldsymbol{p}^i\big).$$

Referring also to equalities (2.3.5), (2.3.6) and (5.1.2), we recall that

$$\boldsymbol{P}(E^i) = \bigcup_{x \in B} \mathcal{H}_0^i(\mathbb{F}, E_x),$$

whereas

$$(\boldsymbol{r}^{ji}, h_0^{ji}, \mathrm{id}_B) \colon \big(\boldsymbol{P}(E^j), \mathcal{H}_0^j(\mathbb{F}), B, \boldsymbol{p}^j\big) \longrightarrow \big(\boldsymbol{P}(E^i), \mathcal{H}_0^i(\mathbb{F}), B, \boldsymbol{p}^i\big)$$

is a morphism of principal bundles with

$$\boldsymbol{r}^{ji}\big(q^1, \ldots, q^j\big) = \big(q^1, \ldots, q^i\big) \quad \text{and} \quad h_0^{ji}\big(g^1, \ldots, g^j\big) = \big(g^1, \ldots, g^i\big).$$

With the notion of related connections in mind (see §1.7.5), we first examine the effect that the presence of a linear plb-connection on E has on the projective system $\{\boldsymbol{P}(E^i)\}_{i \in \mathbb{N}}$:

Lemma 7.3.1 *A linear plb-connection $K = \varprojlim K^i$ on E determines a family of $\big(\boldsymbol{r}^{ji}, h_0^{ji}, \mathrm{id}_B\big)$-related connections $(\theta^i)_{i \in \mathbb{N}}$ on the Banach principal bundles $\boldsymbol{P}(E^i)$, respectively.*

Proof As described in §1.7.4, each linear connection K^i on E^i induces a connection form

$$\omega^i \in \Lambda^1\big(P(E^i), \mathcal{GL}(E^i) \equiv \mathcal{L}(E^i)\big)$$

on the ordinary frame bundle $P(E^i)$ of E^i. In fact, ω^i is completely determined by the corresponding local connection forms $\omega_\alpha^i \in \Lambda^1(U_\alpha, \mathcal{L}(E^i))$, $\alpha \in I$, defined in turn by the Christoffel symbols $\{\Gamma_\alpha^i\}_{\alpha \in I}$ of K^i by the analog of equality (1.7.16), namely

$$\omega_{\alpha,x}^i(v) = \Gamma_\alpha^i(\phi_\alpha(x))\big(\overline{\phi}_\alpha(v)\big), \qquad x \in U_\alpha, v \in T_x B.$$

Here (U_α, ϕ_α) are the charts inducing local trivializations of the involved bundles. As a result, for each $\alpha \in I$, we may define the differential forms $\theta_\alpha^i \in \Lambda^1(U_\alpha, \mathcal{H}^i(\mathbb{F}))$ by setting

(7.3.1) $$\theta_\alpha^i = \big(\omega_\alpha^1, \ldots, \omega_\alpha^i\big);$$

more explicitly,

$$(\theta_\alpha^i)_x(v) := \big((\omega_\alpha^1)_x(v), \ldots, (\omega_\alpha^i)_x(v)\big); \qquad x \in U_\alpha, \ v \in T_x B,$$

for every $i \in \mathbb{N}$. We shall show that $\left(\theta_\alpha^i\right)_{i \in \mathbb{N}}$ determine a connection on the principal bundle $\boldsymbol{P}(E^i)$.

Towards this end, first we readily check that the adjoint representation of $\mathcal{H}_0^i(\mathbb{F})$,

$$\mathrm{Ad}^i \colon \mathcal{H}_0^i(\mathbb{F}) \longrightarrow \mathrm{Aut}(\mathcal{H}^i(\mathbb{F})),$$

is given by

$$\mathrm{Ad}^i\left(g^1, \ldots, g^i\right) = \mathrm{Ad}^1\left(g^1\right) \times \cdots \times \mathrm{Ad}^i\left(g^i\right); \qquad \left(g^1, \ldots, g^i\right) \in \mathcal{H}_0^i(\mathbb{F}),$$

where the operator Ad^i on the right-hand side denotes the usual adjoint representation of $\mathrm{GL}(\mathbb{E}^i)$, for any Banach space $\mathbb{E}^i$; hence,

$$\left[\mathrm{Ad}^i\left(g^1, \ldots, g^i\right)\right]\left(q^1, \ldots, q^i\right) = \left(g^1 \circ q^1 \circ \left(g^1\right)^{-1}, \ldots, g^i \circ q^i \circ \left(g^i\right)^{-1}\right),$$

for every $\left(q^1, \ldots, q^i\right) \in \mathcal{H}^i(\mathbb{F})$.

Similarly, the left Maurer-Cartan differential of $\mathcal{H}_0^i(\mathbb{F})$-valued maps on B (see § 1.2.6) is given by

$$\left(H^1, \ldots, H^i\right)^{-1} d\left(H^1, \ldots, H^i\right) = \left(\left(H^1\right)^{-1} dH^1, \ldots, \left(H^i\right)^{-1} dH^i\right),$$

for every $\left(H^1, \ldots, H^i\right) \in \mathcal{C}^\infty\left(B, \mathcal{H}_0^i(\mathbb{F})\right)$, with $\left(H^k\right)^{-1} dH^k$ denoting the left Maurer-Cartan differential of $H^k \in \mathcal{C}^\infty\left(B, \mathrm{GL}(\mathbb{E}^k)\right)$, $k = 1, \ldots, i$.

With the previous notations, the compatibility condition (1.7.10), adapted to the case of $\boldsymbol{P}(E^i)$, along with (6.5.3) concerning the transition functions of the latter bundle, leads directly to

$$(7.3.2) \qquad \theta_\beta^i = \mathrm{Ad}^i\left(g_{\alpha\beta}^{-1}\right).\theta_\alpha^i + g_{\alpha\beta}^{-1} \cdot dg_{\alpha\beta}$$

on $U_{\alpha\beta}$. Consequently, for each index $i \in \mathbb{N}$, the forms $\{\theta_\alpha\}_{\alpha \in I}$ indeed determine a connection form θ^i on $\boldsymbol{P}(E^i)$, with $\theta^i \in \Lambda^1\left(\boldsymbol{P}(E^i), \mathcal{H}^i(\mathbb{F})\right)$ and local connection forms $\{\theta_\alpha\}_{\alpha \in I}$.

Finally, to prove that θ^j and θ^i are $\left(\boldsymbol{r}^{ji}, h_0^{ji}, \mathrm{id}_B\right)$-related $(j \geq i)$, it is sufficient (by Proposition 1.7.1) to verify the analog of (1.7.20), i.e. equality

$$(7.3.3) \qquad \overline{h_0^{ji}}.\theta_\alpha^j = \mathrm{Ad}^i\left((h_\alpha^i)^{-1}\right)\theta_\alpha^j + (h_\alpha^i)^{-1} dh_\alpha^i,$$

over U_α. Before proving this claim, recall that $\overline{h_0^{ji}} \colon \mathcal{H}^j(\mathbb{F}) \to \mathcal{H}^i(\mathbb{F})$ is the Lie algebra morphism induced by the Lie group homomorphism h_0^{ji}, and $h_\alpha^i \colon U_\alpha \to \mathcal{H}_0^i(\mathbb{F})$ is defined by equality

$$(7.3.4) \qquad \boldsymbol{r}^{ji}(\sigma_\alpha^j(x)) = \sigma_\alpha^i(x) \cdot h_\alpha^i(x),$$

where σ_α^k ($k = j, i; j \geq i$) are the natural local sections of $P(E^k)$. Now, because the local structure of $P(E^k)$ (see Proposition 6.5.1) implies that

$$\sigma_\alpha^k(x) = \left(\left(\tau_x^1\right)^{-1}, \ldots, \left(\tau_x^k\right)^{-1} \right),$$

it follows from (7.3.4) that

$$h_\alpha^i(x) = \left(\mathrm{id}_{\mathbb{E}^1}, \ldots, \mathrm{id}_{\mathbb{E}^i} \right),$$

for all $x \in U_\alpha$; in other words, σ_α^i is constant, thus (7.3.3) reduces to

$$(7.3.5) \qquad\qquad \overline{h_0^{ji}}.\theta_\alpha^j = \theta_\alpha^i.$$

Taking into account that $\overline{h_0^{ji}}$ coincides with

$$h^{ji} \colon \mathcal{H}^j(\mathbb{F}) \longrightarrow \mathcal{H}^i(\mathbb{F}) \colon (g^1, \ldots, g^j) \mapsto (g^1, \ldots, g^i); \qquad j \geq i,$$

(after the identification of the Lie algebra of $\mathcal{H}_0^i(\mathbb{F})$ with $\mathcal{H}^i(\mathbb{F})$), it follows that (7.3.5) is an obvious consequence of (7.3.1). $\qquad\qquad\square$

Based on the preceding result, we define the $\mathcal{H}(\mathbb{F})$-valued 1-form θ on $P(E)$ by setting

$$(7.3.6) \qquad\qquad \theta\left(\left(g^i\right)_{i\in\mathbb{N}} \right) := \varprojlim \left(\theta^i\left(g^1, g^2, \ldots, g^i\right) \right).$$

Note that, after the identification

$$\left(g^i\right)_{i\in\mathbb{N}} \equiv \left(g^1, (g^1, g^2), \ldots (g^1, \ldots, g^i), \ldots\right) = \varprojlim \left((g^1, \ldots, g^i)\right)$$

[see (2.3.9), (2.3.9′)], $(g^1, \ldots, g^i)$ can be thought of as the i-th projection of $(g^i)_{i\in\mathbb{N}}$. Also, according to the comments following Definition 4.2.1, equality (7.3.6) takes the symbolic expression $\theta = \varprojlim \theta^i$.

Besides, using the generalized smooth structure of $P(E)$ discussed in Remarks 6.5.5 (2) and 6.5.5 (3), θ is a generalized smooth form on $P(E)$, so we may write $\theta \in \Lambda^1(P(E), \mathcal{H}(\mathbb{F})))$; therefore, in virtue of Definition 4.2.1 and Theorem 4.2.5, θ may be considered as a connection form on the bundle $P(E)$.

The previous arguments actually prove:

Proposition 7.3.2 *A linear plb-connection* $K = \varprojlim K^i$ *on the plb-vector bundle* $E = \varprojlim E^i$ *determines a (generalized) connection form* θ *on* $P(E)$.

To find the local connection forms of θ, we easily see that the natural sections of $\{\sigma_\alpha\}_{\alpha\in I}$ of $P(E)$, with respect to an open cover $\{U_\alpha\}_{\alpha\in I}$ of B over which all the bundles involved are locally trivial, are given by

$\sigma_\alpha = \varprojlim \sigma_\alpha^i$, where $\{\sigma_\alpha^i\}_{\alpha \in I}$ are the corresponding natural sections of $P(E^i)$, for every $i \in \mathbb{N}$. In particular, for every $x \in U_\alpha$,

$$\sigma_\alpha(x) = \left(\left(\tau_x^i \right)^{-1} \right)_{i \in \mathbb{N}}$$

$$\equiv \left(\left(\tau_x^1 \right)^{-1}, \left(\left(\tau_x^1 \right)^{-1}, \left(\tau_x^2 \right)^{-1} \right), \ldots, \left(\left(\tau_x^1 \right)^{-1}, \ldots, \left(\tau_x^i \right)^{-1} \right), \ldots \right)$$

$$\equiv \left(\sigma_\alpha^1(x), \sigma_\alpha^2(x), \ldots, \sigma_\alpha^i(x), \ldots \right) = \left(\sigma_\alpha^i(x) \right)_{i \in \mathbb{N}}.$$

Accordingly, the local connection forms $\theta_\alpha \in \Lambda^1(U_\alpha, \mathcal{H}(\mathbb{F}))$ ($\alpha \in I$) satisfy, as expected, equality

$$(7.3.7) \qquad\qquad \theta_\alpha = \varprojlim \theta_\alpha^i.$$

Indeed, for every $x \in U_\alpha$, we have:

$$\theta_\alpha(x) = \theta_{\alpha, x} = \left(\sigma_\alpha^* \theta \right)_x = \theta_{\sigma_\alpha(x)} \circ T_x \sigma_\alpha$$

$$= \theta\left(\left(\sigma_\alpha^i(x) \right) \right)_{i \in \mathbb{N}} \circ T_x \sigma_\alpha$$

or, by (7.3.6) and Proposition 3.2.5,

$$= \varprojlim \theta^i \left(\left(\tau_x^1 \right)^{-1}, \ldots, \left(\tau_x^i \right)^{-1} \right) \circ \varprojlim T_x \sigma^i$$

$$= \varprojlim \left(\theta_{\sigma_\alpha^i(x)}^i \circ T_x \sigma^i \right) = \varprojlim \left(\left((\sigma_\alpha^i)^* \theta^i \right)_x \right)$$

$$= \varprojlim \left(\theta_\alpha^i, x \right) = \varprojlim \left(\theta_\alpha^i(x) \right) = \left(\varprojlim \theta_\alpha^i \right)(x),$$

which yields (7.3.7).

Completing Proposition 7.3.2, we prove the following theorem, generalizing the usual association of linear connections on vector bundles with connections on the bundle of frames discussed in § 1.7.4.

Theorem 7.3.3   *There is a bijective correspondence between linear plb-connections on a plb-vector bundle $E = \varprojlim E^i$ and connections on the generalized bundle of frames $P(E)$.*

Proof   We have already seen that a connection $K = \varprojlim K^i$ on E induces a connection form θ on $P(E)$.

Conversely, a connection form $\theta \in \Lambda^1(P(E), \mathcal{H}(\mathbb{F}))$ determines a plb-connection on E: Clearly, $\theta = \varprojlim \theta^i$, where $\theta^i \in \Lambda^1(P(E), \mathcal{H}^i(\mathbb{F}))$). By the general theory of connections on plb-principal bundles, we may write $\theta_\alpha = \varprojlim \theta_\alpha^i$, where $\theta_\alpha = \sigma_\alpha^* \theta$ and $\theta_\alpha^i = (\sigma_\alpha^i)^* \theta^i$, for every $\alpha \in I$ and $i \in \mathbb{N}$. Recall from previous proofs that σ_α and σ_α^i are natural sections of $P(E)$ and $P(E^i)$, respectively.

As a first step to our goal, we check that, for a fixed $i \in \mathbb{N}$, $\left(\theta_\alpha^i \right)_{\alpha \in I}$

are the local connection forms of a connection on $P(E^i)$, inducing in their turn a connection on the ordinary bundle of frames. Indeed, since θ is a connection form on $P(E)$, its local connection forms satisfy the compatibility condition

$$(7.3.8) \qquad \theta_\beta = \mathrm{Ad}\big(g_{\alpha\beta}^{-1}\big).\theta_\alpha + g_{\alpha\beta}^{-1}dg_{\alpha\beta}.$$

Therefore, taking into account the equalities $\theta_\alpha = \varprojlim \theta_\alpha^i$ ($\alpha \in I$) and $g_{\alpha\beta} = \varprojlim g_{\alpha\beta}^i$ (see Proposition 4.1.8), the i-th projection of (7.3.8) leads to the equality

$$(7.3.9) \qquad \theta_\beta^i = \mathrm{Ad}^i\big((g_{\alpha\beta}^i)^{-1}\big).\theta_\alpha^i + \big(g_{\alpha\beta}^i\big)^{-1}dg_{\alpha\beta}^i,$$

which proves that $\big(\theta_\alpha^i\big)_{\alpha\in I}$ determine a connection on $P(E^i)$. On the other hand, for every $x \in U_\alpha$ and every $v \in T_xB$, we have that $\theta_{\alpha,x}^i(v) \in \mathcal{H}^i(\mathbb{F})$; hence, we may write

$$\theta_{\alpha,x}^i(v) = \big(\omega_{\alpha,x}^1(v),\ldots,\omega_{\alpha,x}^i(v)\big),$$

which implies that $\omega_\alpha^k \in \Lambda^1(U_\alpha,\mathcal{L}(\mathbb{E}^k))$ ($k = 1,\ldots,i$) and

$$(7.3.10) \qquad \rho^{jk} \circ \omega_{\alpha,x}^j(v) = \omega_{\alpha,x}^k(v) \circ \rho^{jk}, \qquad j,k = 1,\ldots,i; \, j \geq k,$$

as a consequence of the structure of $\mathcal{H}^i(\mathbb{F})$ [see (5.1.1) in conjunction with (2.3.5)]. Since

$$\omega_{\alpha,x}^i(v) = \mathrm{Pr}_i\big(\theta_{\alpha,x}^i(v)\big); \quad \mathrm{Pr}_i \colon \mathcal{L}(\mathbb{E}^1) \times \cdots \times \mathcal{L}(\mathbb{E}^i) \longrightarrow \mathcal{L}(\mathbb{E}^i),$$

equality (7.3.9) leads to

$$\omega_\beta^i = \mathrm{Ad}\big(g_{\alpha\beta}^i\big).\omega_\alpha^i + \big(g_{\alpha\beta}^i\big)^{-1}dg_{\alpha\beta}^i;$$

more explicitly,

$$\omega_{\beta,x}^i(v) = \big(g_{\alpha\beta}^i(x)\big)^{-1} \circ \omega_\alpha^i \circ g_{\alpha\beta}^i(x) + \big(g_{\alpha\beta}^i\big)^{-1} \circ T_xg_{\alpha\beta}^i(v),$$

for every $x \in U_\alpha$ and $v \in T_xB$. This means that $\{\omega_\alpha^i\}_{\alpha\in I}$ define a connection on the ordinary bundle of frames $P(E^i)$ of E^i.

Now, following the procedure of §1.7.4, and setting

$$(7.3.11) \qquad \big(\Gamma_\alpha^i(y).\lambda\big)(h) := \big(\psi_\alpha^*\omega_\alpha^i\big)_y(h).\lambda,$$

for every $y \in \phi_\alpha(U_\alpha), \lambda \in \mathbb{E}^i, h \in \mathbb{B}$, with $\psi_\alpha = \phi_\alpha^{-1}$, we obtain the family $\big\{\Gamma_\alpha^i \colon \phi_\alpha(U_\alpha) \to \mathcal{L}(\mathbb{E}^i,\mathcal{L}(\mathbb{B},\mathbb{E}^i))\big\}_{\alpha\in I}$. Then (7.3.9) implies the compatibility condition of $\big\{\Gamma_\alpha^i\big\}_{\alpha\in I}$ (for each $i \in \mathbb{N}$), thus the latter are the Christoffel symbols of a linear connection K^i on E^i. Furthermore, translating (7.3.10) in terms of Christoffel symbols, we see that

$$\rho^{ji}\big((\Gamma_\alpha^j(y).\lambda)(h)\big) = \big(\Gamma_\alpha^i(y).\rho^{ji}(\lambda)\big)(h); \quad y \in \phi_\alpha(U_\alpha), \lambda \in \mathbb{E}^i, h \in \mathbb{B},$$

or, equivalently [in virtue of (1.5.8′) and (7.1.3)],

$$\rho^{ji} \circ \Gamma_\alpha^j(y)(\lambda) = \Gamma_\alpha^i(y)\big(\rho^{ji}(\lambda)\big); \quad y \in \phi_\alpha(U_\alpha), \ \lambda \in \mathbb{E}^i,$$

$$\Leftrightarrow \quad r^{ji}\big(\kappa_\alpha^j(y,\lambda)\big) = \kappa_\alpha^i\big(y, \rho^{ji}(\lambda)\big); \quad y \in \phi_\alpha(U_\alpha), \ \lambda \in \mathbb{E}^i,$$

$$\Leftrightarrow \quad r^{ji} \circ \kappa_\alpha^j = \kappa_\alpha^i \circ \big(\mathrm{id}_{\phi_\alpha(U_\alpha)} \times \rho^{ji}\big);$$

that is, we obtain (7.1.6) which guarantees that the linear connections $K^j \equiv \big(\Gamma_\alpha^j\big)_{\alpha \in I}$ and $K^i \equiv \big(\Gamma_\alpha^i\big)_{\alpha \in I}$ are (f^{ji}, id_B)-related. Hence, (7.1.3) is fulfilled and $K := \varprojlim K^i$ is a plb-linear connection on E.

The desired bijectivity is a direct consequence of the association of linear connections with connections on principal bundles of frames, and vice-versa, by relating Christoffel symbols with local connection forms as in (7.3.11) and its inverse in the proof of Lemma 7.3.1.   □

Corollary 7.3.4 *Let $K \equiv \{\Gamma_\alpha\}_{\alpha \in I}$ be an arbitrary linear connection on $E = \varprojlim E^i$. If the Christoffel symbols of K are related with the forms $\{\omega_\alpha^i\}_{\alpha \in I}$ [derived from $\theta \equiv (\theta_\alpha)_{\alpha \in I}$] by (7.3.11), then K is necessarily a plb-linear connection.*

Proof The connection form $\theta \equiv (\theta_\alpha)_{\alpha \in I}$ determines a plb-linear connection $\widetilde{K} = \varprojlim K^i$, with K^i determined by $(\theta_\alpha^i)_{\alpha \in I}$. Since the Christoffel symbols of $\widetilde{K}$ satisfy also (7.3.11), we conclude that $K = \widetilde{K}$. This proves the assertion.   □

8

Geometry of second order tangent bundles

The second order tangent bundle T^2B of a smooth manifold B consists of the equivalence classes of curves in B that agree up to their acceleration, and arises in a natural way in several problems of theoretical physics and differential geometry (cf., for instance, [DG05], [DR82]). However, the vector bundle structure on T^2B is not as straightforward as that of the ordinary (viz. first order) tangent bundle TB of B; in fact, it relies on the choice of a linear connection on B.

Aiming at the reader's convenience, in §§ 8.1.1–8.1.3 we specialize to the ordinary tangent bundle a few facts from the theory of linear connections on vector bundles, exhibited in Chapter 1. In § 8.2 we proceed to the details of the structure of T^2B, for a Banach manifold B. Our next target is to find out the extent of the dependence of the vector bundle structure of T^2B on the choice of the linear connection on B. This naturally leads us to the notion of second order differentials (§ 8.3). With their help, we prove (in § 8.4) that related (or conjugate) connections induce—up to isomorphism—the same vector bundle structure on T^2B.

The last two sections are devoted to the projective limits of second order tangent bundles (§ 8.5), and the generalized second order frame bundle (§ 8.6). Note that, whereas the tangent vectors of curves and velocities of particles naturally form vector bundles, their derivatives, which yield curvatures and accelerations, do not. In order to cover this gap, second order vector bundle structures are constructed for projective limits of Banach modelled manifolds.

8.1 The (first order) tangent bundle in brief

We transcribe a few features of the general theory of linear connections to the particular case of the tangent bundle of a Banach manifold. This will pave the way to the main topics of the present chapter.

8.1.1 Linear connections on manifolds

Let B be a Banach manifold with atlas $\{(U_\alpha, \phi_\alpha)\}_{\alpha \in I}$. We have seen in §1.1.4, in conjunction with §1.4.1, that the structure of the tangent bundle (TB, B, τ_B) of B is determined by the vb-charts $\{(U_\alpha, \phi_\alpha, \Phi_\alpha)\}_{\alpha \in I}$, where the (trivializing) diffeomorphism

$$\Phi_\alpha : TB|_{U_\alpha} \equiv \tau_B^{-1}(U_\alpha) \longrightarrow \phi_\alpha(U_\alpha) \times \mathbb{B}$$

is determined by

$$(8.1.1) \quad \begin{aligned} \Phi_\alpha(v) &= \big(\phi_\alpha(x), \overline{\phi}_\alpha(v)\big) = (\phi_\alpha(x), \Phi_{\alpha,x}(v)) \\ &= \big(\phi_\alpha(x), (\phi_\alpha \circ \gamma)'(0)\big), \end{aligned}$$

for every $x \in B$ and every $v = [(\gamma, x)] \in T_x B$, where γ is a smooth curve in B passing through x.

Analogously, the *double* tangent bundle $(T(TB), TB, \tau_{TB})$ has a local structure induced by the vb-charts $\big(\tau_{TB}^{-1}(TB|_{U_\alpha}), \Phi_\alpha, \widetilde{\Phi}_\alpha\big)$, the diffeomorphism

$$\widetilde{\Phi}_\alpha : \tau_{TB}^{-1}(TB|_{U_\alpha}) \longrightarrow \phi_\alpha(U_\alpha) \times \mathbb{B} \times \mathbb{B} \times \mathbb{B}$$

being given by the analog of (1.5.5), namely

$$(8.1.2) \quad \begin{aligned} \widetilde{\Phi}_\alpha(X) &= \big(\Phi_\alpha(\tau_{TB}(X)), \overline{\Phi}_\alpha(X)\big) = \big(\Phi_\alpha(v), \overline{\Phi}_\alpha(X)\big) = \\ &= \big(\phi_\alpha(x), \overline{\phi}_\alpha(v), \overline{\Phi}_\alpha(X)\big) = (\phi_\alpha(x), (\phi_\alpha \circ \gamma)'(0), (\Phi_\alpha \circ c)'(0)), \end{aligned}$$

for every $X = [(c, v)] \in T_v(TB)$, $v = [(\gamma, x)] \in T_x B$, where c is a smooth curve in TB through v.

For the sake of brevity we shall write

$$(8.1.3) \qquad \Phi_\alpha \equiv (U_\alpha, \phi_\alpha, \Phi_\alpha), \quad \widetilde{\Phi}_\alpha \equiv \big(\tau_{TB}^{-1}(TB|_{U_\alpha}), \Phi_\alpha, \widetilde{\Phi}_\alpha\big).$$

In the same vein, specializing the material of §§1.5.1–1.5.3 to the case of the tangent bundle, we see that a linear connection on TB is a vb-morphism $K : T(TB) \to TB$ whose local representation (with respect to $\widetilde{\Phi}_\alpha$ and Φ_α),

$$K_\alpha = \Phi_\alpha \circ K \circ \widetilde{\Phi}_\alpha^{-1} : \phi_\alpha(U_\alpha) \times \mathbb{B} \times \mathbb{B} \times \mathbb{B} \longrightarrow \phi_\alpha(U_\alpha) \times \mathbb{B},$$

has the form $K_\alpha(x, h, y, k) = (x, k + \kappa_\alpha(x, h).y)$, where the local component

$$\kappa_\alpha \colon \phi_\alpha(U_\alpha) \times \mathbb{B} \to \mathcal{L}(\mathbb{B})$$

is linear with respect to the second variable. As is the custom, K is briefly called a **linear connection on** B.

The Christoffel symbols $\{\Gamma_\alpha\}_{\alpha \in I}$ of K are given by the general formulas of § 1.5.3, according to the form of their range. The compatibility condition of the symbols $\{\Gamma_\alpha \colon \phi_\alpha(U_\alpha) \to \mathcal{L}_2(\mathbb{B}, \mathbb{B}; \mathbb{B})\}_{\alpha \in I}$ reduces to the following variant of (1.5.12):

$$(8.1.4) \quad \begin{aligned} D\phi_{\alpha\beta}(x) \circ \Gamma_\beta(x) &= D^2\phi_{\alpha\beta}(x) + \\ &\quad + \Gamma_\alpha(\phi_{\alpha\beta}(x)) \circ \big(D\phi_{\alpha\beta}(x) \times D\phi_{\alpha\beta}(x)\big), \end{aligned}$$

for every $x \in \phi_\beta(U_{\alpha\beta})$; $\alpha, \beta \in I$, where $\phi_{\alpha\beta} = \phi_\alpha \circ \phi_\beta^{-1}$. This is the case, because now

$$G_{\alpha\beta}(x) = \overline{\phi}_{\alpha,b} \circ \overline{\phi}_{\beta,b}^{-1} = D\phi_{\alpha\beta}(x); \qquad b = \phi_\beta^{-1}(x) \in U_{\alpha\beta}.$$

Analogous conditions hold for the other expressions of the Christoffel symbols.

8.1.2 First order differentials

We fix two Banach manifolds B and B', with respective atlases

$$\{(U_\alpha, \phi_\alpha) \equiv (U_\alpha, \phi_\alpha, \mathbb{B})\}_{\alpha \in I}, \quad \text{and} \quad \{(V_\beta, \phi_\beta) \equiv (V_\beta, \phi_\beta, \mathbb{B}')\}_{\beta \in J}.$$

Let $f \colon B \to B'$ be a smooth map. As in (1.5.21), $f_{\beta\alpha} = \phi_\beta \circ f \circ \phi_\alpha^{-1}$ denotes the local representation of f with respect to the charts (U_α, ϕ_α) and (V_β, ϕ_β) such that $f(U_\alpha) \subseteq V_\beta$. Then the local representation of the ordinary (first order) differential (or tangent map) $Tf \colon TB \to TB'$,

$$\big(\Psi_\beta \circ f \circ \Phi_\alpha^{-1}\big) \colon \phi_\alpha(U_\alpha) \times \mathbb{B} \longrightarrow \phi_\beta(V_\beta) \times \mathbb{B}',$$

relative to the charts Φ_α and Ψ_β [see convention (8.1.3)], is given by the analog of (1.5.20), which now is

$$(8.1.5) \quad \big(\Psi_\beta \circ f \circ \Phi_\alpha^{-1}\big)(x, h) = \big(f_{\beta\alpha}(x), Df_{\beta\alpha}(x).h\big),$$

for every $(x, h) \in \phi_\alpha(U_\alpha) \times \mathbb{B}$.

For a smooth map f as before, the pair of differentials $(T(Tf), Tf)$, with $T(Tf) \colon T(TB) \to T(TB')$, is a vb-morphism between the double tangent bundles $(T(TB), TB, \tau_{TB})$ and $(T(TB'), TB', \tau_{TB'})$. The local representation of $T(Tf)$ [in terms of the vb-charts $\widetilde{\Phi}_\alpha$ and $\widetilde{\Psi}_\alpha$ as in (8.1.2)] is the map

$$\widetilde{\Psi}_\beta \circ T(Tf) \circ \widetilde{\Phi}_\alpha^{-1} \colon \phi_\alpha(U_\alpha) \times \mathbb{B} \times \mathbb{B} \times \mathbb{B} \longrightarrow \phi_\beta(V_\beta) \times \mathbb{B}' \times \mathbb{B}' \times \mathbb{B}',$$

given by [see also (1.5.25)]

$$\left(\widetilde{\Psi}_\beta \circ Tf \circ \widetilde{\Phi}_\alpha^{-1}\right)(x, h, y, k) =$$

(8.1.6)
$$= \left(f_{\beta\alpha}(x), Df_{\beta\alpha}(x).h, Df_{\beta\alpha}(x).y, Df_{\beta\alpha}(x).k + \right.$$
$$\left. + D^2 f_{\beta\alpha}(x)(y, h)\right),$$

for every $(x, h, y, k) \in \phi_\alpha(U_\alpha) \times \mathbb{B} \times \mathbb{B} \times \mathbb{B}$.

8.1.3 Related linear connections on manifolds

Let B and B' be Banach manifolds and $f \colon B \to B'$ a smooth map. Assume that B, B' are equipped with the connection K and K', respectively. We say that K and K' are **f-related** if they are (Tf, f)-related in the sense of § 1.5.6.

The general formula (1.5.27), expressing related connections in terms of local components, now becomes

(8.1.7)
$$\kappa'_\beta\big(f_{\beta\alpha}(x), Df_{\beta\alpha}(x).h\big)\big(Df_{\beta\alpha}(x).y\big) =$$
$$= Df_{\beta\alpha}\big(\kappa_\alpha(x, h).y\big) - D^2 f_{\beta\alpha}(x)(y, h),$$

for every $(x, h, y, k) \in \phi_\alpha(U_\alpha) \times \mathbb{B} \times \mathbb{B} \times \mathbb{B}$. Therefore, if K and K' are *linear* connections, then, in virtue of (1.5.30), they are f-related if and only if

(8.1.8) $\Gamma'_\beta(f_{\beta\alpha}(x)) \circ \big(Df_{\beta\alpha}(x) \times Df_{\beta\alpha}(x)\big) = Df_{\beta\alpha} \circ \Gamma_\alpha(x) - D^2 f_{\beta\alpha}(x),$

for every charts (U_α, ϕ_α) and (V_β, ϕ_β) with $f(U_\alpha) \subseteq V_\beta$, and every $x \in \phi_\alpha(U_\alpha)$. In the preceding equality we have considered Christoffel symbols of the form $\Gamma_\alpha \colon \phi_\alpha(U_\alpha) \to \mathcal{L}_2(\mathbb{B}, \mathbb{B}; \mathbb{B})$ and $\Gamma'_\beta \colon \phi_\beta(V_\beta) \to \mathcal{L}_2(\mathbb{B}', \mathbb{B}'; \mathbb{B}')$.

8.2 Second order tangent bundles

We fix throughout this section a smooth manifold B, modelled on a Banach space $\mathbb{B}$, with atlas $\{(U_\alpha, \phi_\alpha)\}_{\alpha \in I}$. For every $x \in B$,

$$C_x = \{\gamma \colon (-\varepsilon, \varepsilon) \to B \mid \gamma \text{ smooth with } \gamma(0) = x; \ \varepsilon > 0\},$$

obviously denotes the set of smooth curves through x. We define the following equivalence relation in C_x:

(8.2.1) $\gamma_1 \approx_x \gamma_2 \;\Leftrightarrow\; \dot{\gamma}_1(0) = \dot{\gamma}_2(0)$ and $\ddot{\gamma}_1(0) = \ddot{\gamma}_2(0)$;

that is, the curves are **tangent of second order** or **equivalent up to acceleration**

Here, for an arbitrary $\gamma \in C_x$, the curves $\dot{\gamma}$ and $\ddot{\gamma}$ are, respectively, the first and second derivatives (or **velocity** and **acceleration**) of γ, defined by

$$\dot{\gamma} \colon (-\varepsilon, \varepsilon) \longrightarrow TB \colon t \longmapsto T_t\gamma(\partial_t),$$
$$\ddot{\gamma} \colon (-\varepsilon, \varepsilon) \longrightarrow T(TB) \colon t \longmapsto T_t\dot{\gamma}(\partial_t),$$

where $\partial = d/dt$ is the basic vector field of $\mathbb{R}$, thus ∂_t coincides with $1 \in \mathbb{R}$ under the natural identification $T_t\mathbb{R} \equiv \mathbb{R}$.

In accordance with the first order equivalence of curves given in § 1.1.4, we also have that

$$\gamma_1 \approx_x \gamma_2 \;\Leftrightarrow\; \text{there is a chart } (U, \phi) \text{ of } B \text{ such that:}$$
$$(\phi \circ \gamma_1)'(0) = (\phi \circ \gamma_2)'(0) \quad \text{and} \quad (\phi \circ \gamma_1)''(0) = (\phi \circ \gamma_2)''(0), \quad \text{or}$$
$$D(\phi \circ \gamma_1)(0) = D(\phi \circ \gamma_2)(0) \quad \text{and} \quad D^2(\phi \circ \gamma_1)(0) = D^2(\phi \circ \gamma_2)(0).$$

It is a matter of routine checking to see that the latter conditions are independent of the choice of charts at x.

In analogy to the ordinary (first order) tangent space, the **second order tangent space** or **tangent space of order two** at $x \in B$ is defined by

$$T_x^2 B := C_x / \approx_x,$$

while the **second order tangent bundle** or **tangent bundle of order two** of B is

$$T^2 B := \bigcup_{x \in B} T_x^2 B.$$

It is worth noting here that $T_x^2 B$ can be always thought of as a topological vector space isomorphic to $\mathbb{B} \times \mathbb{B}$ via the bijection

$$\overline{\phi}_\alpha^2 \colon T_x^2 B \xrightarrow{\;\cong\;} \mathbb{B} \times \mathbb{B} \colon [(\gamma, x)]_2 \longmapsto \big((\phi_\alpha \circ \gamma)'(0), (\phi_\alpha \circ \gamma)'(0)\big),$$

where $[(\gamma, x)]_2$ stands for the equivalence class of γ with respect to $\approx_x$. However, this structure depends on the choice of the chart (U_α, ϕ_α); hence, we cannot define a vector bundle structure on $T^2 B$ based on the aforementioned bijections, for all $x \in B$. A convenient way to overcome this obstacle is to assume that B is endowed with the additional structure of a linear connection (see §8.1.1).

Theorem 8.2.1 *If B admits a linear connection K, then T^2B becomes a Banach vector bundle with structure group $GL(\mathbb{B} \times \mathbb{B})$.*

Proof Let us denote by $\pi_2 \colon T^2B \to B$ the natural projection given by $\pi_2([(\gamma,x)]_2) = x$. If $\{\Gamma_\alpha \colon \phi_\alpha(U_\alpha) \to \mathcal{L}_2(\mathbb{B}, \mathbb{B}; \mathbb{B})\}_{a \in I}$ are the Christoffel symbols of K with respect to the atlas $\{(U_a, \phi_a)\}_{a \in I}$ of B, then, for each $\alpha \in I$, we define the map $\tau_\alpha^2 \colon \pi_2^{-1}(U_\alpha) \to U_\alpha \times \mathbb{B} \times \mathbb{B}$ with

$$(8.2.2) \quad \begin{aligned} \tau_\alpha^2([(\gamma,x)]_2) := \big(&x, (\phi_\alpha \circ \gamma)'(0), (\phi_\alpha \circ \gamma)''(0)+ \\ &+ \Gamma_\alpha(\phi_\alpha(x))\big((\phi_\alpha \circ \gamma)'(0), (\phi_\alpha \circ \gamma)'(0))\big), \end{aligned}$$

for every $x \in \pi_2^{-1}(U_\alpha)$. The maps $\{\tau_\alpha^2\}_{a \in I}$ are obviously well-defined and injective. They are also surjective, since any element $(x, h, k) \in U_\alpha \times \mathbb{B} \times \mathbb{B}$ can be obtained, via τ_α^2, as the image of the equivalence class $[(\gamma, x)]_2$ of the smooth curve $\gamma = \phi_\alpha^{-1} \circ \sigma \colon (-\varepsilon, \varepsilon) \to B$, where

$$(8.2.3) \qquad \sigma(t) := \phi_\alpha(x) + ht + \frac{1}{2}\big(k - \Gamma_\alpha(\phi_\alpha(x))(h,h)\big)t^2,$$

with ε small enough so that $\sigma\big((-\varepsilon, \varepsilon)\big) \subset \phi_\alpha(U_\alpha)$.

On the other hand, since the diagram

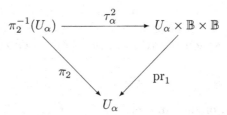

is commutative, it is clear that the pairs $(U_\alpha, \tau_\alpha^2)$, for all $\alpha \in I$, determine on T^2B the structure of a locally trivial fibre bundle, such that the restrictions of τ_α^2 to the fibres, $\tau_{\alpha,x}^2 \colon T_xB \to \mathbb{B} \times \mathbb{B}$, are linear isomorphisms, for every $x \in B$.

The question now is whether the previous structure is that of a vector bundle. The answer will follow from the behaviour of τ_α^2 on overlappings. In this respect, let us consider two trivializations $(U_\alpha, \tau_\alpha^2)$ and (U_β, τ_β^2) of T^2B with $U_{\alpha\beta} = U_\alpha \cap U_\beta \neq \emptyset$. For every $x \in U_{\alpha\beta}$ we check that

$$\big(\tau_{\alpha,x}^2 \circ (\tau_{\beta,x}^2)^{-1}\big)(h,k) = \tau_{\alpha,x}([(\gamma,x)]_2),$$

where, by the analog of (8.2.3), $\gamma \colon (-\varepsilon, \varepsilon) \to U_{\alpha\beta}$ is the smooth curve trough x with

$$\gamma(t) = \phi_\beta^{-1}\left(\phi_\beta(x) + ht + \frac{1}{2}(k - \Gamma_\beta(\phi_\beta(x))(h,h))t^2\right).$$

As a result, setting $\phi_{\alpha\beta} = \phi_\alpha \circ \phi_\beta^{-1}$, (8.2.2) implies that

$$\left(\tau_{\alpha,x}^2 \circ (\tau_{\beta,x}^2)^{-1}\right)(h,k) = \tau_{\alpha,x}([(\gamma,x)]_2) =$$
$$= \left((\phi_\alpha \circ \gamma)'(0), (\phi_\alpha \circ \gamma)''(0) + \Gamma_\alpha(\phi_\alpha(x))((\phi_\alpha \circ \gamma)'(0), (\phi_\alpha \circ \gamma)'(0))\right)$$
$$= \left(D\phi_{\alpha\beta}(\phi_\beta(x)).h, D\phi_{\alpha\beta}(\phi_\beta(x)).k - D\phi_{\alpha\beta}(\phi_\beta(x)) \circ \Gamma_\beta(\phi_\beta(x)).(h,h)\right.$$
$$\left. + D^2\phi_{\alpha\beta}(\phi_\beta)(x)(h,h)\right),$$

or, taking into account the compatibility condition of the Christoffel symbols (8.1.4) (and noting also the difference between the present x and that in the aforementioned equality),

$$(8.2.4) \qquad \left(\tau_{\alpha,x}^2 \circ (\tau_{\beta,x}^2)^{-1}\right)(h,k) = \left(D\phi_{\alpha\beta}(\phi_\beta(x)).h, D\phi_{\alpha\beta}(\phi_\beta(x)).k\right).$$

Therefore, $\tau_{\alpha,x}^2 \circ (\tau_{\beta,x}^2)^{-1} \in \mathcal{L}is(\mathbb{B} \times \mathbb{B})$, and the transition maps

$$T_{\alpha\beta}^2 \colon U_{\alpha\beta} \longrightarrow \mathcal{L}is(\mathbb{B} \times \mathbb{B}) \colon x \longmapsto \tau_{\alpha,x}^2 \circ (\tau_{\beta,x}^2)^{-1}$$

are smooth, because

$$(8.2.5) \qquad T_{\alpha\beta}^2 = \left(D\phi_{\alpha\beta} \circ \phi_\beta\right) \times \left(D\phi_{\alpha\beta} \circ \phi_\beta\right) = T_{\alpha\beta} \times T_{\alpha\beta},$$

where $\{T_{\alpha\beta} \colon U_{\alpha\beta} \to \mathcal{L}is(\mathbb{B})\}_{\alpha,\beta \in I}$ are the transition functrions of TB. Consequently, (T^2B, B, π_2) is indeed a vector bundle over B, of fibre type $\mathbb{B} \times \mathbb{B}$, with structure group $\mathrm{GL}(\mathbb{B} \times \mathbb{B})$. $\qquad\square$

The vb-charts of T^2B are the triplets $(U_\alpha, \phi_\alpha, \Phi_\alpha^2)$, where the diffeomorphisms $\Phi_\alpha^2 \colon \pi_2^{-1}(U_\alpha) \to \phi_\alpha(U_\alpha) \times \mathbb{B} \times \mathbb{B}$ are given by

$$(8.2.6) \qquad \Phi_\alpha^2 := (\phi_\alpha \times \mathrm{id}_\mathbb{B} \times \mathrm{id}_\mathbb{B}) \circ \tau_\alpha^2.$$

A byproduct of the preceding proof is the following:

Corollary 8.2.2 *The second order tangent bundle T^2B is vb-isomorphic to $TB \times_B TB$.*

Proof In virtue of (8.2.4), the cocycles of both bundles coincide. $\qquad\square$

From the preceding constructions it is also clear that the vector space structure of T_x^2B is given by

$$[(\gamma_1,x)]_2 + \lambda\,[(\gamma_2,x)]_2 = \left(\tau_{\alpha,x}^2\right)^{-1}\left(\tau_{\alpha,x}^2([(\gamma_1,x)]_2) + \lambda\,\tau_{\alpha,x}^2[(\gamma_2,x)]_2\right),$$

which is independent of the choice of charts in virtue of (8.2.4)

We conclude this section by proving the converse of Theorem 8.2.1 if the isomorphism of Corollary 8.2.2 has an explicit expression. More precisely, we consider the trivializations $\{(\pi_2^{-1}(U_\alpha), \tau_\alpha^2)\}_{\alpha \in I}$ of T^2B (see the proof of Theorem 8.2.1)), and assume that their restrictions to each fibre over $x \in B$, $\pi_2^{-1}(x) \cong \tau_B^{-1}(x) \times \tau_B^{-1}(x)$, is written in the form

$$(8.2.7) \qquad\qquad \tau_{\alpha,x}^2 \equiv \tau_{\alpha,x} \times \tau_{\alpha,x}',$$

such that

$$(8.2.8) \qquad \tau_{\alpha,x}^2\big([[(\gamma,x)]_2\big) = \big(\tau_{\alpha,x}([[(\gamma,x)]]), \tau_{\alpha,x}'([[(\gamma,x)]])\big),$$

for every smooth curve γ through x. We recall that $\tau_B \colon TB \to B$ is the projection of the tangent bundle, $\tau_\alpha, \tau_\alpha' \colon \tau_B^{-1} \xrightarrow{\ \cong\ } U_\alpha \times \mathbb{B}$ are the trivializations of TB over U_α and $\tau_{\alpha,x}, \tau_{\alpha,x}' \colon T_xB \xrightarrow{\ \cong\ } \mathbb{B}$ the induced isomorphisms on the fibres.

Theorem 8.2.3 *Let B be a smooth manifold modelled on the Banach space $\mathbb{B}$, and assume that the second order tangent bundle T^2B of B has a vector bundle structure of fibre type $\mathbb{B} \times \mathbb{B}$, isomorphic to the fibre product $TB \times_B TB$. Then, under the conditions (8.2.7) and (8.2.7), B admits a linear connection.*

Proof Let $\{(\pi_2^{-1}(U_\alpha), \tau_\alpha^2)\}_{\alpha \in I}$ be the trivializations of T^2B satisfying (8.2.7) and (8.2.8). Then, we may construct a chart (U, ϕ_α) of B such that

$$T_x\phi_\alpha(\dot\gamma(0)) = \big(\mathrm{pr}_1 \circ \tau_\alpha^2\big)([\gamma, x]_2) = \tau_\alpha([[(\gamma,x)]]).$$

Indeed, if (U, ϕ) is an arbitrarily chosen chart of B with $U \subseteq U_\alpha$, we may take $\phi_\alpha := \tau_{\alpha,x} \circ (T_x\phi)^{-1} \circ \phi$. Using charts of the previous form, we define the Christoffel symbols $\Gamma_\alpha \colon \phi_\alpha(U_\alpha) \to \mathcal{L}_2(\mathbb{B}, \mathbb{B}; \mathbb{B})$ of the desired connection by setting

$$\Gamma_\alpha(y)(u, u) := \tau_{\alpha,x}^2([[(\gamma, x)]_2) - (\phi_\alpha \circ \gamma)''(0); \qquad y \in \phi_\alpha(U_\alpha),$$

where γ is a curve of B representing the tangent vector u, with respect to the chart (U_α, ϕ_α). The remaining values of $\Gamma_\alpha(y)$ on elements of the form (u, v), with $u \neq v$, are defined by demanding $\Gamma_\alpha(y)$ to be symmetric bilinear maps. $\{\Gamma_\alpha\}_{\alpha \in I}$ satisfy the necessary compatibility condition (8.1.4) since the trivializations $\{(\pi_2^{-1}(U_\alpha), \tau_\alpha^2)\}_{\alpha \in I}$ agree, via the transition functions of T^2B, on all common areas of their domains. Hence, $\{\Gamma_\alpha\}_{\alpha \in I}$ give rise to a linear connection on B. $\qquad\square$

8.3 Second order differentials

To examine the dependence of the vector bundle structure of the second order tangent bundle on the choice of linear connections on the base space, we need the notion of the differential of second order. To define it, we consider two smooth manifolds B and B' modelled on the Banach spaces $\mathbb{B}$ and $\mathbb{B}'$, respectively, with corresponding atlases $\mathcal{A} = \{(U_\alpha, \phi_\alpha)\}_{\alpha \in I}$ and $\mathcal{B} = \{(V_\beta, \psi_\beta)\}_{\beta \in J}$. We fix two linear connections K and K' on B and B', respectively, with Christoffel symbols $\{\Gamma_\alpha \colon \phi_\alpha(U_\alpha) \to \mathcal{L}_2(\mathbb{B}, \mathbb{B}; \mathbb{B})\}_{\alpha \in I}$ and $\{\Gamma_\beta \colon \phi_\beta(U_\beta) \to \mathcal{L}_2(\mathbb{B}', \mathbb{B}'; \mathbb{B}')\}_{\beta \in J}$. As proved in the preceding section, the pairs (B, K) and (B', K') induce the second order tangent bundles $T^2 B$ and $T^2 B'$ with vector bundle atlases $\mathcal{A}^2 = \{(\pi_{B,2}^{-1}(U_\alpha), \phi_\alpha, \Phi_\alpha)\}_{\alpha \in I}$ and $\mathcal{B}^2 = \{\pi_{B',2}^{-1}(V_\beta), \psi_\beta, \Psi_\beta)\}_{\beta \in J}$.

Definition 8.3.1 If $f \colon B \to B'$ is a smooth map, the **second order differential of** f is the map $T^2 f \colon T^2 B \to T^2 B'$, with $T^2 f([(\gamma, x)]_2) := [(f \circ \gamma, f(x))]_2$.

$T^2 f$ is well-defined. Indeed, if $\gamma_1, \gamma_2 \in C_x$ such that $\gamma_1 \approx_x \gamma_2$, then taking any charts $(U_\alpha, \phi_\alpha) \in \mathcal{A}^2$ and $(V_\beta, \phi_\beta) \in \mathcal{B}^2$ with $x \in U_\alpha$ and $f(U_\alpha) \subseteq V_\beta$ (provided by the smoothness of f at x), we check that

$$(f \circ \gamma_1)(0) = (f \circ \gamma_2)(0) = f(x),$$
$$(f \circ \gamma_1)^\cdot(0) = T_0(f \circ \gamma_1)(\partial_0) = T_x f(\dot\gamma_1(0))$$
$$= T_x f(\dot\gamma_2(0)) = (f \circ \gamma_2)^\cdot(0),$$
$$(f \circ \gamma_1)^{\cdot\cdot}(0) = T_{\dot\gamma_1(0)}(T_{\gamma_1(0)}f)(\ddot\gamma_1(0))$$
$$= T_{\dot\gamma_2(0)}(T_{\gamma_2(0)}f)(\ddot\gamma_2(0))$$
$$= (f \circ \gamma_2)^{\cdot\cdot}(0).$$

The same is true if we use the definition of equivalence classes by means of local charts.

Lemma 8.3.2 *The pair* $(T^2 f, f)$ *determines a fibre bundle morphism between* $(T^2 B, B, \pi_2)$ *and* $(T^2 B', B', \pi_2')$.

Proof Equality $\pi_2' \circ T^2 f = f \circ \pi_2$ is immediately verified. It remains to ascertain the differentiability of $T^2 f$. For an arbitrary $[(\gamma_0, x_0)]_2 \in T^2 B$, the smoothness of f implies the existence of two charts $(U_\alpha, \phi_\alpha) \in \mathcal{A}$ and $(V_\beta, \psi_\beta) \in \mathcal{B}$ with $x_0 \in U_\alpha$ and $f(U_\alpha) \subseteq V_\beta$, inducing the smooth local representation $f_{\beta\alpha} := \psi_\beta \circ f \circ \phi_\alpha^{-1}$ of f. Considering now the corresponding charts $\pi_2^{-1}(U_\alpha) \equiv T^2 B|_{U_\alpha}$ and $(\pi_2')^{-1}(V_\beta) \equiv T^2 B'|_{V_\beta}$ of

T^2B and T^2B', we check that the following diagram is commutative:

$$
\begin{array}{ccc}
T^2B|_{U_\alpha} & \xrightarrow{\quad T^2f \quad} & T^2B'|_{V_\beta} \\
\Big\downarrow{\Phi_\alpha^2} & & \Big\downarrow{\Psi_\beta^2} \\
\phi_\alpha(U_\alpha) \times \mathbb{B} \times \mathbb{B} & \xrightarrow[\Psi_\beta^2 \circ T^2f \circ (\Phi_\alpha^2)^{-1}]{} & \psi_\beta(V_\beta) \times \mathbb{B}' \times \mathbb{B}'
\end{array}
$$

Therefore, for every $(y, h, k) \in \phi_\alpha(U_\alpha) \times \mathbb{B} \times \mathbb{B}$,

$$
\left(\Psi_\beta^2 \circ T^2f \circ (\Phi_\alpha^2)^{-1} \right)(y, h, k) = \left(\Psi_\beta \circ T^2f \right)([(\gamma, x)]_2),
$$

with $[(\gamma, x)]_2$ determined by the smooth curve $\gamma(t) = \phi_\alpha(\sigma(t))$, where

$$
x = \phi_\alpha^{-1}(y); \quad \sigma(t) = y + ht + \frac{1}{2}\left(k - \Gamma_\alpha(k, k)\right)t^2, \quad t \in (\epsilon, \epsilon).
$$

This means that

$$
(\phi_\alpha \circ \gamma)(0) = y, \quad (\phi_\alpha \circ \gamma)'(0) = h, \quad (\phi_\alpha \circ \gamma)''(0) = k - \Gamma_\alpha(k, k).
$$

As a result, the local representation of T^2f turns into

$$
\left(\Psi_\beta^2 \circ T^2f \circ (\Phi_\alpha^2)^{-1} \right)(y, h, k) = \Psi_\beta\left([[(f \circ \gamma, f(x))]]\right) =
$$

$$
= \left(\psi_\beta(f(x)), (\psi_\beta \circ f \circ \gamma)'(0)), (\psi_\beta \circ f \circ \gamma)''(0)) + \right.
$$

$$
\left. + \Gamma_\beta'(\psi_\beta(f(x)))\left((\psi_\beta \circ f \circ \gamma)'(0), (\psi_\beta \circ f \circ \gamma)'(0)\right) \right)
$$

$$
= \left((f_{\beta\alpha} \circ (\phi_\alpha \circ \gamma))(x), Df_{\beta\alpha}(\phi_\alpha(x)).(\phi_\alpha \circ \gamma)'(0) + \right.
$$

$$
+ Df_{\beta\alpha}(\phi_\alpha(x)).(\phi_\alpha \circ \gamma)''(0) +
$$

$$
\left. + D^2 f_{\beta\alpha}(\phi_\alpha(x))\left((\phi_\alpha \circ \gamma)'(0), (\phi_\alpha \circ \gamma)'(0)\right) \right)
$$

$$
= \left(f_{\beta\alpha}(y), Df_{\beta\alpha}(y).h, Df_{\beta\alpha}(y).k - Df_{\beta\alpha}(y).\Gamma_\alpha(y)(h, h) + \right.
$$

$$
\left. + D^2 f_{\beta\alpha}(y).(h, h) + \Gamma_\beta'(f_{\beta\alpha}(y))\left(Df_{\beta\alpha}(y).h, Df_{\beta\alpha}(y).h\right) \right),
$$

which proves the smoothness of T^2f at an arbitrary element of T^2B and concludes the proof. $\qquad\square$

From the previous computations and equality

$$
\left(\Psi_\beta^2 \circ T^2f \circ (\Phi_\alpha^2)^{-1} \right)(y, h, k) =
$$

$$
= \left(f_{\beta\alpha}(y), \left(\Psi_{\beta, f(x)}^2 \circ T_x^2 f \circ (\Phi_{\alpha, x}^2)^{-1} \right)(h, k) \right),
$$

with $x = \phi_\alpha^{-1}(y)$, it follows that

$$\left(\Psi^2_{\beta, f(x)} \circ T^2_x f \circ (\Phi^2_{\alpha, x})^{-1}\right)(h, k) =$$

(8.3.1) $$= \left(Df_{\beta\alpha}(y).h, Df_{\beta\alpha}(y).k - Df_{\beta\alpha}(y).\Gamma_\alpha(y)(h, h) +\right.$$

$$\left. + D^2 f_{\beta\alpha}(y).(h, h) + \Gamma'_\beta(f_{\beta\alpha}(y))\big(Df_{\beta\alpha}(y).h, Df_{\beta\alpha}(y).h\big)\right)$$

Hence, despite the fact that $T^2 f$ is a fibre bundle morphism, the presence of the Christoffel symbols and the derivatives of second order prevents it from being necessarily linear on the fibres, in contrast to the case of ordinary (first order) differentials. Here, related (or conjugate) connections provide an efficient way to surmount this obstacle.

Indeed, if K and K' are f-related, then (8.1.8) transforms (8.3.1) into

(8.3.2) $$\left(\Psi^2_{\beta, f(x)} \circ T^2_x f \circ (\Phi^2_{\alpha, x})^{-1}\right)(h, k) =$$

$$= \big(Df_{\beta\alpha}(\phi_\alpha(x)).h, Df_{\beta\alpha}(\phi_\alpha(x)).k\big),$$

for every $(x, h, k) \in U_\alpha \times \mathbb{B} \times \mathbb{B}$. Therefore, one infers the following:

Corollary 8.3.3   *If the connections K and K' on B and B', respectively, are f-related, then the second order differential $T^2 f \colon T^2 B \to T^2 B'$ is linear on the fibres*

Remark 8.3.4   The assumption that K and K' are f-related, employed in Corollary 8.3.3, is a sufficient, but not necessary, condition ensuring the linearity of $T^2 f$ on the fibres. The optimal (necessary and sufficient) choice would be to assume that the "problematic" part appearing in (8.3.1), namely

$$\mathbb{B} \ni u \longmapsto - Df_{\beta\alpha}(\phi_\alpha(x)).\Gamma_\alpha(\phi_\alpha(x))(u, u) + D^2 f_{\beta\alpha}(\phi_\alpha(x))(u, u) +$$

$$+ \Gamma'_\beta(f_{\beta\alpha}(\phi_\alpha(x)))\big(Df_{\beta\alpha}(\phi_\alpha(x))(u), Df_{\beta\alpha}(\phi_\alpha(x))(u)\big)) \in \mathbb{B}$$

is a linear, not necessarily zero, map. The possible geometric consequences of such an assumption remain an open problem.

Example 8.3.5   We give here some particular examples of related connections in order to clarify a bit the preceding remark.

1) In the case of a constant map f, equality (8.1.8) collapses to a trivial identification of zero quantities, since $f_{\beta\alpha}$ is constant. As a result, all linear connections are related through constant maps.

2) If we consider the map $f = \mathrm{id}_B$, then necessarily $K = K'$. This

agrees with the fact that, in this case, equality (8.1.8) yields

$$D\phi_{\alpha\beta}(\phi_\alpha(x)).\Gamma_\alpha(\phi_\alpha(x))(h,k)) =$$
$$D^2\phi_{\beta\alpha}(\phi_\alpha(x))(h,h) + \Gamma'_\beta(\phi_\beta(x))\big(D\phi_{\beta\alpha}(\phi_\alpha(x))(h), D\phi_{\beta\alpha}(\phi_\alpha(x))(h)\big),$$

which is precisely the compatibility condition (1.8.4) of the Christoffel symbols of a connection on B.

Completing Lemma 8.3.2 and Corollary 8.3.3, we obtain the first main result of this section:

Theorem 8.3.6 *Let T^2B, T^2B' be the second order tangent bundles determined by the pairs (B,K), (B',K'), and let $f\colon B \to B'$ be a smooth map. If the connections K and K' are f-related, then (T^2f, f) is a vector bundle morphism.*

Proof The only thing we need to check is (see, e.g., [Lan99]) the smoothness of the map

$$U_\alpha \longrightarrow \mathcal{L}(\mathbb{B}\times\mathbb{B}, \mathbb{B}'\times\mathbb{B'})\colon x \mapsto \Psi^2_{\beta,f(x)} \circ T^2_x f \circ (\Phi^2_{\alpha,x})^{-1},$$

which is an obvious consequence of (8.3.2)   □

8.4 Connection dependence of second order tangent bundles

As we have seen in §8.2, the vector bundle structure of T^2B depends heavily on the choice of a linear connection K on the base manifold B. The results of the preceding section allow us to estimate the extent of this dependence. In fact, we obtain:

Theorem 8.4.1 *Let K, K' be two linear connections on a Banach manifold B. If f is a diffeomorphism of B such that K and K' are f-related, then the vector bundle structures on T^2B, determined by K and K', are isomorphic.*

Proof In virtue of Theorem 8.3.6, (T^2f, f) is a vb-isomorphism.   □

We introduce the following convenient terminology: Two pairs (B,K) and (B',K'), where the linear connections K, K' are f-related with respect to a diffeomorphism f of B, as in Theorem 8.4.1, are called **equivalent**. The corresponding equivalence class of (B,K) is denoted by $[(B,K)]_f$. Then we have:

Corollary 8.4.2 *The elements of* $[(B, K)]_f$ *determine, up to isomorphism, the same vector bundle structure on* T^2B. *Consequently, the latter structure depends not only on a pair* (B, K) *but also on the entire class* $[(B, K)]_f$.

From the preceding discussion, it is now clear that the vb-identification of two bundles T^2B, T^2B' is not ensured by the existence of a mere diffeomorphism $f\colon B \xrightarrow{\sim} B$. One has to take into account the geometry of B and B', as it is expressed by the linear connections on them.

An interesting question is whether it is possible to characterize the isomorphism classes of second order tangents using *systems of connections*. A brief description of these systems and relevant comments are given in the Appendix (item 6, p. 273).

8.5 Second order Fréchet tangent bundles

As in § 3.1, we consider a plb-manifold M, realized as the projective system of Banach manifolds $\{M^i; \mu^{ji}\}_{i,j\in\mathbb{N}}$ and modelled on the Fréchet space $\mathbb{F}$, where $\mathbb{F} \equiv \varprojlim\{\mathbb{E}^i; \rho^{ji}\}_{i,j\in\mathbb{N}}$, with $\mathbb{E}^i$ being the Banach space models of M^i, respectively, for all $i \in \mathbb{N}$. We shall prove that the corresponding second order tangent bundle of M belongs to the category of plb-vector bundles. The first result towards this direction is:

Proposition 8.5.1 *The second order tangent bundles* $\{T^2M^i\}_{i\in\mathbb{N}}$ *form a projective system with limit set-theoretically isomorphic to* T^2M.

Proof For any pair of indices (i, j) with $j \geq i$, we define the map

$$\mu_2^{ji}\colon T^2M^j \longrightarrow T^2M^i\colon [(\gamma, x)]_2^j \mapsto \left[\left(\mu^{ji} \circ \gamma, \mu^{ji}(x)\right)\right]_2^i,$$

where the brackets $[\cdot, \cdot]_2^j$ and $[\cdot, \cdot]_2^i$ denote the second order equivalence classes of curves in M^j and M^i, respectively, defined in § 8.2. These maps are well-defined, since two equivalent curves γ_1, γ_2 on M^j [in the sense of (8.2.1)] yield

$$T^{(n)}\mu_2^{ji}\left(\gamma_1^{(n)}(0)\right) = \left(\mu_2^{ji} \circ \gamma_1\right)^{(n)}(0) = \left(\mu_2^{ji} \circ \gamma_2\right)^{(n)}(0) = T^{(n)}\mu_2^{ji}\left(\gamma_2^{(n)}(0)\right),$$

for every $n = 0, 1, 2$, where

$$T^{(1)}\mu_2^{ji}\colon TM^j \longrightarrow TM^i \quad \text{and} \quad T^{(2)}\mu_2^{ji}\colon T(TM^j) \longrightarrow T(TM^i)$$

are the first (ordinary) and second order differential of μ_2^{ji}, respectively, and

$$\gamma^{(n)}(0) = \begin{cases} \gamma(0), & n = 0 \\ \dot{\gamma}(0), & n = 1 \\ \ddot{\gamma}(0), & n = 2 \end{cases}$$

The family $\{T^2M^i; \mu_2^{ji}\}_{i,j \in \mathbb{N}}$ determines a projective system, since the necessary conditions

$$\mu_2^{ik} \circ \mu_2^{ji} = \mu_2^{jk}, \qquad j \geq i \geq k$$

are immediately verified from the analogous conditions of the initial connecting morphisms $\{\mu^{ji}\}_{i,j \in \mathbb{N}}$.

On the other hand, the second order differentials of the canonical projections $\mu^i \colon M \to M^i$ of M, namely

$$T^2\mu^i \colon T^2M \longrightarrow T^2M^i \colon [(\gamma, x)]_2 \mapsto \left[(\mu^i \circ \gamma, \mu^i(x))\right]_2^i; \qquad i \in \mathbb{N},$$

can be taken as the canonical projections μ_2^i of the limit space $\varprojlim T^2M^i$; that is, $\mu_2^i := T^2\mu^i$, since the equality $\mu_2^{ji} \circ \mu_2^j = \mu_2^i$ holds for any $j \geq i$. As a result, we obtain the map

$$F := \varprojlim T^2\mu^i \colon T^2M \longrightarrow \varprojlim(T^2M^i):$$

$$[(\gamma, x)]_2 \longmapsto \left(\left[(\mu^i \circ \gamma, \mu^i(x))\right]_2^i\right)_{i \in \mathbb{N}}.$$

This is an injection because $F([(\gamma_1, x)]) = F([(\gamma_2, x)])$ implies

$$T^{(n)}\mu^i\big(\gamma_1^{(n)}(0)\big) = \big(\mu^i \circ \gamma_1\big)^{(n)}(0) = \big(\mu^i \circ \gamma_2\big)^{(n)}(0) = T^{(n)}\mu^i\big(\gamma_2^{(n)}(0)\big),$$

for every $n = 0, 1, 2$; therefore, $\gamma_1^{(n)}(0) = \gamma_2^{(n)}(0)$ $(n = 0, 1, 2)$, since

$$TM \equiv \varprojlim TM^i \quad \text{and} \quad T(TM) \equiv \varprojlim T(TM^i),$$

in virtue of Theorem 3.2.8.

The surjectivity of F is a bit more complicated and goes as follows: Given any element

$$a = \left(\left[(\gamma^i, x^i)\right]_2^i\right)_{i \in \mathbb{N}} \in \varprojlim(T^2M^i),$$

the definition of μ_2^{ji} implies that

$$(8.5.1) \qquad \left[(\mu^{ji} \circ \gamma^j, \mu^{ji}(x^j))\right]_2^i = \left[(\gamma^i, x^i)\right]_2^i; \qquad j \geq i,$$

thus $x = (x^i) \in M = \varprojlim M^i$. Moreover, if $\left(U = \varprojlim U^i, \phi = \varprojlim \phi^i\right)$ is a projective limit chart of M at x, and

$$\left(\pi_M^{-1}(U) = \varprojlim \pi_{M^i}^{-1}(U^i), \Phi = T\phi = \varprojlim T\phi^i\right),$$

$$\left(\pi_{TM}^{-1}\left(\pi_M^{-1}(U) = \varprojlim \pi_{TM^i}^{-1}(\pi_{M^i}^{-1}(U^i))\right), \widetilde{\Phi} = T(T\phi) = \varprojlim T(T\phi^i)\right)$$

are the corresponding charts of TM and $T(TM)$, respectively, it turns out that the equality

$$\left((\phi^i \circ \mu^{ji} \circ \gamma^j)(0), T\phi^i\left((\mu^{ji} \circ \gamma^j)^{\cdot}(0)\right)\right) = \left((\phi^i \circ \gamma^i)(0), T\phi^i\left((\gamma^i)^{\cdot}(0)\right)\right)$$

implies that

$$\rho^{ji}\left((\phi^j \circ \gamma^j)(0)\right), T\phi^i\left(T\mu^{ji}((\gamma^j)^{\cdot}(0))\right)) =$$
$$= \left((\phi^i \circ \gamma^i)(0), T\phi^i((\gamma^i)^{\cdot}(0))\right).$$

Therefore, the vectors $u = \left((\phi^i \circ \gamma^i)(0)\right)_{i \in \mathbb{N}}$ and $v = \left((\phi^i \circ \gamma^i)^{\cdot}(0)\right)_{i \in \mathbb{N}}$ belong to $\mathbb{F} \cong \varprojlim \mathbb{E}^i$.

Similarly, relations (8.5.1) ensure that $(\mu^{ji} \circ \gamma^j)^{\cdot\cdot}(0) = (\gamma^i)^{\cdot\cdot}(0)$ which, by means of the charts of $T(TM)$ defined above, yields

$$T(T\phi^i)\left((\mu^{ji} \circ \gamma^j)^{\cdot\cdot}(0)\right) = T(T\phi^i)\left((\gamma^i)^{\cdot\cdot}(0)\right)$$

or, equivalently,

$$\rho^{ji}\left((\phi^j \circ \gamma^j)^{\cdot\cdot}(0)\right) = (\phi^i \circ \gamma^i)^{\cdot\cdot}(0),$$

for every $j \geq i$. Hence,

$$w = \left((\phi^i \circ \gamma^i)^{\cdot\cdot}(0)\right)_{i \in \mathbb{N}} \in \mathbb{F} \cong \varprojlim \mathbb{E}^i.$$

Considering now the curve h in $\mathbb{F}$ with

$$h(t) = u + tv + \frac{t^2}{2}w; \qquad t \in \mathbb{R},$$

and the curve $\gamma := \phi^{-1} \circ h$ of M (with an appropriate restriction of the domain of h, if necessary), we easily check that

$$(\mu^i \circ \gamma)(0) = \mu^i(x) = x^i = \gamma^i(0),$$

$$(\mu^i \circ \gamma)^{\cdot}(0) = \left((\phi^i)^{-1} \circ \rho^i \circ h\right)^{\cdot}(0) = T((\phi^i)^{-1})\left((\rho^i \circ h)^{\cdot}(0)\right)$$
$$= T((\phi^i)^{-1})(\rho^i(v)) = T((\phi^i)^{-1})\left((\phi^i \circ \gamma^i)^{\cdot}(0)\right)$$
$$= (\gamma^i)^{\cdot}(0),$$

$$(\mu^i \circ \gamma)^{\cdot\cdot}(0) = \left((\phi^i)^{-1} \circ \rho^i \circ h\right)^{\cdot\cdot}(0) = T(T((\phi^i)^{-1}))\left((\rho^i \circ h)^{\cdot\cdot}(0)\right)$$
$$= T(T((\phi^i)^{-1}))(\rho^i(w)) = T(T((\phi^i)^{-1}))\left((\phi^i \circ \gamma^i)^{\cdot\cdot}(0)\right)$$
$$= (\gamma^i)^{\cdot\cdot}(0),$$

for all indices i, j with $j \geq i$. Consequently, the curves $\mu^i \circ \gamma$ and γ^i are (second order) equivalent, thus $F([(\gamma, x)]_2) = \left([(\gamma^i, x^i)]_2^i \right)_{i \in \mathbb{N}} = a$. This completes the surjectivity of F and establishes the desired set-theoretical isomorphism between $T^2 M$ and $\varprojlim(T^2 M^i)$. $\qquad\qquad\square$

Based on the preceding identification of $T^2 M$ and $\varprojlim(T^2 M^i)$, we may define a Fréchet vector bundle structure on $T^2 M$ by means of an appropriate linear connection on M. The problems concerning the structure group of this bundle are surmounted by replacing, once again, the pathological $\mathrm{GL}(\mathbb{F} \times \mathbb{F})$ by the topological group (see also §5.2)

$$\mathcal{H}_0(\mathbb{F} \times \mathbb{F}) := \left\{ (l^i)_{i \in \mathbb{N}} \in \prod_{i=1}^{\infty} \mathrm{GL}(\mathbb{E}^i \times \mathbb{E}^i) : \varprojlim l^i \text{ exists} \right\}.$$

More precisely, we prove the following main result.

Theorem 8.5.2 *If a Fréchet manifold $M = \varprojlim M^i$ is endowed with a linear plb-connection $K = \varprojlim K^i$, then $T^2 M$ is a Fréchet vector bundle over M with structural group $\mathcal{H}_0(\mathbb{F} \times \mathbb{F})$.*

Proof Let $\left\{ (U_\alpha = \varprojlim U_\alpha^i, \phi_\alpha = \varprojlim \phi_\alpha^i) \right\}_{\alpha \in I}$ be a plb-atlas of M and let $\left\{ \Gamma_\alpha^i : \phi_\alpha^i(U_\alpha^i) \to \mathcal{L}_2(\mathbb{E}^i, \mathbb{E}^i; \mathbb{E}^i) \right\}_{\alpha \in I}$ be the Christoffel symbols of each factor linear connection K^i on M^i ($i \in \mathbb{N}$). Then, as proved in Theorem 8.2.1, each $T^2 M^i$ is a Banach vector bundle over M^i of fibre type $\mathbb{E}^i$. The corresponding local trivializations

$$\tau_\alpha^i : \left(\pi_2^i \right)^{-1} (U_\alpha^i) \longrightarrow U_\alpha^i \times \mathbb{E}^i \times \mathbb{E}^i; \qquad \alpha \in I,$$

are given by

$$\tau_\alpha^i([\gamma, x]_2^i) = \left((x, (\phi_\alpha^i \circ \gamma)'(0), (\phi_\alpha^i \circ \gamma)''(0) + \right.$$
$$\left. + \Gamma_\alpha^i(\phi_\alpha^i(x)) \left((\phi_\alpha^i \circ \gamma)'(0) \right), (\phi_\alpha^i \circ \gamma)'(0) \right).$$

Taking into account that the families $\{\mu_2^{ji}\}_{i,j \in \mathbb{N}}$, $\{\mu^{ji}\}_{i,j \in \mathbb{N}}$ and $\{\rho^{ji}\}_{i,j \in \mathbb{N}}$ are connecting morphisms of the projective systems $T^2 M = \varprojlim(T^2 M^i)$, $M = \varprojlim M^i$ and $\mathbb{F} = \varprojlim \mathbb{E}^i$, respectively, we check that the projections $\left\{ \pi_2^i : T^2 M^i \to M^i \right\}_{i \in \mathbb{N}}$ satisfy the equality

$$\mu^{ji} \circ \pi_2^j = \pi_2^i \circ \mu_2^{ji}; \qquad j \geq i,$$

while the trivializations $\{\tau_\alpha^i\}_{i \in \mathbb{N}}$ satisfy

$$(\mu^{ji} \times \rho^{ji} \times \rho^{ji}) \circ \tau_\alpha^j = \tau_\alpha^i \circ \mu_2^{ji}, \qquad j \geq i.$$

As a consequence, conditions (PVB. 1) and (PVB. 2) of Definition 5.2.1

are fulfilled and the family $\{T^2M^i; \mu_2^{ji}\}_{i,j\in\mathbb{N}}$ is a projective system of Banach vector bundles, thus the derived limit $T^2M = \varprojlim(T^2M^i)$ is a (Fréchet) plb-vector bundle. In particular, its projection to the base is

$$\pi_2 = \varprojlim \pi_2^i \colon T^2M \longrightarrow M,$$

and the local trivializations have the form

$$(8.5.2) \qquad \tau_\alpha = \varprojlim \tau_\alpha^i \colon \pi_2^{-1}(U_\alpha) \longrightarrow U_\alpha \times \mathbb{F} \times \mathbb{F},$$

for every $\alpha \in I$. The corresponding transition functions

$$(8.5.3) \qquad T_{\alpha\beta} = \tau_{\alpha,x} \circ \tau_{\beta,x}^{-1}; \qquad \alpha, \beta \in I,$$

can be considered as taking values in the generalized Lie group $\mathcal{H}_0(\mathbb{F}\times\mathbb{F})$, since $T_{\alpha\beta} = \varepsilon \circ T_{\alpha\beta}^*$, where $\{T_{\alpha\beta}^*\}_{\alpha,\beta\in I}$ are the smooth maps

$$T_{\alpha\beta}^* \colon U_\alpha \cap U_\beta \longrightarrow \mathcal{H}_0(\mathbb{F} \times \mathbb{F}) \colon x \mapsto \left(\tau_{\alpha,x}^i \circ \left(\tau_{\beta,x}^i\right)^{-1}\right)_{i\in\mathbb{N}}$$

and ε is the natural inclusion

$$\varepsilon \colon \mathcal{H}_0(\mathbb{F} \times \mathbb{F}) \longrightarrow \mathcal{L}(\mathbb{F} \times \mathbb{F}) \colon (l^i)_{i\in\mathbb{N}} \mapsto \varprojlim l^i.$$

In this way, T^2M is endowed with a vector bundle structure over M, with fibres of type $\mathbb{F} \times \mathbb{F}$ and structure group $\mathcal{H}_0(\mathbb{F} \times \mathbb{F})$. This bundle is isomorphic to $TM \times_M TM$ since both bundles have identical transition functions:

$$T_{\alpha\beta}(x) = \tau_{\alpha,x} \circ \tau_{\beta,x}^{-1} =$$
$$= \left(D(\psi_a \circ \psi_\beta^{-1}) \circ \psi_\beta\right)(x) \times \left(D(\psi_a \circ \psi_\beta^{-1}) \circ \psi_\beta\right)(x). \qquad \square$$

In analogy to Theorem 8.2.3), we obtain the following converse of Theorem 8.5.2.

Theorem 8.5.3 *Let T^2M be a Fréchet plb-vector bundle over M, with structure group $\mathcal{H}_0(\mathbb{F} \times \mathbb{F})$, as in Theorem 8.5.2. If T^2M is isomorphic to $TM \times_M TM$ under the analogs of (8.2.7) and (8.2.8), then M admits a linear connection which can be realized as a projective limit of linear connections.*

Proof We have seen that the vector bundle structure on T^2M is defined by a family of trivializations $\{\tau_\alpha \colon \pi_2^{-1}(U_\alpha) \to U_\alpha \times \mathbb{F} \times \mathbb{F}\}_{\alpha\in I}$, realized as projective limits of the trivializations $\tau_\alpha^i \colon (\pi_2^i)^{-1}(U_\alpha^i) \to U_\alpha^i \times \mathbb{E}^i \times \mathbb{E}^i$ of T^2M^i ($i \in \mathbb{N}$). Consequently, every factor T^2M^i is a vector bundle isomorphic to $TM^i \times_{M^i} TM^i$, and, following the proof of Theorem 8.2.3, M^i admits a linear connection K^i, with Christoffel symbols given by

$$\Gamma^i_\alpha(y)(u^i, u^i) = \tau^{\alpha,i}_{\alpha,x} \left([\gamma^i, x]^i_2\right) - (\phi^i_\alpha \circ \gamma^i)''(0),$$

where γ^i is the curve of M^i representing the vector u^i, with respect to the chart $(U^i_\alpha, \phi^i_\alpha)$. It is now readily checked that $\varprojlim \left(\Gamma^i_\alpha(y^i)(u^i, u^i)\right)$ exists, for every $y = (y^i) \in \phi(U) = \varprojlim \phi^i(U_i)$ and $(u^i) \in \mathbb{F} = \varprojlim \mathbb{E}^i$. This ensures that the connections $\{K^i\}_{i \in \mathbb{N}}$ form a projective system with projective limit the desired linear connection $K = \varprojlim K^i$ on M. $\square$

Regarding the structure group of T^2M and $TM \times_M TM$, we note that, in virtue of Remarks 5.2.6 and Definition 5.2.7 (for $E = TM$), we have the identification

$$\mathcal{H}_0(\mathbb{F}) \times \mathcal{H}_0(\mathbb{F}) \equiv \mathcal{H}_0(\mathbb{F} \times \mathbb{F}).$$

8.6 Second order frame bundles

As already discussed in §§1.6 and 6.5, several geometric properties of vector bundles can be studied by using the corresponding bundle of linear frames. The purpose of the present section is to exhibit the structure of the second order frame bundles associated with Banach and plb-manifolds.

We first consider a Banach manifold B with model $\mathbb{B}$. We further assume that B is endowed with a linear connection K, thus T^2B admits a corresponding vector bundle structure (see Theorem 8.2.1). Then, the **second order frame bundle** of B is defined by

$$(8.6.1) \qquad P^2(B) := \bigcup_{x \in B} \mathcal{L}is(\mathbb{B} \times \mathbb{B}, T^2_x B).$$

(Compare with the ordinary frame bundle defined in §1.6.5.)

Proposition 8.6.1 $P^2(B)$ *is principal bundle over B, with structure group* $\mathrm{GL}(\mathbb{B} \times \mathbb{B})$.

Proof Let $\{(U_\alpha, \phi_\alpha)\}_{\alpha \in I}$ be a smooth atlas of B and the corresponding trivializations $\{\tau^2_\alpha \colon \pi_2^{-1}(U_\alpha) \to U_\alpha \times \mathbb{B} \times \mathbb{B})\}_{\alpha \in I}$ of T^2B (see Theorem 8.2.1). We denote by $\tau^2_{\alpha,x} \in \mathcal{L}is(T^2_x B, \mathbb{B} \times \mathbb{B})$ the restriction of τ^2_α to the fibre over $x \in B$, and by $p \colon P^2(B) \to B$ the projection given by $p(h) = x$, for every $h \in \mathcal{L}is(\mathbb{B} \times \mathbb{B}, T^2_x B)$.

The group $\mathrm{GL}(\mathbb{B} \times \mathbb{B})$ acts on (the right of) $P^2(B)$ in a natural way;

namely, $h \cdot g := h \circ g$, for every $(h, g) \in P^2(B) \times \mathrm{GL}(\mathbb{B} \times \mathbb{B})$. Then the local structure of $P^2(B)$ is obtained by the obvious bijections

$$F_\alpha \colon p^{-1}(U_\alpha) \longrightarrow U_\alpha \times \mathrm{GL}(\mathbb{B} \times \mathbb{B}) \colon h \mapsto \left(p(h), \tau^2_{\alpha, p(h)} \circ h\right), \qquad \alpha \in I.$$

Indeed, each $X_\alpha := p^{-1}(U_\alpha)$, $a \in I$, can be endowed with a smooth manifold structure modelled on the Banach space $\mathbb{B} \times \mathrm{GL}(\mathbb{B} \times \mathbb{B})$. Since $F_\alpha(X_\alpha \cap X_\beta) = U_{\alpha\beta} \times \mathrm{GL}(\mathbb{B} \times \mathbb{B})$ is an open subset of $F_\alpha(X_\alpha)$, it follows that $X_\alpha \cap X_\beta$ is open in X_α. Moreover, the differential structure of $X_\alpha \cap X_\beta$ as a submanifold of X_α coincides with that obtained via X_β because $F_\beta \circ F_\alpha^{-1}$ is the diffeomorphism

$$(F_\beta \circ F_\alpha^{-1})(x, g) = \left(x, \left(\mathrm{comp} \circ (T^2_{\beta\alpha} \times \mathrm{id}_{\mathcal{L}(\mathbb{B} \times \mathbb{B})})\right)(x, g)\right),$$

for every $(x, g) \in U_{\alpha\beta} \times \mathrm{GL}(\mathbb{B} \times \mathbb{B})$. Here,

$$\mathrm{comp} \colon \mathcal{L}(\mathbb{B} \times \mathbb{B}) \times \mathcal{L}(\mathbb{B} \times \mathbb{B}) \longrightarrow \mathcal{L}(\mathbb{B} \times \mathbb{B}) \colon (f, g) \mapsto f \circ g$$

denotes the composition map, and $\{T^2_{\alpha\beta}\}_{\alpha, \beta \in I}$ are the transition functions of $T^2 B$ [see (8.2.5)]. Therefore, by the gluing Lemma ([Bou67, N° 5.2.4]), $P^2(B)$ turns to be a principal Banach bundle, with transition functions

$$G^2_{\alpha\beta} \colon U_{\alpha\beta} \longrightarrow \mathrm{GL}(\mathbb{B} \times \mathbb{B}) \colon x \mapsto F_{\alpha, x} \circ F_{\beta, x}^{-1}.$$

They are related with the transition functions of $T^2 B$ by

$$G^2_{\alpha\beta}(x)(g) = \left(F_{\alpha, x} \circ F_{\beta, x}^{-1}\right)(g) = T^2_{\alpha\beta}(x) \circ g. \qquad \square$$

The bundle $T^2 B$ is associated with $P^2(B)$ by means of the action of $\mathrm{GL}(\mathbb{B} \times \mathbb{B})$ on the right of $P^2(B) \times \mathbb{B} \times \mathbb{B}$,

$$\left(h, (u, v)\right) \cdot g = \left((h \circ g), g^{-1}(u, v)\right).$$

More precisely, we prove the following:

Theorem 8.6.2 *The quotient* $\widetilde{E} = \left(P^2(B) \times \mathbb{B} \times \mathbb{B}\right) / \mathrm{GL}(\mathbb{B} \times \mathbb{B})$ *co-incides, up to isomorphism, with* $T^2 B$.

Proof With the notations of the proof of Proposition 8.6.1, we define the projection

$$\tilde{\pi} \colon \widetilde{E} \longrightarrow B \colon [h, (u, v)] \mapsto p(h),$$

and the local trivializations

$$\widetilde{\Phi}_\alpha \colon \tilde{\pi}^{-1}(U_\alpha) \longrightarrow U_\alpha \times \mathbb{B} \times \mathbb{B} \colon$$

$$[(h, (u, v))] \longmapsto \left(p(h), \left(\mathrm{pr}_2 \circ F_\alpha\right)(h).(u, v)\right) = \left(p(h), \tau^2_{\alpha, p(h)} \circ h\right),$$

for all $\alpha \in I$, where pr_2 denotes the projection of $U_\alpha \times \mathrm{GL}(\mathbb{B}, \mathbb{B})$ to the second factor.

Each $\widetilde{\Phi}_\alpha$ is injective, because

$$\widetilde{\Phi}_\alpha([(h, (u, v))]) = \widetilde{\Phi}_\alpha([(h_1, (u_1, v_1))])$$
$$\Rightarrow \quad p(h) = p(h_1) := x \in B, \ (\tau_{\alpha,x}^2 \circ h)(u, v) = (\tau_{\alpha,x}^2 \circ h_1)(u_1, v_1).$$

Therefore, $h(u, v) = h_1(u_1, v_1)$ and the classes $[h, (u, v)]$, $[h_1, (u_1, v_1)]$ coincide via the isomorphism $g := h_1^{-1} \circ h$.

Also, $\widetilde{\Phi}_\alpha$ is surjective: If $(x, (u, v))$ is an arbitrary element of $U_\alpha \times \mathbb{B} \times \mathbb{B}$, then

$$\widetilde{\Phi}_\alpha([(h, (u, v))]) = (x, (\tau_{\alpha,x}^2 \circ h)(u, v)) = (x, (u, v)),$$

where $h := (\tau_{\alpha,x}^2)^{-1} \in \mathcal{L}is(\mathbb{B} \times \mathbb{B}, T_x^2 B)$.

The restrictions of the previous trivializations to the fibres, namely $\widetilde{\Phi}_{\alpha,x} := \mathrm{pr}_2 \circ \widetilde{\Phi}_\alpha|_{\tilde{\pi}^{-1}(x)}$, $x \in B$, imply that $\widetilde{\Phi}_{\alpha,x} \circ \widetilde{\Phi}_{\beta,x}^{-1} = \tau_{\alpha,x}^2 \circ (\tau_{\beta,x}^2)^{-1}$. Hence, it is a matter of routine checking to verify that $\widetilde{E}$ admits a vector bundle structure with corresponding transition functions

$$\widetilde{T}_{\alpha\beta}(x) = \widetilde{\Phi}_{\alpha,x} \circ \widetilde{\Phi}_{\beta,x}^{-1} = \tau_{\alpha,x}^2 \circ (\tau_{\beta,x}^2)^{-1} = T_{\alpha\beta}^2(x), \qquad x \in U_{\alpha\beta}.$$

The preceding identification of the transition functions of $\widetilde{E}$ and $T^2 B$ establishes a vector bundle isomorphism. $\qquad\square$

Remark 8.6.3 For the sake of completeness, we describe a concrete isomorphism $G: \widetilde{E} \to T^2 B$, by setting $G([(h, (u, v))]) := h(u, v)$.

G is well-defined: If $[(h, (u, v))] = [(h_1, (u_1, v_1))]$, then there exists a $g \in \mathrm{GL}(\mathbb{B} \times \mathbb{B})$ such that $h \circ g = h_1$ and $g(u_1, v_1) = (u, v)$, thus $h(u, v) = h_1(u_1, v_1)$.

G is injective: Equality $G([(h, (u, v))]) = G([(h_1, (u_1, v_1))])$ implies that $h(u, v) = h_1(u_1, v_1)$; therefore, for $g := h_1^{-1} \circ h$,

$$(h_1, (u_1, v_1)) \cdot g = (h_1 \circ g, g^{-1}(u_1, v_1)) = (h, (u, v)).$$

G is surjective: Indeed, for an arbitrary $w \in T_x^2 B$, if (U_α, ϕ_α) is a chart of B at x, and τ_α^2 is the corresponding trivialization of $T^2 B$, then, for

$$h := (\tau_{\alpha,x}^2)^{-1} \in \mathcal{L}is(\mathbb{B} \times \mathbb{B}, T_x^2 B) = P^2(B)_x, \ (u, v) := \tau_{\alpha,x}^2 \in \mathbb{B} \times \mathbb{B},$$

it follows that $G([(h, (u, v))]) = w$.

Finally, we verify that

$$\tau_\alpha^2 \circ G \circ \widetilde{\Phi}_\alpha^{-1} = \mathrm{id}\big|_{U_\alpha \times \mathbb{B} \times \mathbb{B}}; \qquad \alpha \in I,$$

while, for each $\alpha \in I$,

$$\tau^2_{\alpha,x} \circ G_x \circ \widetilde{\Phi}^{-1}_{\alpha,x} = \mathrm{id}_{\mathbb{B} \times \mathbb{B}}, \qquad x \in U_\alpha.$$

The preceding equalities imply immediately conditions (VBM. 1) and (VBM. 2) of § 1.4.3 and ensure that G is a vb-isomorphism.

We want to study now the frame bundle of a plb-manifold

$$M = \varprojlim \{M^i; \mu^{ji}\}_{i,j \in \mathbb{N}},$$

modelled on the Fréchet space $\mathbb{F} \cong \varprojlim \{\mathbb{E}^i; \rho^{ji}\}_{i,j \in \mathbb{N}}$, where $\mathbb{E}^i$ are the Banach space models of M^i ($i \in \mathbb{N}$). To this end, we further assume that

(8.6.2) *the canonical projections $\mu^i \colon M \to M^i (i \in \mathbb{N})$ are surjective.*

(Recall that $\rho^i \colon \mathbb{F} \to \mathbb{E}^i$ are surjective maps, see Remarks 2.3.9.) Then we set

$$\boldsymbol{P}^2(M^i) := \bigcup_{x^i \in M^i} \left\{ (h^k)_{1 \leq k \leq i} \right\},$$

where the maps $h^k \in \mathcal{L}is(\mathbb{E}^k \times \mathbb{E}^k, T^2_{\mu^{ik}(x^i)} M^k)$ satisfy the equalities

$$\mu^{mk}_2 \circ h^m = h^k \circ (\rho^{mk} \times \rho^{mk}), \qquad i \geq m \geq k.$$

(Compare with the bundles $\boldsymbol{P}(E^i)$ of Proposition 6.5.1.)

Proposition 8.6.4 *Each $\boldsymbol{P}^2(M^i)$ ($i \in \mathbb{N}$) is a principal fibre bundle over M^i, with structure group the Banach-Lie group $\mathcal{H}^i_0(\mathbb{F} \times \mathbb{F})$.*

Proof By (8.6.2), for every $x^i \in M^i$ there is an $x \in M$ such that $\mu^i(x) = x^i$. Let $\left(U_\alpha = \varprojlim U^i_\alpha, \phi_\alpha = \varprojlim \phi^i_\alpha \right)$, $a \in I$, be a family of plb-charts of M that cover all the possible selections of elements x^i and x with $x \in U_\alpha$, and let $\left(\pi_2^{-1}(U_\alpha) = \varprojlim (\pi^i_2)^{-1}(U^i_\alpha), \tau_\alpha = \varprojlim \tau^i_\alpha \right)$ be the corresponding trivialization of $T^2 M$ as defined by (8.5.2). Recall that the diffeomorphisms $\tau_\alpha \colon \pi_2^{-1}(U_\alpha) \longrightarrow U_\alpha \times \mathbb{F} \times \mathbb{F}$ induce fibre-wise the linear isomorphisms $\tau_{\alpha,x} \colon T^2_x M \to \mathbb{F} \times \mathbb{F}$, with

$$\tau_{\alpha,x} = \mathrm{pr}_2 \circ \tau_\alpha|_{\pi_2^{-1}(x)} = \varprojlim \tau^i_{\alpha,x} = \varprojlim \left(\mathrm{pr}_2 \circ \tau^i_\alpha|_{(\pi^i_2)^{-1}(x^i)} \right),$$

where now $\mathrm{pr}_2 \colon U_\alpha \times \mathbb{F}^2 \to \mathbb{F}^2$ denotes the projection to the second factor. Next, we define the projections

$$p^i \colon \boldsymbol{P}^2(M^i) \longrightarrow M^i \colon \left(h^1, h^2, \dots, h^i \right) \mapsto x^i,$$

for all $(h^1, h^2, \ldots, h^i)$ with $h^i \in \mathcal{L}is\left(\mathbb{E}^i \times \mathbb{E}^i, T^2_{x^i} M^i\right)$, as well as the action of $\mathcal{H}^i_0(\mathbb{F} \times \mathbb{F})$ on the right of $\boldsymbol{P}^2(M^i)$:

$$\left(h^1, h^2, \ldots, h^i\right) \cdot \left(g^1, g^2, \ldots, g^i\right) := \left(h^1 \circ g^1, h^2 \circ g^2, \ldots, h^i \circ g^i\right).$$

Following the general pattern of the proof of Proposition 8.6.1 (with the appropriate modifications), we define the bijections

$$(8.6.3) \qquad \begin{aligned} \Phi^i_\alpha \colon \left(\boldsymbol{p}^i\right)^{-1}(U^i_\alpha) &\longrightarrow U^i_\alpha \times \mathcal{H}^i_0(\mathbb{F} \times \mathbb{F}) \colon \\ \left(h^1, \ldots, h^i\right) &\longmapsto \left(\boldsymbol{p}^i\left(h^1, \ldots, h^i\right), \tau^1_{\alpha, x^i} \circ h^1, \ldots, \tau^i_{\alpha, x^i} \circ h^i\right), \end{aligned}$$

if $\boldsymbol{p}^i\left(h^1, \ldots, h^i\right) = x^i$.

The injectivity of Φ^i_α ($a \in I$) is obvious, while any $(x^i, g^1, \ldots, g^i) \in U^i_\alpha \times \mathcal{H}^i_0(\mathbb{F} \times \mathbb{F})$ can be written as

$$\Phi^i_\alpha \left(\left(\tau^1_{\alpha, x^i}\right)^{-1} \circ g^1, \ldots, \left(\tau^i_{\alpha, x^i}\right)^{-1} \circ g^i\right),$$

thus showing that Φ^i_α is a surjection. As a result, each $X_\alpha := (\boldsymbol{p}^i)^{-1}(U^i_\alpha)$ can be endowed with the structure of a Banach manifold modelled on $\mathbb{E}^i \times \mathcal{H}^i_0(\mathbb{F} \times \mathbb{F})$. Moreover, for every (α, β),

$$\Phi^i_\alpha(X_\alpha \cap X_\beta) = (U^i_\alpha \cap U^i_b) \times H^i_0(\mathbb{F} \times \mathbb{F})$$

is open in $\Phi^i_\alpha(X_\alpha)$, thus $X_\alpha \cap X_\beta$ is open in X_α.

The differential structure of $X_a \cap X_b$, as an (open) submanifold of X_α, coincides with the one induced by X_β, since $\Phi^i_\beta \circ (\Phi^i_\alpha)^{-1}$ is a diffeomorphism of $(U^i_\alpha \cap U^i_\beta) \times \mathcal{H}^i_0(\mathbb{F} \times \mathbb{F})$. Indeed,

$$\left(\Phi^i_\beta \circ (\Phi^i_\alpha)^{-1}\right)(x^i, g^1, \ldots, g^i) =$$
$$= \left(x^i, \left(\left(\mathrm{comp} \circ (T^k_\beta \times \mathrm{pr}_k)\right)(x^i, g^1, \ldots, g^i)\right)_{k=1,2,\ldots,i}\right),$$

where

$$\mathrm{comp} \colon \mathcal{L}(\mathbb{E}^k) \times \mathcal{L}(\mathbb{E}^k) \longrightarrow \mathcal{L}(\mathbb{E}^k) \colon (f, g) \mapsto f \circ g,$$

$\left(T^k_{\alpha\beta}\right)_{a,b \in I}$ are the transition functions of $T^2 M^k$, and

$$\mathrm{pr}_k \colon \prod_{k=1}^{i} \mathcal{L}(\mathbb{E}^k) \longrightarrow \mathcal{L}(\mathbb{E}^k)$$

is the projection to the k-th factor. Hence, by the gluing Lemma,

$$\boldsymbol{P}^2(M^i) = \bigcup_{a \in I} X_a,$$

is a Banach principal bundle with local trivializations determine by the equality (8.6.3). □

The principal bundles $\{P^2(M^i)\}_{i\in\mathbb{N}}$ form a projective system. More precisely, for every $i, j \in \mathbb{N}^2$ with $j \geq i$, the connecting morphisms are given by

$$r^{ji}\colon P^2(M^j) \longrightarrow P^2(M^i)\colon \left(h^1, h^2, \ldots, h^j\right) \mapsto \left(h^1, h^2, \ldots, h^i\right),$$

obviously satisfying the necessary relations $r^{ik} \circ r^{ji} = r^{jk}$ $(j \geq i \geq k)$. Similarly, the connecting morphisms

$$h_0^{ji}\colon \mathcal{H}_0^j(\mathbb{F} \times \mathbb{F}) \longrightarrow \mathcal{H}_0^i(\mathbb{F} \times \mathbb{F})\colon \left(g^1, g^2, \ldots, g^j\right) \mapsto \left(g^1, g^2, \ldots, g^i\right),$$

satisfy $h_0^{ik} \circ h_0^{ji} = h_0^{jk}$ $(j \geq i \geq k)$. Consequently, $\varprojlim P^2(M^i)$ exists and can be endowed with a principal bundle structure, as asserted by the following result.

Theorem 8.6.5 $P^2(M) := \varprojlim P^2(M^i)$ *is a (Fréchet) plb-principal bundle over M, with structure group $\mathcal{H}_0(\mathbb{F} \times \mathbb{F})$.*

Proof The trivializations (8.6.3) of $P^2(M^i)$, $i \in \mathbb{N}$, form a projective system, because

$$\left(\mu^{ji} \times h^{ji}\right) \circ \Phi_\alpha^j = \Phi_\alpha^i \circ r^{ji}, \qquad j \geq i.$$

Taking into account that

$$\varprojlim U_\alpha^i = U_\alpha \quad \text{and} \quad \varprojlim \left(\mathcal{H}_0^i(\mathbb{F} \times \mathbb{F})\right) = \mathcal{H}_0(\mathbb{F} \times \mathbb{F}),$$

we see that the isomorphisms

(8.6.4) $\Phi_\alpha := \varprojlim \Phi_\alpha^i \colon p^{-1}(U_\alpha) \to U_\alpha \times \mathcal{H}_0(\mathbb{F} \times \mathbb{F}); \qquad a \in I,$

are well-defined, for $p = \varprojlim p^i$. These isomorphisms provide local topological trivializations on $P^2(M)$, which can be also thought of as differential ones under the conventions of §5.1, regarding the generalized differential structure of $\mathcal{H}_0(\mathbb{F} \times \mathbb{F})$. Moreover, each Φ_α is equivariant with respect to the action of $\mathcal{H}_0(\mathbb{F} \times \mathbb{F})$ on the right of $P^2(M) = \varprojlim P^2(M^i)$, induced by their counterpart actions on the factors. Therefore, $P^2(M)$ becomes a plb-principal bundle over M, with structure group $\mathcal{H}_0(\mathbb{F} \times \mathbb{F})$ and transition functions given by

$$g_{\alpha\beta}^2 = \Phi_\alpha \circ \Phi_\beta^{-1} = \varprojlim \left(\Phi_\alpha^i \circ \left(\Phi_\beta^i\right)^{-1}\right); \qquad \alpha, \beta \in I. \qquad \square$$

The principal bundle $\boldsymbol{P}^2(M) := \varprojlim \boldsymbol{P}^2(M^i)$ is called the **generalized second order frame bundle** of the Fréchet manifold $M = \varprojlim M^i$.

The preceding definition is a natural generalization of the standard second order frame bundle within the framework of Fréchet manifolds. As a matter of fact, if M is a *Banach* manifold, modelled on $\mathbb{E}$, then the projective systems $\{M^i; \mu^{ji}\}$ and $\{\boldsymbol{P}^2(M^i); r^{ji}\}$ reduce to the trivial ones $\{M; \mathrm{id}_M\}$ and $\{\boldsymbol{P}^2(M); \mathrm{id}_{P^2(M)}\}$, respectively, where $P^2(M)$ is the second order frame bundle of M defined by (8.6.1). Analogously, $\mathcal{H}_0(\mathbb{F} \times \mathbb{F})$ coincides with $\mathrm{GL}(\mathbb{E} \times \mathbb{E})$. Therefore, in this case, $\boldsymbol{P}^2(M)$ reduces precisely to the second order frame bundle of M, i.e. $\boldsymbol{P}^2(M) = \varprojlim \boldsymbol{P}^2(M^i) = P^2(M)$.

It is worth noticing that the generalized second order frame bundle $\boldsymbol{P}^2(M)$ of $M = \varprojlim M^i$, apart from being a projective limit of Banach principal bundles, it can be expressed also in a form analogous to that of its factors; namely,

$$\boldsymbol{P}^2 M \equiv \bigcup_{x \in M} \left\{ (h^i)_{i \in \mathbb{N}} \big| h^i \in \mathcal{L}is(\mathbb{E}^i \times \mathbb{E}^i, T^2_{\mu^i(x)} M^i) : \varprojlim h^i \text{ exists} \right\},$$

in view of the identification $((h^1, h^2, ..., h^i))_{i \in \mathbb{N}} \equiv (h^i)_{i \in \mathbb{N}}$ [see also equalities (2.3.9) and (2.3.9')].

Analogously to the case of ordinary frame bundles, the second order tangent bundle $T^2 M$ of a plb-manifold $M = \varprojlim M^i$ is associated with the generalized bundle $\boldsymbol{P}^2 M$. More precisely, we prove the following:

Theorem 8.6.6 *The quotient space* $\widetilde{\boldsymbol{E}} := (\boldsymbol{P}^2 M \times (\mathbb{F} \times \mathbb{F})) / \mathcal{H}_0(\mathbb{F} \times \mathbb{F})$, *derived from the action of* $\mathcal{H}_0(\mathbb{F} \times \mathbb{F})$ *on the right of* $\boldsymbol{P}^2 M \times (\mathbb{F} \times \mathbb{F})$,

$$\big((h^i), (u^i, v^i) \big)_{i \in \mathbb{N}} \cdot (g^i)_{i \in \mathbb{N}} := \left((h^i \circ g^i), (g^i)^{-1} (u^i, v^i) \right)_{i \in \mathbb{N}},$$

is isomorphic to the second order tangent bundle $T^2 M$.

Proof Let $\widetilde{\pi}$ be the natural projection of $\widetilde{\boldsymbol{E}}$ to the base manifold M,

$$\widetilde{\pi} \colon \widetilde{\boldsymbol{E}} \longrightarrow M \colon [((h^i), (u^i, v^i))] \mapsto \boldsymbol{p}((h^i)) := \varprojlim (\boldsymbol{p}^i(h^i)),$$

if $\boldsymbol{p}$ and $\boldsymbol{p}^i$ are the projections of the bundles $\boldsymbol{P}^2(M)$ and $\boldsymbol{P}^2(M^i)$, respectively (see the relative notations in the proofs of Proposition 8.6.4 and Theorem 8.6.5). Working with an open plb-covering

$$\left\{ \left(U_\alpha = \varprojlim U^i_\alpha, \phi_\alpha = \varprojlim \phi^i_\alpha \right) \right\}_{\alpha \in I}$$

of M and the corresponding trivializations [see also (8.6.2)]

$$\left\{ \Phi_\alpha = \varprojlim \Phi^i_\alpha \colon \boldsymbol{p}^{-1}(U_\alpha) \to U_\alpha \times \mathcal{H}_0(\mathbb{F} \times \mathbb{F}) \right\}_{\alpha \in I}$$

of $P^2(M)$, we define the maps

$$\widetilde{\Phi}_\alpha : \widetilde{\pi}^{-1}(U_\alpha) \longrightarrow U_\alpha \times \mathbb{F} \times \mathbb{F} :$$

$$[((h^i), (u^i, v^i))] \longmapsto \Big(p((h^i)), \Phi_{\alpha,2}((h^i))((u^i, v^i)) \Big),$$

for all $\alpha \in I$, where $\Phi_{\alpha,2}$ denotes the projection of Φ_α to $\mathcal{H}_0(\mathbb{F} \times \mathbb{F})$.
Each $\widetilde{\Phi}_\alpha$ is injective: First observe that

$$\widetilde{\Phi}_\alpha([((h^i), (u^i, v^i))]) = \widetilde{\Phi}_\alpha([((\underline{h}^i), (\underline{u}^i, \underline{v}^i))]) =$$

$$\Rightarrow p((h^i)) = p((\underline{h}^i)) = (x^i) \in M = \varprojlim M^i.$$

Since (h^i), $(\underline{h}^i) \in P^2 M$, the limits $\varprojlim h^i$ and $\varprojlim \underline{h}^i$ can be defined.
Moreover, in virtue of (8.6.3),

$$\widetilde{\Phi}_\alpha([((h^i), (u^i, v^i))]) = \widetilde{\Phi}_\alpha([((\underline{h}^i), (\underline{u}^i, \underline{v}^i))])$$

$$\Rightarrow \Phi_{\alpha,2}([((h^i), (u^i, v^i))]) = \Phi_{\alpha,2}([((\underline{h}^i), (\underline{u}^i, \underline{v}^i))])$$

$$\Rightarrow \Big((\tau^i_{\alpha,x^i} \circ h^i)(u^i, v^i) \Big)_{i\in\mathbb{N}} = \Big((\tau^i_{\alpha,x^i} \circ \underline{h}^i)(\underline{u}^i, \underline{v}^i) \Big)_{i\in\mathbb{N}}.$$

As a result,

$$\Big(h^i(u^i, v^i) \Big)_{i\in\mathbb{N}} = \Big(\underline{h}^i(\underline{u}^i, \underline{v}^i) \Big)_{i\in\mathbb{N}}.$$

Considering now the isomorphisms $g^i := (\underline{h}^i)^{-1} \circ h^i \in \mathrm{GL}(\mathbb{E}^i \times \mathbb{E}^i)$, we
obtain their limit $\varprojlim g^i$. Thus, with respect to the action of $g := (g^i) \in \mathcal{H}_0(\mathbb{F} \times \mathbb{F})$, we conclude that $[((h^i), (u^i, v^i))] = [((\underline{h}^i), (\underline{u}^i, \underline{v}^i))]$.

Also, $\widetilde{\Phi}_\alpha$ is surjective. Indeed, for any $((x^i), (u^i, v^i))_{i\in\mathbb{N}} \in U_\alpha \times \mathbb{F} \times \mathbb{F}$,
each linear isomorphism $h^i := (\tau^i_{\alpha,x^i})^{-1}$ belongs to $\mathcal{L}is(\mathbb{E}^i \times \mathbb{E}^i, T^2_{x^i} M^i)$.
It is now readily checked that $(h^i)_{i\in\mathbb{N}} \in P^2(M)$. Therefore,

$$\widetilde{\Phi}_\alpha([((h^i), (u^i, v^i))]) = ((x^i), ((\tau^i_\alpha \circ h^i)(u^i, v^i))) = ((x^i), (u^i, v^i)),$$

thus proving the desired surjectivity.

Since, for every $\alpha, \beta \in I$, $\widetilde{\Phi}_{\alpha,x} \circ \widetilde{\Phi}_{\beta,x}^{-1} = \tau_{\alpha,x} \circ \tau_{\beta,x}^{-1} \in \mathcal{L}is(\mathbb{F} \times \mathbb{F})$, where
$\tau_{\alpha,x} = \varprojlim \tau^i_{\alpha,x^i}$, with $x = (x^i)$, we easily verify that $\widetilde{E}$ is indeed a vector
bundle with local trivializations $\{(U_\alpha, \widetilde{\Phi}_\alpha)\}_{\alpha \in I}$.

On the other hand, if we denote by $\{T_{\alpha\beta}\}_{\alpha\beta \in I}$ the transition functions
of $\widetilde{E}$, equality (8.5.3) implies that

$$T_{\alpha\beta}(x) = \widetilde{\Phi}_{\alpha,x} \circ \widetilde{\Phi}_{\beta,x}^{-1} = \tau_{\alpha,x} \circ \tau_{\beta,x}^{-1} = T^2_{\alpha\beta}(x); \qquad x \in U_{\alpha\beta}$$

where now $\{T^2_{\alpha\beta}\}_{\alpha,\beta \in I}$ are the transition functions of $T^2 M$. Therefore,,
$\widetilde{E}$ is isomorphic to $T^2 M$. $\qquad\square$

As in the case of Theorem 8.6.2 (see also Remark 8.6.3), we can describe a concrete vb-isomorphism between $\widetilde{E}$ and T^2M, namely

$$G: \widetilde{E} \longrightarrow T^2M: \big[((h^i),(u^i,v^i))\big] \mapsto \big(h^i(u^i,v^i)\big).$$

The range of G is indeed T^2M because, for every $\big[((h^i),(u^i,v^i))\big] \in \widetilde{E}$, the family $\big\{h^i: \mathbb{E}^i \times \mathbb{E}^i \to T^2_{x^i}M^i\big\}_{i\in\mathbb{N}}$ is a projective system, and $(u^i),(v^i)$ belong to $\mathbb{F} = \varprojlim \mathbb{E}^i$; therefore, $\big(h^i(u^i,v^i)\big)_{i\in\mathbb{N}} \in T^2M = \varprojlim T^2M^i$. G is well-defined, since $\big[((h^i),(u^i,v^i))\big] = \big[((\underline{h}^i),(\underline{u}^i,\underline{v}^i))\big]$ implies the existence of an element $(g^i) \in \mathcal{H}_0(\mathbb{F} \times \mathbb{F})$, such that

$$h^i \circ g^i = \underline{h}^i, g^i(\underline{u}^i,\underline{v}^i) = (u^i,v^i); \qquad i \in \mathbb{N},$$

$$\big(h^i(u^i,v^i)\big)_{i\in\mathbb{N}} = \big(h^i(g^i(\underline{u}^i,\underline{v}^i))\big)_{i\in\mathbb{N}} = \big(\underline{h}^i(\underline{u}^i,\underline{v}^i)\big)_{i\in\mathbb{N}}.$$

The next step is to show that G is a bijection. First we see that $G\big(\big[((h^i),(u^i,v^i))\big]\big) = G\big(\big[((\underline{h}^i),(\underline{u}^i,\underline{v}^i))\big]\big)$ yields $h^i(u^i,v^i) = \underline{h}^i(\underline{u}^i,\underline{v}^i)$, for all $i \in \mathbb{N}$. Also, the isomorphisms $g^i := (\underline{h}^i)^{-1} \circ h^i$ $(i \in \mathbb{N})$ define a projective limit, as a consequence of the existence of the projective limits of both families (h^i) and $(\underline{h}^i)$. Because

$$((\underline{h}^i),(\underline{u}^i,\underline{v}^i)) \cdot (g^i) = ((\underline{h}^i \circ g^i),(g^i)^{-1}(\underline{u}^i,\underline{v}^i)) = ((h^i),(u^i,v^i)),$$

we conclude that $\big[((h^i),(u^i,v^i)\big] = \big[((\underline{h}^i),(\underline{u}^i,\underline{v}^i))\big]$, which proves that G is injective.

The surjectivity of G goes as follows: If $(w^i) \in T^2M = \varprojlim T^2M^i$, where $w^i \in T^2_{x^i}M^i$ and $x = (x^i) \in M = \varprojlim M^i$, then, taking a plb-chart $\big(U_\alpha = \varprojlim U^i_\alpha, \phi_\alpha = \varprojlim \phi^i_\alpha\big)$ of M and the corresponding chart

$$\big(U_\alpha = \varprojlim U^i_\alpha, \Phi_\alpha = \varprojlim \Phi^i_\alpha\big)$$

of T^2M, the linear isomorphism [see also (8.5.2)]

$$\tau_{\alpha,x} := \mathrm{pr}_2 \circ \Phi_\alpha\big|_{\pi_2^{-1}(x)} = \varprojlim \tau^i_{\alpha,x^i}: T^2_xM \xrightarrow{\simeq} \mathbb{F} \times \mathbb{F}$$

is defined. Hence,

$$\big((\tau^i_{\alpha,x^i})^{-1}\big)_{i\in\mathbb{N}} \in P^2M, \quad \big(\tau^i_{\alpha,x^i}(w^i)\big)_{i\in\mathbb{N}} = (u^i,v^i)_{i\in\mathbb{N}} \in \mathbb{F} \times \mathbb{F},$$

and

$$G\left(\left[\big((\tau^i_{\alpha,x^i})^{-1}\big)_{i\in\mathbb{N}},(u^i,v^i)_{i\in\mathbb{N}}\right]\right) =$$
$$= \left(\big(\tau^i_{\alpha,x^i}\big)^{-1}(u^i,v^i)\right)_{i\in\mathbb{N}} = (w^i)_{i\in\mathbb{N}}.$$

To prove that G is a vector bundle isomorphism it suffices to ensure that it preserves the trivializations of the two bundles involved. This is verified by the following equalities, for every $\alpha \in I$:

$$(\Phi_\alpha \circ G)\left(\left[\left((h^i)_{i\in\mathbb{N}}, (u^i, v^i)_{i\in\mathbb{N}}\right)\right]\right) =$$

$$= \Phi_\alpha\left(\left(h^i(u^i, v^i)\right)_{i\in\mathbb{N}}\right)$$

$$= \left(\Phi_\alpha^i\left(h^i(u^i, v^i)\right)\right)_{i\in\mathbb{N}}$$

$$= \left(p\left((h^i)_{i\in\mathbb{N}}\right), \left((\tau_\alpha^i \circ h^i)(u^i, v^i)\right)_{i\in\mathbb{N}}\right)$$

$$= \widetilde{\Phi}_\alpha\left(\left[\left((h^i)_{i\in\mathbb{N}}, (u^i, v^i)_{i\in\mathbb{N}}\right)\right]\right).$$

Appendix: Further study

In the following list we select a few problems presenting a research interest, naturally complementing the main ideas and methods expounded in the course of this work.

1. State and prove a Chern-Weil theorem in the framework of projective limit principal and vector bundles. The later case may be of particular interest because the ordinary general linear group $\mathrm{GL}(\mathbb{F})$ of the Fréchet fibre type of the bundle should be replaced by the generalized Lie group $\mathcal{H}_0(\mathbb{F})$ with Lie algebra $\mathcal{H}(\mathbb{F})$.

2. Investigate the possibility to prove the analog of the holonomy theorem (concerning the algebra of the holonomy group). The Banach case of the classical result of W. Ambrose and I.M. Singer is studied in [Mag04] and [Vas78(b)].

3. Many aspects of the ordinary geometry of projective limits of Lie groupoids and Lie algebroids can be extended to the Fréchet framework by using the methods of this book. A first attempt towards this direction appears in [Cab12].

4. Investigate and develop an approach to infinite-dimensional symplectic geometry within the projective limit framework.

5. The development of a general theory of G-structures, where G is a projective limit Fréchet-Lie group, also may be interesting. Applications of this approach would provide a Fréchet bundle with the analogs of many classical structures.

6. Another point of view of the totality of linear connections on a smooth finite dimensional manifold M that is worth noting here, is that of *system of connections* devised by Mangiarotti and Modugno ([MM83], [Mod87]). Namely, whereas the function space of all linear connections is infinite dimensional, even in the case of finite dimensional M, it is

possible to obtain a finite dimensional bundle-representation of all linear connections on M in terms of such a connection system. Indeed, there exists a unique universal connection of which every connection in the system of connections is a pullback. A similar relation holds between the corresponding universal curvature and the curvatures of the connections of the system (in this respect see Cordero, Dodson and deLeon [CDL89]). This is a different representation of an object similar to that introduced by Narasimhan and Ramanan [NR61, NR63] for G-bundles, also allowing a proof of Chern-Weil's theorem (cf. [CDL89], [Gar72], [KN69]).

The system of all linear connections on a finite dimensional manifold M has a representation on the tangent bundle via the **system space**

$$C_T = \{\alpha \otimes j\gamma \in T^*M \otimes_M JTM \mid j\gamma\colon TM \to TTM \text{ projects onto } \mathrm{id}_{TM}\}.$$

Here we view id_{TM} as a section of $T^*M \otimes TM$, which is a subbundle of $T^*M \otimes TTM$, with local expression $dx^\lambda \otimes \partial_\lambda$.

The fibred morphism for the system C_T is

$$\xi_T\colon C_T \times_M TM \longrightarrow JTM \subset T^*M \otimes_{TM} TTM,$$
$$(\alpha \otimes j\gamma, \nu) \longmapsto \alpha(\nu)j\gamma.$$

In coordinates (x^λ) on M and (y^λ) on TM,

$$\xi_T = dx^\lambda \otimes (\partial_\lambda - \gamma^i_\lambda\, \partial_i) = dx^\lambda \otimes (\partial_\lambda - y^j\, \Gamma^i_{j\lambda}\, \partial_i).$$

Each section of $C_T \to M$, such as $\tilde{\Gamma}\colon M \to C_T\colon (x^\lambda) \to (x^\lambda, \gamma_{\mu\vartheta})$, determines the unique linear connection $\Gamma = \xi_T \circ (\tilde{\Gamma} \circ \pi_T, \mathrm{id}_{TM})$ with Christoffel symbols $\Gamma^\lambda_{\mu\vartheta}$.

On the fibred manifold $\pi_1\colon C_T \times_M TM \to C_T$, the universal connection is given by:

$$\Lambda_T : C_T \times_M TM \longrightarrow J(C_T \times_M TM) \subset T^*C_T \otimes T(C_T \times_M TM)$$
$$(x^\lambda, v^\lambda_{\mu\nu}, y^\lambda) \longmapsto [(X^\lambda, V^\lambda_{\mu\nu}) \to (X^\lambda, V^\lambda_{\mu\nu}, Y^\mu V^\lambda_{\mu\nu} X^\nu)].$$

In coordinates,

$$\Lambda_T = dx^\lambda \otimes \partial_\lambda + dv^a \otimes \partial_a + y^\mu v^i_{\mu\nu}\, dx^\nu \otimes \partial_i.$$

Explicitly, each $\tilde{\Gamma} \in Sec(C_T/M)$ gives an injection $(\tilde{\Gamma} \circ \pi_T, \mathrm{id}_{TM})$, of TM into $C_T \times TM$, which is a section of π_1, Γ coincides with the restriction of Λ_T to this section:

$$\Lambda_{T\,|(\tilde{\Gamma}\circ\pi_T, I_{TM})TM} = \Gamma,$$

and the universal curvature of the connection Λ is given by:

$$\Omega_T = d_{\Lambda_T} \Lambda_T : C_T \times_M TM \to \wedge^2(T^* C_T) \otimes_{TM} V(TM).$$

So, here the universal curvature Ω_T has the coordinate expression:

$$\Omega_T = \frac{1}{2} \left(y^k v_{k\lambda}^j \, \partial_j y^m v_{m\mu}^i \, dx^\lambda \wedge dx^\mu + 2 \, \partial_a y^m v_{m\mu}^i \, dx^a \wedge dx^\mu \right) \otimes \partial_i.$$

For more details of the corresponding universal calculus see Dodson and Modugno [DM86]. In the case of Riemannian and pseudo-Riemannian manifolds, Canarutto and Dodson [CD85] used systems of principal connections to establish certain incompleteness stability properties; Del Riego and Dodson [DD88] established certain topological and universal properties of sprays and Lie algebras, obtaining associated completeness criteria.

The system of linear connections provides a bundle framework in which choices of linear connection may be made, and hence vector bundle structures on T^2M are determined. It would be interesting to extend to infinite dimensional Banach and even Fréchet manifolds the systems of connections approach of [MM83] and its associated universal connections [CDL89]. That might make it possible to characterize further the isomorphism classes of second order tangent bundles. Specifically, since all connections are pullbacks of the universal connection, in what way are these pullbacks characterized through the conjugacy classes?

7. A natural question is the study of tangent bundles of higher order, extending the properties and results of Chapter 8.

8. The possibility to obtain metrics on the projective limits of Hilbert or Finsler bundles could give many important results related with the metric and leading to a wealth of applications.

9. It would be interesting to extend the finite-dimensional results of Dodson and Vazquez-Abal [DV90, DV92] for bundle projection and lifting of harmonicity. This could apply to the infinite dimensional case of a projective limit Hilbert manifold $\mathbb{E} = \lim_{\infty \leftarrow s} \mathbb{E}^s$, of a projective system of smooth Hilbert manifolds $\mathbb{E}^s$, consisting of sections of a tensor bundle over a smooth compact finite dimensional Riemannian manifold (M, g). Such spaces arise in geometry and physical field theory and they have many desirable properties but it is necessary to establish existence of the projective limits for various geometric objects. Smolentsev [Smo07] gives a detailed account of the underlying theory we need—that paper is particularly concerned with the manifold of

sections of the bundle of smooth symmetric 2-forms on M and its critical points for important geometric functionals. We may mention the work of Bellomonte and Trapani [BT11]who investigated directed systems of Hilbert spaces whose extreme spaces are the projective and the inductive limit of a directed contractive family of Hilbert spaces. Via the volume form on (n-dimensional compact) (M, g) a weak induced metric on the space of tensor fields is $\int_M g(X, Y)$ but there is a stronger family [Smo07] of inner products on $\mathbb{E}^s$, the completion Hilbert space of sections. For sections X, Y of the given tensor bundle over M we put

$$(X, Y)_{g,s} = \sum_{i=0}^{s} \int_M g(\nabla^{(i)} X, \nabla^{(i)} Y) \quad s \geq 0.$$

Then the limit $\mathbb{E} = \lim_{\infty \leftarrow s} \mathbb{E}^s$ with limiting inner product $g_{\mathbb{E}}$ is a Fréchet space with topology independent of the choice of metric g on M. In particular it is known, for example see Omori [Omo70, Omo97] and Smolentsev [Smo07], that the smooth diffeomorphisms $f : (M, g) \to (M, g)$ form a strong projective limit Lie group Diff(M) modelled on the projective limit manifold

$$\Gamma(TM) = \lim_{\infty \leftarrow s} \Gamma^s(TM)$$

of smooth sections of the tangent bundle. Moreover, the curvature and Ricci tensors are equivariant under the action of Diff(M) which yields the Bianchi identities as consequences.

10. A large body of work has concerned the properties of operators, particularly linear ones, on infinite dimensional spaces, because of their importance in representing ordinary differential equations on function spaces. The fibred equivalent of such problems yields partial differential equations on manifolds. A common problem in applications of linear models is the characterization and solution of continuous linear operator equations on Hilbert, Banach and Fréchet spaces. However, there are many open problems. For example, it is known that for a continuous linear operator T on a separable Banach space $\mathbb{E}$ there may be no non-trivial closed subspace nor non-trivial closed subset such that $A \subset \mathbb{E}$ with $TA \subset A$ ([Rea88, AAB94, Enf87]). Atzmon [Atz83] provided what turned out to be an example of an operator on a somewhat artificial space but Goliński [Gol12, Gol13] gave an operator without invariant subset and showed that operators without nontrivial invariant subspaces exist on $C^\infty(K)$ for an arbitrary smooth compact manifold K.

There has been substantial interest from differential geometry in hypercyclic operators, whose iterations generate dense subsets, a review is given in [Dod12]. A continuous linear operator T on a topological vector space $\mathbb{E}$ is *cyclic* if for some $f \in \mathbb{E}$ the span of $\{T^n f, n \geq 0\}$ is dense in $\mathbb{E}$. On finite-dimensional spaces there are many cyclic operators but no hypercyclic operators. The operator T is called *chaotic* [GEM11] if it is hypercyclic and its set of periodic points is dense in $\mathbb{E}$. Each operator on the Fréchet space of analytic functions on $\mathbb{C}^N$, which commutes with all translations and is not a scalar multiple of the identity, is chaotic. On the Fréchet space $\mathbb{H}(\mathbb{C})$ of functions analytic on $\mathbb{C}$, the translation by a fixed nonzero $\alpha \in \mathbb{C}$ is hypercyclic and so is the differentiation operator $f \mapsto f'$. All infinite-dimensional separable Banach spaces admit hypercyclic operators but finite dimensional spaces do not. In particular a Fréchet space admits a hypercyclic operator if and only if it is separable and infinite-dimensional, Ansari [Ans97], and the spectrum of a weakly hypercyclic operator must meet the unit circle, eg. Dilworth and Troitsky [Dil03]. Such contexts indicate a number of areas of potential application of the projective limit approach to study the corresponding Fréchet differential geometry and its operators.

11. The quantum completion $\bar{\mathcal{A}}$ of the *space of connections* in a manifold can be viewed as the set of morphisms from the groupoid of the edges to the compact gauge group $\bar{\mathcal{G}}$ and Velhinho [Vel02] used this to generalize the description of the gauge-invariant quantum configuration space $\overline{\mathcal{A}/\mathcal{G}}$. The definition of functional calculus on $\overline{\mathcal{A}/\mathcal{G}}$ relies on the representation of $\bar{\mathcal{A}}$ and $\overline{\mathcal{A}/\mathcal{G}}$ as *projective limits* of families of finite-dimensional compact manifolds, offering means to construct measures and vector fields. This groupoid approach is applied in [Vel02] to show that the quotient of $\bar{\mathcal{A}}$ by the gauge group is homeomorphic to $\overline{\mathcal{A}/\mathcal{G}}$, clarifying the relation between the two spaces. See also Thiemann [Thi07] for a detailed discussion of $\overline{\mathcal{A}/\mathcal{G}}$, the development of measures and functional calculus on projective limits and their role in loop quantum gravity.

12. In his thesis of 1967, Ebin [Ebin67] gave a detailed study of the space of Riemannian metrics and in particular those on a compact smooth manifold M; he gave a summary of those results in [Ebin68]. The Riemannian metrics $\mathcal{M} \subseteq C^\infty(S^2 T^*)$, being sections of the bundle of smooth symmetric covariant 2-tensors on M induce positive definite bilinear forms on the tangent spaces of M. The group $\mathcal{D}$ of diffeomorphisms of M with C^∞ topology acts on the right of $C^\infty(S^2 T^*)$ by

pull-back

$$A : C^\infty(S^2 T^*) \times \mathcal{D} \to C^\infty(S^2 T^*) : (\gamma, \eta) \mapsto \eta^*(\gamma)$$

and $\mathcal{M}$ is invariant under this action. The restriction of this action is a right action on $\mathcal{M}$

$$A : \mathcal{M} \times \mathcal{D} \to \mathcal{M} : (\gamma, \eta) \mapsto \eta^*(\gamma).$$

because $(\xi\eta)^*\gamma = \eta^*\xi^*(\gamma)$. For $\lambda \in \mathcal{M}$ the isotropy group of λ is

$$I_\lambda = \{\eta \in \mathcal{D} | \eta^*(\lambda) = \lambda\}.$$

Theorem ([[Ebin67, Ebin68]]). *A induces a homeomorphism of* $\mathcal{D}/I_\gamma$ *onto the orbit* O_γ *of* $\mathcal{D}$ *through* γ *by* $\eta I_\gamma \mapsto \eta^*(\gamma)$. *Then there is a subspace* $S \subseteq \mathcal{M}$ *containing* γ *with the following properties:*
(1) $A(I_\gamma, S) = S$,
(2) If $\eta \in \mathcal{D}$ *with* $\eta^*(S) \bigcap S \neq \emptyset$, *then* $\eta \in I_\gamma$,
(3) There exists a neighbourhood U *of the identity coset in* $\mathcal{D}/I_\gamma$ *and a local section* $\chi : U \to \mathcal{D}$ *such that*

$$F : U \times S \to \mathcal{M} : (u, s) \mapsto \chi(u)^*(s)$$

is a homeomorphism onto a neighbourhood of γ.

Then for all $\lambda \in \mathcal{M}$ sufficiently near γ there exists $\eta \in \mathcal{D}$ such that $I_\lambda \subseteq \eta I_\gamma \eta^{-1}$. Now, $\mathcal{M}$ is locally like a Fréchet space and does not directly admit a manifold structure so in order to prove the theorem Ebin ([Ebin67]) enlarged it to belong to the Sobolev space $H^s(S^2 T^*)$ used by Palais [Pa65], denoted the enlarged space by $\mathcal{M}^s$. Next he enlarged $\mathcal{D}$ to $\mathcal{D}^{s+1}$, the H^{s+1} maps $M \to M$ with H^{s+1} inverses, as in Palais [Pa68]. A $\mathcal{D}^{s+1}$-invariant Riemannian metric was constructed on $\mathcal{M}^s$. Then using the normal bundle to O_γ^s and the exponential map on $\mathcal{M}^s$ the properties (3) of S in the above theorem were established.

We know that the space $\mathcal{M}$ can be represented as a projective limit of Banach manifolds from the earlier papers [Gal96, Gal98], as described in Chapter 4. Ghahremani-Gol and Razavi [GGR13] used this projective limit of Banach manifolds to represent the infinite dimensional space of Riemannian metrics on a compact manifold. Using the work of Galanis and coworkers on existence and uniqueness of integral curves of a projective system of vector fields, described here in Chapter 2, they applied it to the parabolic partial differential equations for the Ricci flow and its integral curves. They found short-time solutions that are locally unique and, in particular, showed that the Ricci flow curve starting from an Einstein metric is not a geodesic. This work has a number of potential lines of further development because of the importance of spaces of metrics in many physical applications.

13. Information geometry and in particular quantum information theory increasingly make use of infinite-dimensional spaces of probability density functions and geometrical constructions thereon. Current research in this context may be found in [NBh12, NBa13, Nil14] and there are many applications in the sciences. Banach manifolds can be used to represent an infinite dimensional family of probability density functions of exponential type, however, unfortunately the all-important likelihood function is not continuous on this manifold and Fukumizu [Fuk05] turned to the weaker topology arising from reproducing kernel Hilbert space structure, cf. also [SM94]. It seems likely that these developments may benefit from the projective limit approach to geometric structures on infinite-dimensional manifolds.

References

[AA96] P.L. Antonelli and M. Anastasiei: *The Differential Geometry of La-grangians which Generate Sprays*. Kluwer, Dordrecht, 1996.

[AAB94] Y.A. Abramovich, C.D. Aliprantis and O. Burkinshaw: *Invariant Ssubspace theorems for positive operators*. J. Functional Analysis **14**(1994), 95-111.

[ABB09] S. Agethen, K.D. Bierstedt and J. Bonet: *Projective limits of weighted (LB)-spaces of continuous functions*. Arch. Math. (Basel) **92** (2009), 384-398.

[ADG07] M. Aghasi, C.T.J. Dodson, G.N. Galanis and A. Suri: *Infinite dimensional second order ordinary differential equations via T^2M*. Nonlinear Analysis **67** (2007), 2829–2838.

[ADG08] M. Aghasi, C.T.J. Dodson, G.N. Galanis and A. Suri: *Conjugate connections and differential equations on infinite dimensional manifolds*. VIII International Colloquium on Differential Geometry, Santiago de Compostela, 7–11 July 2008. World Scientific, Hackensack, NJ, 227–236, 2009.

[AI92] A. Ashtekar and C.J. Isham: *Representations of the holonomy algebras of gravity and non-abelian gauge theories*. Class. Quantum Grav. **9** (1992), 1433–1467.

[AIM93] P.L. Antonelli, R.S. Ingarden and M.S. Matsumoto: *The Theory of Sprays and Finsler Spaces with Applications in Physics and Biology*. Kluwer, Dordrecht, 1993.

[AL94] A. Ashtekar and J. Lewandowski: *Representation Theory of Analytic Holonomy C^*-algebras, Knots and Quantum Gravity*. J.C. Baez ed., Oxford University Press, Oxford, 1994.

[AJP97] S. Albeverio, J. Jost, S. Paycha, S. Scarlatti: *A mathematical introduction to string theory. Variational problems, geometric and probabilistic methods*. London Mathematical Society Lecture Note Series **225**. Cambridge University Press, Cambridge, 1997.

[AL95] A. Ashtekar and J. Lewandowski: *Differential geometry on the space of connections via graphs and projective limits*. J. Geom. Phys **17** (1995), 191–230.

[AM99] M. C. Abbati and A. Manià: *On differential structure for projective limits of manifolds*. J. Geom. Phys. **29** (1999), 35-63.

[AMR88] R. Abraham, J.E. Marsden and T. Ratiu: *Manifolds, Tensor Analysis, and Applications* (2nd edition). Springer, New York, 1988.

[AR67] R. Abraham and J. Robbin: *Transversal Mappings and Flows*. Benjamin, New York, 1967.

[AO09] R.P. Agarwal and D. O'Regan: *Fixed point theory for various classes of permissible maps via index theory*. Commun. Korean Math. Soc. **24** (2009), 247-263.

[APS60] W. Ambrose, R.S. Palais and I.M. Singer: *Sprays*. Anais da Academia Brasieira de Ciencias **32** (1960), 1–15.

[Ans97] S.I. Ansari. *Existence of hypercyclic operators on topological vector spaces*. J. Funct. Anal. **148** (1997), 384-390.

[Atz83] A. Atzmon: *An operator without invariant subspaces on a nuclear Fréchet space*. Ann. of Math. **117** (1983), 669–694.

[Bae93] J.C. Baez: *Diffeomorphism-invariant generalized measures on the space of connections modulo gauge transformations*. Proceeding of the Conference on Quantum Topology, Manhattan, Kansas, March 24-28, 1993.

[BB03] K.D. Bierstedt and J. Bonet: *Some aspects of the modern theory of Fréchet spaces*. RACSAM. Rev. R. Acad. Cienc.Exactas Fís. Nat. Ser. A Mat. **97** (2003), 159–188.

[BDH86] E. Behrends, S. Dierolf, and P. Harmand: *On a problem of Bellenot and Dubinsky*. Math. Ann. **275** (1986), 337–339.

[Ble81] D. Bleecker: *Gauge Theory and Variational Principles*. Addison-Wesley, Reading, Massachusetts, 1981.

[BM09] F. Bayart and E. Matheron: *Dynamics of Linear Operators*. Cambridge Tracts in Mathematics 179, Cambridge University Press, Cambridge, 2009.

[BP75] C. Bessaga and A. Pełczyński: *Selected topics in infinite dimensional topology*. PWN, Warszawa 1975.

[BT11] G. Bellomonte and C. Trapani: *Rigged Hilbert spaces and contractive families of Hilbert spaces*. Monatsh. Math. **164** (2011), 271-285.

[BMM89] J. Bonet, G. Metafune, M. Maestre, V.B. Moscatelli and D. Vogt: *Every quojection is the quotient of a countable product of Banach spaces* (Istanbul, 1988), 355–356. NATO Adv. Sci. Inst. Ser. C Math. Phys. Sci. 287, Kluwer, Dordrecht, 1989.

[Bou67] N. Bourbaki: *Varietés différentielles et analytiques. Fascicule de résultats*, §§1–7. Hermann, Paris, 1967.

[Bou71] N. Bourbaki: *Varietés différentielles et analytiques. Fascicule de résultats*, §§8–15. Hermann, Paris, 1971.

[Bou72] N. Bourbaki: *Groupes et algèbres de Lie*. Chapitres 2–3, Paris, 1972.

[Cab12] P. Cabau: *Strong projective limits of Banach Lie algebroids*. Portugal. Math. **69**,1 (2012), 1–21.

[Car67(a)] H. Cartan: *Calcul Différentiel*. Hermann, Paris, 1971.

[Car67(b)] H. Cartan: *Formes Différentielles*. Hermann, Paris, 1967.

[CD85] D. Canarutto and C.T.J. Dodson: *On the bundle of principal connections and the stability of b-incompleteness of manifolds*. Math. Proc. Camb. Phil. Soc. **98** (1985), 51–59.

[CDL89] L.A. Cordero, C.T.J. Dodson and M.de Leon: *Differential Geometry of Frame Bundles*. Kluwer, Dordrecht, 1989.

[CEO09] R. Choukri, A. El Kinani, and M. Oudadess: *On some von Neumann topological algebras*. Banach J. Math. Anal. **3** (2009), 55-63.

[CK03] A. Constantin and B. Kolev: *Geodesic flow on the diffeomorphism group of the circle*. Comm. Math. Helv. **78** (2003), 787–804.

[Dal00] H.G. Dales: *Banach algebras and automatic continuity*. London Mathematical Society Monographs, New Series **24**. Oxford Science Publications, The Clarendon Press, Oxford University Press, New York, 2000.

[DD88] L. Del Riego and C.T.J. Dodson: *Sprays, universality and stability*. Math. Proc. Camb. Phil. Soc. **103** (1988), 515–534.

[DEF99] P. Deligne, P. Etingof, D.S. Freed, L.C. Jeffrey, D. Kazhdan, J.W. Morgan, D.R. Morrison, E. Witten (Editors): *Quantum fields and strings: a course for mathematicians*, Vol. 1, 2. Material from the Special Year on Quantum Field Theory held at the Institute for Advanced Study, Princeton NJ, 1996–1997. AMS, Providence RI, 1999.

[Die72] J. Dieudonné: *Treatise on Analysis, Vol. III*. Academic Press, New York, 1972.

[Dil03] S.J. Dilworth and V.G. Troitsky: *Spectrum of a weakly hypercyclic operator meets the unit circle*. Contemporary Mathematics **321** (2003), 67-69.

[DG04] C.T.J. Dodson and G.N. Galanis: *Second order tangent bundles of infinite dimensional manifolds*. J. Geom. Phys. **52** (2004), 127–136.

[DG05] C.T.J. Dodson and G.N. Galanis: *Bundles of acceleration on Banach manifolds*. Nonlinear Analysis **63** (2005), 465-471.

[DGV05] C.T.J. Dodson, G.N. Galanis and E. Vassiliou: *A generalized second order frame bundle for Fréchet manifolds*. J. Geom. Phys. **55** (2005), 291–305.

[DGV06] C.T.J. Dodson, G.N. Galanis and E. Vassiliou: *Isomorphism classes for Banach vector bundle structures of second tangents*. Math. Proc. Camb. Phil. Soc. **141** (2006), 489–496.

[DM86] C.T.J. Dodson and M. Modugno: *Connections over connections and universal calculus*. Proc. VI Convegno Nazionale di Relativita General e Fisica Della Gravitazione Florence, 10-13 October 1984, 89–97, Eds. R. Fabbri and M. Modugno, Pitagora Editrice, Bologna, 1986.

[DP97] C.T.J. Dodson and P.E. Parker: *A User's Guide to Algebraic Topology*. Kluwer, Dordrecht, 1997.

[Dod88] C.T.J. Dodson: *Categories, Bundles and Spacetime Topology* (2nd edition). Kluwer, Dordrecht, 1988.

[Dod12] C.T.J. Dodson: *A review of some recent work on hypercyclicity*. Invited paper, Workshop celebrating the 65 birthday of L.A. Cordero, Santiago de Compostela, June 27-29, 2012. Balkan J. Geom. App. (2014), in press.

[Dom62] P. Dombrowski: *On the geometry of the tangent bundle*. J. Reine und Angewante Math. **210** (1962), 73–88.

[Dow62] C.H. Dowker: *Lectures on Sheaf Theory*. Tata Inst. Fund. Research, Bombay, 1962.

[DR82] C.T.J. Dodson and M.S. Radivoiovici: *Tangent and Frame bundles of order two*. Anal. Ştiint. Univ. "Al. I. Cuza" **28** (1982), 63-71.

[DRP95] L. Del Riego and P.E. Parker: *Pseudoconvex and disprisoning homogeneus sprays*, Geom. Dedicata **55** (1995), no. 2, 211–220.

[Dub79] E. Dubinsky: *The structure of nuclear Fréchet spaces*. Lecture Notes in Mathematics **720**, Springer-Verlag, Heidelberg, 1979.

[Dug75] J. Dugundji: *Topology*. Allyn and Bacon, Boston, 1975.

[Dup78] J.L. Dupont: *Curvature and Characteristic Classes*. Lecture Notes in Mathematics **640**, Springer-Verlag, Heidelberg, 1978.

[DV90] C.T.J. Dodson and M.E. Vazquez-Abal: *Harmonic fibrations of the tangent bundle of order two*. Boll. Un. Mat. Ital. 7 4-B (1990) 943-952.

[DV92] C.T.J. Dodson and M.E. Vazquez-Abal: *Tangent and frame bundle harmonic lifts*. Mathematicheskie Zametki of Acad. Sciences of USSR **50**, 3, (1991), 27-37 (Russian). Translation in *Math. Notes* **3-4** (1992), 902908.
http://www.maths.manchester.ac.uk/kd/PREPRINTS/91MatZemat.pdf

[DZ84] S. Dierolf and D. N. Zarnadze: *A note on strictly regular Fréchet spaces*. Arch. Math. **42** (1984), 549–556.

[Ebin67] D.G. Ebin: *On the space of Riemannian metrics*. Doctoral Thesis, Massachusetts Institute of Technology, Cambridge, Mass., 1967.

[Ebin68] D.G. Ebin: On the space of Riemannian metrics. *Bull. Amer. Math. Soc.* **74** (1968), 1001-1003.

[EE67] C.J. Earle and J. Eells Jr.: *Foliations and fibrations*. J. Diff. Geom. 1 (1967), 33–41.

[Eel66] J. Eells Jr.: *A setting for global analysis*. Bull. A.M.S **72** (1966), 751–807.

[Eli67] H.I. Eliasson: *Geometry of manifolds of maps*. J. Diff. Geom. 1 (1967), 169–174.

[EM70] D.G. Ebin and J. Marsden: *Groups of diffeomorphisms and the motion of an incompressible fluid*. Ann. of Math. **92** (1970), 101–162.

[Enf87] P. Enflo: *On the invariant subspace problem for Banach spaces*. Acta Mathematica (1987), 213–313.

[FK72] P. Flaschel and W. Klingenberg: *Riemannsche Hilbert-mannigfaltigkeiten. Periodische Geodatische*. Lecture Notes in Mathematics **282**, Springer-Verlag, Heidelberg, 1972.

[Fuk05] K. Fukumizu: *Infinite dimensional exponential families by reproducing kernel Hilbert spaces*. Proc. 2nd International Symposium on Information Geometry and its Applications, December 12-16, 2005, Tokyo, pp. 324-333.

[FW96] L. Frerick and J. Wengenroth: *A sufficient condition for vanishing of the derived projective limit functor*. Archiv der Mathematik **67** (1996), 296–301.

[Gal96] G. Galanis: *Projective limits of Banach-Lie groups*. Period. Math. Hungar. **32** (1996), 179–191.

[Gal97(a)] G. Galanis: *On a type of linear differential equations in Fréchet spaces*. Ann. Scuola Norm. Sup. Pisa **24** (1997), 501–510.

[Gal97(b)] G. Galanis: *On a type of Fréchet principal bundles over Banach bases.* Period. Math. Hungar. **35** (1997), 15–30.

[Gal98] G. Galanis: *Projective limits of Banach vector bundles.* Portugal. Math. **55** (1998), 11-24.

[Gal04] G. Galanis: *Differential and geometric structure for the tangent bundle of a projective limit manifold.* Rend. Seminario Matem. Padova **112** (2004), 104–115.

[Gal07] G. Galanis: *Universal connections in Fréchet principal bundles.* Period. Math. Hungar. **54** (2007), 1–13.

[GP05] G. Galanis and P. Palamides: *Nonlinear differential equations in Fréchet spaces and continuum cross-sections.* Anal. Ştiint. Univ. "Al. I. Cuza" **51** (2005), 41–54.

[Gar72] P. L. Garcia: *Connections and 1-jet fibre bundles.* Rend. Sem. Mat. Univ. Padova **47** (1972), 227–242.

[GEM11] K-G. Grosse-Erdmann and A.P. Manguillot: *Linear Chaos.* Universitext, Springer, London, 2011.

[GGR13] H. Ghahremani-Gol, A. Razavi: *Ricci flow and the manifold of Riemannian metrics.* Balkan J. Geom. App. **18** (2013,) 20-30.

[GHV73] W. Greub, S. Halperin and R. Vanstone: *Connections, Curvature and Cohomology, Vol. II.* Academic Press, N. York, 1973.

[God73] R. Godement: *Topologie Algébrique et Théorie des Faisceaux* (3ème édition). Hermann, Paris, 1973.

[Gol12] M. Goliński: *Invariant subspace problem for classical spaces of functions* J. Funct. Anal. **262** (2012), 1251–1273.

[Gol13] M. Goliński: *Operator on the space rapidly decreasing functions with all non-zero vectors hypercyclic* Adv. Math. **244** (2013), 663–677.

[Gro58] A. Grothendieck: *A general theory of fibre spaces with structural sheaf* (2nd edition). Kansas Univ., 1958.

[GV98] G. Galanis and E. Vassiliou: *A Floquet-Liapunov theorem in Fréchet spaces.* Ann. Scuola Norm. Sup. Pisa **27** (1998), 427–436.

[Ham82] R.S. Hamilton: *The inverse function theorem of Nash and Moser.* Bull. Amer. Math. Soc. **7** (1982), 65–222.

[Har64] P. Hartman: *Ordinary Differential Equations.* Wiley, New York, 1964.

[Hir66] F. Hirzebruch: *Topological Methods in Algebraic Geometry.* Springer-Verlag, New York, 1966.

[Hye45] D.H. Hyers: *Linear topological spaces.* Bull. Amer. Math. Soc. **51** (1945), 1–24.

[Jar81] H. Jarchow: *Locally Convex Spaces.* Teubner, Stuttgart, 1981.

[KJ80] S.G. Kreĭn and N.I. Yatskin: *Linear Differential Equations on Manifolds.* Voronezh Gos. Univ., Voronezh, 1980 (in Russian).

[KLT09] J. Kakol, M.P. Lopez Pellicer and A.R. Todd: *A topological vector space is Fréchet-Urysohn if and only if it has bounded tightness.* Bull. Belg. Math. Soc. Simon Stevin **16** (2009), 313-317.

[KM90] A. Kriegl and P.W. Michor: *The Convenient Setting of Global Analysis.* Mathematical Surveys and Monographs **53** (1997), American Mathematical Society.

[KM97] A. Kriegl and P.W. Michor: *A convenient setting for real analytic mappings*. Acta Math. **165** (1990),105–159.

[KN68] S. Kobayashi and K. Nomizu: *Foundations of Differential Geometry, Vol. I.* Interscience, New York, 1968.

[KN69] S. Kobayashi and K. Nomizu: *Foundations of Differential Geometry, Vol. II.* Interscience, New York, 1969.

[Kos60] J.L. Koszul: *Lectures on Fibre Bundles and Differential Geometry.* Tata Institute, Bombay, 1960.

[KS09] A. Kogasaka and K. Sakai: *A Hilbert cube compactification of the function space with the compact-open topology.* Cent. Eur. J. Math. **7** (2009), 670-682.

[Kur68] K. Kuratowski: *Topology.* Halner, New York, 1968.

[Lan99] S. Lang: *Fundamentals of Differential Geometry.* Springer, New York, 1999.

[Laz65] M. Lazard: *Groupes Différentiables.* Notes, Institut H. Poincaré, Paris, 1965.

[Lem86] R. Lemmert: *On ordinary differential equations in locally convex spaces.* Nonlinear Analysis, Theory, Methods and Applications **10** (1986), 1385–1390.

[Les67] J.A. Leslie: *On a differential structure for the group of diffeomorphisms.* Topology **46** (1967), 263–271.

[Les68] J.A. Leslie: *Some Frobenious theorems in Global Analysis.* J. Diff. Geom. **42** (1968), 279–297.

[LGV] V. Lakshmikantham, T. Gnana Bhaskar, J. Vasundhara Devi: *Theory of Set Differential Equations in a Metric Space* (to appear).

[LT09] A.T-M. Lau and W. Takahashi: *Fixed point properties for semigroup of nonexpansive mappings on Fréchet spaces.* Nonlinear Anal. **70** (2009), 3837–3841.

[Lob92] S.G. Lobanov: *Picard's theorem for ordinary differential equations in locally convex spaces.* Izv. Ross.Akad. Nauk Ser. Mat. **56** (1992), 1217–1243; English translation in Russian Acad. Sci. Izv. Math. **41** (1993), 465–487.

[Mag04] J.-P. Magnot: *Structure groups and holonomy in infinite dimensions.* Bull. Sci. Math. **128** (2004), 513–529.

[Mai62] B. Maissen: *Lie Gruppen mit Banachräumen als Parameterräume.* Acta Mathem. **108** (1962), 229–270.

[Man98] P. Manoharan: *Characterization for spaces of sections.* Proc. Amer. Math. Soc. **126** (1998), 1205–1210.

[Man02] P. Manoharan: *On the geometry of free loop spaces.* Int. J. Math. Math. Sci. **30** (2002), 15–23.

[Max72] L. Maxim: *Connections compatible with Fredholm structures on Banach manifolds.* Anal. Ştiint. Univ. "Al. I. Cuza" Iaşi **18** (1972), 384–400.

[Mil58] J. Milnor: *On the existence of a connection with curvature zero.* Com. Math. Helvetici **32** (1958), 215–223.

[MM83] L. Mangiarotti and M. Modugno: *Fibred spaces, jet spaces and connections for field theories.* Proc. International Meeting on Geometry and

Physics, Florence, 12-15 October 1982. Ed. M. Modugno, Pitagora Editrice, Bologna, 1983, 135–165.

[MV85] R. Meise and D. Vogt: *A characterization of the quasinormable Fréchet spaces.* Math. Nachr. **122** (1985), 141–150.

[MV97] R. Meise and D. Vogt: *Introduction to Functional Analysis.* Oxford Graduate Texts in Mathematics 2, Clarendon Press, Oxford University Press, New York, 1997.

[Mod87] M. Modugno: *Systems of vector valued forms on a fibred manifold and applications to gauge theories.* Proc. Conference Differential Geometric Methods in Mathematical Physics, Salamanca 1985. Lecture Notes in Mathematics **1251**, Springer-Verlag, Heidelberg, 1987, 238–264.

[Nab00] G.L. Naber: *Topology, Geometry, and Gauge Fields. Interactions.* Springer, New York, 2000.

[NBa13] F. Nielsen and F. Barbaresco (Eds.): *Geometric Science of Information*, Proceedings GSI 2013. Lecture Notes in Computer Science 8085, Springer, Heidelberg (2013).

[NBh12] F. Nielsen and R. Bhatia (Eds.): *Matrix Information Geometry.* Springer-Verlag, Heidelberg, 2012. http://www.springer.com/engineering/signals/book/978-3-642-30231-2

[Nee06] K-H. Neeb: *Infinite Dimensional Lie Groups*, 2005 Monastir Summer School Lectures, Lecture Notes, January 2006. http://www.math.uni-hamburg.de/home/wockel/data/monastir.pdf

[Nee09] K-H. Neeb and C. Wockel, *Central extensions of groups of sections.* Ann. Global Anal. Geom. **36** (2009), 381-418.

[Nic95] L.I. Nicolaescu: *Lecture Notes on the Geometry of Manifolds.* World Scientific, Singapore, 1996.

[Nil14] F. Nielsen (Ed.): *Geometric Theory of Information.* Springer, Heidelberg (2014) in press.

[NR61] M.S. Narasimhan and S. Ramanan: *Existence of universal connections I.* Amer. J. Math. **83** (1961), 563–572.

[NR63] M. S. Narasimhan and S. Ramanan: *Existence of universal connections II.* Amer. J. Math. **85** (1963), 223–231.

[NS95] S. Nag and D. Sullivan: *Teichmuller theory and the universal period mapping via quantum calculus and the $H^{1/2}$ space on the circle.* Osaka Journal Math. **32** (1995), 1–34.

[Omo70] H. Omori: *On the group of diffeomorphisms on a compact manifold.* Proc. Symp. Pure Appl. Math. AMS **XV** (1970), 167–183.

[Omo74] H. Omori: *Infinite Dimensional Lie Transformation Groups*, Lecture Notes in Mathematics **427**, Springer-Verlag, Heidelberg, 1974.

[Omo78] H. Omori: *On Banach Lie groups acting on finite dimensional manifolds.* Tohoku Math. J. **30** (1978), 223–250.

[Omo97] H. Omori: *Infinite-dimensional Lie groups.* Translations of Mathematical Monographs **158**, Amer. Math. Soc., 1997.

[Pa65] R.S. Palais: *Seminar on the Atiyah-Singer index theormem.* Ann. Math. Studies **57**, Princeton Univ. Press, Princeton NJ, 1965.

[Pa68] R.S. Palais: *Foundations of global non-linear analysis.* W.A. Benjamin, New York, 1968.

[Pal68] V.P. Palamodov: *The projective limit functor in the category of linear topological spaces*: Math. USSR-Sbornik **75** (**117**) (1968), 529–559. http://iopscience.iop.org/0025-5734/4/4/A05

[Pap80] N. Papaghiuc: *Équations différentielles linéaires dans les espaces de Fréchet*. Rev. Roumaine Math. Pures Appl. **25** (1980), 83–88.

[Pay01] S. Paycha: *Basic prerequisites in differential geometry and operator theory in view of applications to quantum field theory*. Preprint Université Blaise Pascal, Clermont, France, 2001.

[Pen67] J.-P. Penot: *De submersions en fibrations. Séminaire de Géométrie Différentielle de P. Libermann*. Paris, 1967.

[Pen69] J.-P. Penot: *Connexion linéaire déduite d' une famille de connexions linéaires par un foncteur vectoriel multilinéaire*. C. R. Acad. Sc. Paris **268** (1969), série A, 100–103.

[Pha69] Q.M. Pham: *Introduction à la Géométrie des Variétés Différentiables*. Dunod, Paris, 1969.

[Pir09] A.Yu. Pirkovskii: *Flat cyclic Fréchet modules, amenable Fréchet algebras, and approximate identities*. Homology, Homotopy Appl. **11** (2009), 81-114.

[PV95] M. Poppenberg and D. Vogt: *A tame splitting theorem for exact sequences of Fréchet spaces*. Math. Z. **219** (1995), 141–161.

[Rea88] C.J. Read: *The invariant subspace problem for a class of Banach spaces, 2. Hypercyclic operators*. Israel J. Math. **63** (1988), 1–40.

[Sau87] D.J. Saunders: *Jet fields, connections and second order differential equations*. J. Phys. A: Math. Gen. **20** (1987), 3261–3270.

[Sch80] H.H. Schaeffer: *Topological Vector Spaces*. Springer-Verlag, Heidelberg, 1980.

[SM94] C.G. Small and D.L. McLeash: *Hilbert space methods in probability and statistical inference*. John Wiley, Chichester, 1994, reprinted 2011.

[Smo07] N.K. Smolentsev: *Spaces of Riemannian metrics*. Journal of Mathematical Sciences **142** (2007), 2436-2519.

[SS70] L.A. Steen and J.A. Seebach Jnr.: *Counterexamples in Topology*. Holt, Rinehart and Winston, New York, 1970.

[SW72] R. Sulanke and P. Wintgen: *Differentialgeometrie und Faserbündel*. Birkhäuser Verlag, Basel, 1972.

[Tak79] F. Takens: *A global version of the inverse problem of the calculus of variations*. J. Dif. Geom. **14** (1979), 543–562.

[Thi07] T. Thiemann: *Modern canonical quantum general relativity*. Cambridge University Press, Cambridge UK, 2007.

[Tka10] M. Tkachenko: *Abelian groups admitting a Fréchet-Urysohn pseudocompact topological group topology*. J. Pure Appl. Algebra **214** (2010), 1103-1109.

[Val89] M. Valdivia: *A characterization of totally reflexive Fréchet spaces*. Math. Z. **200** (1989), 327–346.

[Vas78(a)] E. Vassiliou: *(f, φ, h)-related connections and Liapunoff's theorem*. Rend. Circ. Mat. Palermo **27** (1978), 337–346.

[Vas78(b)] E. Vassiliou: *On the infinite dimensional holonomy theorem*. Bull. Soc. Roy. Sc. Liège **9-10** (1978), 223–228.

[Vas81] E. Vassiliou: *On affine transformations of banachable bundles.* Colloq. Math. **44** (1981), 117–123.

[Vas82] E. Vassiliou: *Transformations of linear connections.* Period. Math. Hung. **13** (1982), 289–308.

[Vas83] E. Vassiliou: *Flat bundles and holonomy homomorphisms.* Manuscripta Math. **42** (1983), 161–170.

[Vas86] E. Vassiliou: *Transformations of linear connections II.* Period. Math. Hung. **17** (1986), 1–11.

[Vas13] E. Vassiliou: *Local connection forms revisited.* Rend. Circ. Mat. Palermo **62** (2013), 393–408. http://arxiv.org/pdf/1305.6471.pdf.

[Vel02] J.M. Velhinho: *A groupoid approach to spaces of generalized connections.* J. Geometry and Physics **41** (2002), 166-180.

[VerE83] P. Ver Eecke: *Fondements du Calcul Différentielle.* Presses Universitaires de France, Paris, 1983.

[VerE85] P. Ver Eecke: *Applications du Calcul Différentielle.* Presses Universitaires de France, Paris, 1985.

[Vero74] M.E. Verona: *Maps and forms on generalised manifolds.* St. Cerc. Mat. **26** (1974), 133–143 (in romanian).

[Vero79] M.E. Verona: *A de Rham theorem for generalized manifolds.* Proc. Edinburg Math. Soc. **22** (1979), 127–135.

[VG97] E. Vassiliou and G. Galanis: *A generalized frame bundle for certain Fréchet vector bundles and linear connections.* Tokyo J. Math. **20** (1997), 129–137.

[Vil67] J. Vilms: *Connections on tangent bundles.* J. Diff. Geom. **1** (1967), 235–243.

[Vog77] D. Vogt: *Characterisierung der Unterräume von s.* Math. Z. **155** (1977), 109-117.

[Vog79] D. Vogt: *Sequence space representations of spaces of test functions and distributions.* Functional analysis, holomorphy and approximation theory (Rio de Janeiro, 1979), pp. 405–443. Lecture Notes in Pure and Appl. Math. **83**, Dekker, New York, 1983.

[Vog83] D. Vogt: *Fréchetträume, zwishen denen jede stetige Abbildung beschränkt ist.* J. Reine Angew. Math. **345** (1983), 182-200.

[Vog87] D. Vogt: *On the functors* $\mathrm{Ext}^1(E, F)$ *for Fréchet spaces.* Studia Math. **85** (1987), 163-197.

[Vog10] D. Vogt: *A nuclear Fréchet space consisting of* C^∞*-functions and failing the bounded approximation property.* Proc. Amer. Math. Soc. **138** (2010), 1421-1423.

[VW80] D. Vogt and M.J. Wagner: *Josef Charakterisierung der Quotintenräume von s und eine Vermutung von Martineau.* Studia Math. **67** (1980), 225-240.

[VW81] D. Vogt and M.J. Wagner: *Charakterisierung der Quotintenräume der nuklearen stabilen Potenzreihenräume von unendlichen Typ.* Studia Math. **70** (1981), 63-80.

[War83] F.W. Warner: *Foundations of Differentiable Manifolds and Lie Groups.* GTM **94**, Springer-Verlag, New York, 1983.

[Wen03] J. Wengenroth: *Derived functors in functional analysis.* Lecture Notes in Mathematics **1810**, Springer-verlag, Berlin, 2003.

[Wol09] E. Wolf: *Quasinormable weighted Fréchet spaces of entire functions.* Bull. Belg. Math. Soc. Simon Stevin **16** (2009), 351-360.

List of Notations

The list contains most of notations used in this work,
together with a brief description and the page of their
first appearance.

Chapter 1

$\mathcal{L}(\mathbb{E}, \mathbb{F})$	space of continuous linear maps from $\mathbb{E}$ to $\mathbb{F}$, 1
$\mathcal{L}(\mathbb{E})$	abbreviation of $\mathcal{L}(\mathbb{E}, \mathbb{E})$, 1
$\mathcal{L}is(\mathbb{E}, \mathbb{F})$	space of invertible elements in $\mathcal{L}(\mathbb{E}, \mathbb{F})$, 1
$\mathcal{L}is(\mathbb{E})$	abbreviation of $\mathcal{L}is(\mathbb{E}, \mathbb{E})$, 1
$\mathrm{GL}(\mathbb{E})$	general linear group of $\mathbb{E}$, 1
$Df(x) \in \mathcal{L}(\mathbb{E}, \mathbb{F})$	(Fréchet) derivative of $f\colon U \subseteq \mathbb{E} \to \mathbb{F}$ at $x \in U$, 1
$Df\colon U \to \mathcal{L}(\mathbb{E}, \mathbb{F})$	total (Fréchet) derivative of f, 2
$D^k f$	k-th derivative of f, 2
C^∞	symbol of smoothness, 2
(U, ϕ)	chart of a manifold, 2
$(U, \phi, \mathbb{B})$	chart with specific model $\mathbb{B}$, 2
$\mathcal{A}$	(maximal) atlas of a manifold, 2
$f_{VU},\, f_{\psi\phi},\, f_{\beta\alpha}$	expressions of local representations of $f\colon M \to N$, 2
$[(\alpha, x)]$ or $[\alpha, x]$	equivalent class of tangent curves, 3
$T_x M$	tangent space of M at x, 3
$\overline{\phi}\colon T_x M \to \mathbb{B}$	isomorphism induced by a chart $(U, \phi, \mathbb{B})$, 3
$\tau_M\colon TM \to M$	projection of the tangent bundle, 4

$(\pi^{-1}(U), \Phi)$,	chart of TM induced by (U, ϕ), 4
$T_x f$	differential or tangent map of $f: M \to N$ at x, 4
$\dot{\alpha}(t)$	velocity vector of a smooth curve α at t, 5
Tf	(total) differential or tangent map of $f: M \to N$, 6
$\mathcal{X}(M)$	space of smooth vector fields on M, 6
$X(f)$	map defined by $X(f)(x) := T_x f(X_x)$, 6
$\Phi \circ X \circ \phi^{-1}$	local representation of $X \in \mathcal{X}(M)$, 6
X_ϕ	principal part of $\Phi \circ X \circ \phi^{-1}$, 7
X_α	abbreviation of X_{ϕ_α}, 7
$\gamma: G \times G \to G$	multiplication of a group G, 8
$\alpha: G \to G$	inversion of a group G, 8
$\lambda_g: G \to G$	left translation by $g \in G$, 8
$\rho_g: G \to G$	right translation by $g \in G$, 8
$\mathcal{L}(G)$	Lie algebra of a Lie group G, 8
$h: \mathcal{L}(G) \to T_e G$	identification of $\mathcal{L}(G)$ with $T_e G$, 9
$\mathfrak{g}$	$T_e G$ considered as a Lie algebra via h, 9
$\exp \equiv \exp_G$	exponential map of G, 9
$\mathrm{Ad}: G \to \mathrm{Aut}(\mathfrak{g})$	adjoint representation of G, 9
$\mathcal{A}_k(T_x B, \mathfrak{g})$	space of k-alternating maps $T_x B \to \mathfrak{g}$, 10
$A_k(TB, \mathfrak{g})$	(total space of the) bundle of k-alternating maps, 10, 22
$\Lambda^k(B, \mathfrak{g})$	space of $\mathfrak{g}$-valued k-forms on B, 10
$D^l f \equiv f^{-1} df$	left Maurer-Cartan/logarithmic differential, 11
$D^r f \equiv df.f^{-1}$	right Maurer-Cartan/logarithmic differential, 11
$F_\theta: \widetilde{M} \to G$	fundamental solution of $D^r z = \tilde{\pi}^* \theta$, 12
$\theta^\#: \pi_1(M) \to G$	monodromy homomorphism of equation $D^r x = \theta$, 12
$\delta: M \times G \to M$	action of G on (the right of) M, 12
X^*	fundamental or Killing vector field, 13
$R_g = \delta_g: M \to M$	right translation of M by $g \in G$, 13
$E_x = \pi^{-1}(x)$	fibre, over x, of a vector bundle E, 14
$U_{\alpha\beta}$	abbreviation of $U_\alpha \cap U_\beta$, 14
(U_α, τ_α)	trivialization of a vector bundle, 14

$\tau_{\alpha,x}: \pi^{-1}(x) \to \mathbb{E}$	isomorphism induced by (U_α, τ_α), 14
$\ell = (E, B, \pi)$ or E	vector bundle, 15
$(U_\alpha, \phi_\alpha, \Phi_\alpha)$	vector bundle chart, 15
$\Phi_{\alpha,x}: E_x \to \mathbb{E}$	isomorphism induced by $(U_\alpha, \phi_\alpha, \Phi_\alpha)$, 15
$T_{\alpha\beta}$	transition map (function) of a vector bundle over $U_{\alpha\beta}$, 16
f_x	restriction of f on the fibre over x, 17
$(f, h): \ell_1 \to \ell_2$	morphism of vector bundles, 17
$f: E_1 \to E_2$	another notation for morphisms of vector bundles over the same base, 17
$\mathcal{VB}_B$	category of vector bundles over B, 17
$\mathcal{VB}_B(\mathbb{E})$	category of vector bundles over B, of fibre type $\mathbb{E}$, 17
$\mathfrak{GL}(\mathbb{E})$	sheaf of germs of smooth GL($\mathbb{E}$)-valued maps, 19
$H^1(B, \mathfrak{GL}(\mathbb{E}))$	1st cohomology group of B with coefficients in $\mathfrak{GL}(\mathbb{E})$, 19
$(E_1 \times_B E_2, B, \pi)$	fibre product of vector bundles over B, 19
$(E_1 \oplus E_2, B, \pi)$	direct/Whitney sum of vector bundles over B, 19
$\begin{aligned}f^*(\ell) = \\ = (f^*(E), Y, f^*(\pi))\end{aligned}$	pull-back of $\ell = (E, B, \pi)$ by $f: Y \to B$, 20
$(L(E, E'), B, L)$	linear map bundle, 21
$L_k(E_1 \times \cdots \times E_k, E')$	k-linear bundle map, 22
$A_k(E, E')$	k-alternating map bundle, 22
$A_k(TB, \mathfrak{g})$	k-alternating map bundle with fibres $\mathcal{A}_k(T_x B, \mathfrak{g})$, 22
$\Gamma(E) \equiv \Gamma(B, E)$	$\mathcal{C}^\infty(B, \mathbb{R})$-module of global smooth sections of (E, B, π), 23
$\Gamma(U, E)$	module of smooth sections of E over $U \subseteq B$, 23
$\Phi \circ \xi \circ \phi^{-1}$	local representation of $\xi \in \Gamma(E)$ relative to a vector bundle chart (U, ϕ, Φ), 23
$\xi_\phi: \phi(U) \to \mathbb{E}$	local principal part of $\Phi \circ \xi \circ \phi^{-1}$, 23
$\xi_\alpha: \phi_\alpha(U_\alpha) \to \mathbb{E}$	abbreviation of ξ_{ϕ_α}, 23
$P^k(\mathbb{B}, \mathbb{E})$	space of $\mathbb{E}$-valued k-polynomials on $\mathbb{B}$, 23
$p^k f(a)$	polynomial $(f(a), Df(a), \ldots, D^k f(a))$, 23
$j_x^k \xi$	k-jet of ξ at x, 24
$J^k(\ell) = (J^k E, B, \pi^k)$	k-jet bundle of sections of $\ell = (E, B, \pi)$, 24

$K := r \circ V$	connection map, 28
$K_\alpha \equiv K_{U_\alpha}$	K relative to $(U_\alpha, \phi_\alpha, \Phi_\alpha)$, 28
κ_α	local component of a connection K, 28
Γ_α	Christoffel symbol (map) over (U_α, ϕ_α), 29
∇	covariant derivation, 31
$\Gamma_\gamma(E)$	set of sections of E along γ, 32
$\partial \equiv \dfrac{d}{dt}$	basic vector field of $\mathbb{R}$, 32
$\tau_\gamma \colon E_{\gamma(0)} \to E_{\gamma(1)}$	parallel displacement or translation along γ, 33, 69
$^K\Phi_b,$	holonomy group of (a linear connection) K with reference point b, 33
$^K\Phi_b^0,$	restricted holonomy group of K with reference point b, 33
$f_{\beta\alpha}^{\#}$	local principal part of a vector bundle morphism (f, h), 34
$f^{\#} = f_{\alpha\alpha}^{\#}$	local principal part of a vector bundle morphism (f, id_B), 37
(U, Φ) or (U, Ψ)	local trivialization of a principal bundle (cf. vector bundle charts, p. 15), 38
$\Phi_x \colon \pi^{-1}(x) \to G$	fibre isomorphism induced by (U, Φ) (cf. the vector bundle analog, p. 15), 39
$\ell = (P, G, B, \pi)$ or P	principal bundle, 37
$\Gamma(U, P)$	set of smooth sections of P over $U \subseteq B$, 40
$s_\alpha \in \Gamma(U_\alpha, P)$	(natural) section of P over U_α, 40
$k \colon P \times_B P \to G$	map defined by $q = p \cdot k(p, q)$, 40
$g_{\alpha\beta} \colon U_{\alpha\beta} \to G$	transition map or function of a principal bundle P over $U_{\alpha\beta}$, 41
$h^*(\ell) =$ $= (h^*(P), G, B', \pi^*)$	pull-back of $\ell = (P, G, B, \pi)$ by $h \colon B' \to B$, 43
$\ell(E) \equiv P(E)$	frame bundle of a vector bundle (E, B, π_E), 44
$P \times^G H$	quotient of $P \times H$ induced by a Lie group morphism $\phi \colon G \to H$, 46
$\varphi(\ell)$	principal bundle $(P \times^G H, H, B, \pi_H)$ associated to $(P, G.B, \pi)$ by $\varphi \colon G \to H$, 47
$\nu \colon P \times \mathfrak{g} \to TP$	vector bundle morphism with $\nu(p, X_e) = X_p^*$, 52
VP	vertical subbundle of TP, 52

$\nu_p\colon \mathfrak{g} \to V_pP$	isomorphism identifying $\mathfrak{g}$ with V_pP, 52
HP	horizontal subbundle of TP, 54
u^h	horizontal component of $u \in T_pP$, 55
$\Lambda^1(P, \mathfrak{g})$	space of $\mathfrak{g}$-valued 1-forms on P, 55
$\omega \in \Lambda^1(P, \mathfrak{g})$	connection form on P, 55
u^v	vertical component of $u \in T_pP$, 55
$\omega_\alpha \in \Lambda^1(U_\alpha, \mathfrak{g})$	local connection form over U_α, 56
$g_\alpha\colon \pi^{-1}(U_\alpha) \to G$	the map with $g_\alpha(p) = (\mathrm{pr}_2 \circ \Phi_\alpha)(p)$, 57
$\Omega \in \Lambda^2(P, \mathfrak{g})$	curvature form of ω, 64
$\Omega_\alpha \in \Lambda^2(U_\alpha, \mathfrak{g})$,	local curvature form over U_α, 64
ω^0	canonical flat connection on $B \times G$, 67
$\widehat{\alpha}_p$	horizontal lift of α with $\widehat{\alpha}_p(0) = p$, 68
Φ_x	holonomy group of ω with reference point $x \in B$, 69
Φ_x^0	restricted holonomy group with reference point $x \in B$, 70
C_x	loop group at $x \in B$, 69
C_x^0	group of 0-homotopic loops at $x \in B$, 69
k_p	the map given by $\tau_\alpha(p) = p \cdot k_p(\tau_\alpha)$, 70
Φ_p	holonomy group of ω with reference point $p \in P$, 70
Φ_p^0	restricted holonomy group with reference point $p \in P$, 70
$P[p]$	holonomy bundle of P at p, 70
$H(B, G)$	set of classes of equivalent flat bundles with base B and group G, 71
$h_\omega\colon \pi_1(B) \to G$	holonomy homomorphism of a flat bundle (P, ω), 71
$\mathcal{S}(B, G)$	set of classes of similar homomorphisms $h\colon \pi_1(B) \to G$, 71

Chapter 2

$p\colon \mathbb{F} \to \mathbb{R}$	seminorm, 74
$\Gamma = \{p_\alpha\}_{\alpha \in I}$	family of seminorms, 74
$\mathcal{T}_\Gamma$	topology induced by Γ above, 74
$\mathcal{B}_\Gamma$	neighborhood basis of $\mathcal{T}_\Gamma$, 74
$\mathbb{F}$	often a Fréchet space (mainly from Chapter 2 onwards), 75

$Df \colon U \times \mathbb{F}_1 \to \mathbb{F}_2$	total derivative of f (à la Leslie, in Fréchet spaces), 80
$\{E^i, \rho^{ji}\}_{i,j \in I}$	projective system, 82
$\rho^{ji} \colon E^j \to E^i$	connecting morphisms of the preceding, 82
$\varprojlim E^i$	projective limit of $\{E^i, \rho^{ji}\}_{i,j \in I}$, 82
$\rho^i \colon \varprojlim E^i \to E^i$	i-th canonical projection of the preceding, 82
$(x_i) \equiv (x_i)_{i \in I}$	equivalent expressions of elements in projective limits, 82
$\varprojlim f^i$	projective limit of maps, 84
$\mathcal{H}(\mathbb{F}_1, \mathbb{F}_2)$	Fréchet space of particular continuous linear maps from $\mathbb{F}_1$ to $\mathbb{F}_2$, 88
$\mathcal{H}^i(\mathbb{F}_1, \mathbb{F}_2)$	Banach space of particular continuous linear maps from $\mathbb{F}_1$ to $\mathbb{F}_2$, 88
$\varepsilon \colon \mathcal{H}(\mathbb{F}_1, \mathbb{F}_2) \to \mathcal{L}(\mathbb{F}_1, \mathbb{F}_2)$	the map $(f^i) \mapsto \varprojlim f^i$, 88
$\mathcal{L}_I(\mathbb{F})$	a particular subspace of $\mathcal{L}(\mathbb{F}) = \mathcal{L}(\mathbb{F}, \mathbb{F})$, 95
$\mathcal{H}(\mathbb{F})$	abbreviation of $\mathcal{H}(\mathbb{F}, \mathbb{F})$, 95
comp	the composition map $\mathcal{L}_I(\mathbb{F}) \times \mathcal{L}(\mathbb{F}) \to \mathcal{L}(\mathbb{F})$: $(f, g) \mapsto f \circ g$, 96
ev	the evaluation map $\mathcal{L}(\mathbb{E}, \mathbb{F}) \times \mathbb{E} \to \mathbb{F}$: $(f, a) \mapsto f(a)$, 99

Chapter 3

$\{M^i; \mu^{ji}\}_{i,j \in \mathbb{N}}$	projective system of manifolds, 106
$\mu^i \colon M = \varprojlim M^i \to M^i$	the i-th canonical projection of M, 108
$\left(\varprojlim U^i, \varprojlim \varphi^i \right)$	typical chart of a projective limit manifold, 107
$\{G^i; g^{ji}\}_{i,j \in \mathbb{N}}$	projective system of Lie groups, 122
$g^i \colon G = \varprojlim G^i \to G^i$	i-th canonical projection of G, 121

Chapter 4

$\{\ell^i; F^{ji}\}_{i,j \in \mathbb{N}}$	projective system of principal bundles, with $\ell^i = (P^i, G^i, B, \pi^i)$, $F^{ji} = (p^{ji}, g^{ji}, \mathrm{id}_B)$, 140
$p^{ji} \colon P^j \to P^i$	connecting morphisms of $\{P^i, p^{ji}\}_{i,j \in \mathbb{N}}$, 140
$p^i \colon P = \varprojlim P^i \to P^i$	i-th canonical projection of P, 140
$(U_\alpha, \Phi_\alpha)_{i \in \mathbb{N}}$	abbreviation of a trivializing cover $\left(\varprojlim_{i \in \mathbb{N}} U_\alpha^i, \varprojlim_{i \in \mathbb{N}} \Phi_\alpha^i \right)_{i \in \mathbb{N}}$ of $\varprojlim \ell^i$, 140

$\underline{k}^i : (\pi^i)^{-1} \to G^i$ the map determined by $u = s^i(\pi^i(u)) \cdot \underline{k}^i(u)$, 145

$\underline{k} : \pi^{-1}(U) \to G$ the projective limit of the preceding, 145

$\omega^i(u^i)$ same as $\omega^i_{u^i}$, 151

$\omega^i(u^i)(w)$ same as $\omega^i_{u^i}(w)$, 151

$\omega = \varprojlim \omega^i$ projective limit of connection forms, 152

$\{\omega^i_\alpha\}_{\alpha \in I}$ (for fixed $i \in \mathbb{N}$) the local connection forms of ω^i, 155

Chapter 5

$\mathcal{H}^i(\mathbb{F})$ abbreviation of $\mathcal{H}^i(\mathbb{F}, \mathbb{F})$, 183

$\mathcal{H}^i_0(\mathbb{F})$ the group $\mathcal{H}^i(\mathbb{F}) \cap \prod_{j=1}^i \mathcal{L}is(\mathbb{E}_j)$, 184

$\mathcal{H}_0(F)$ the group $\mathcal{H}(\mathbb{F}) \cap \prod_{j=1}^\infty \mathcal{L}is(\mathbb{E}_j)$, 184

$\{(E^i, B, \pi^i); f^{ji}\}_{i,j \in \mathbb{N}}$ projective system of vector bundles, 185

$f^{ji} : E^j \to E^i$ connecting morphisms of the preceding, 185

E^i_U variant of $(\pi^i)^{-1}(U)$, 185

$f^i : E = \varprojlim E^i \to E^i$ i-th canonical projection of E, 193

$(T^*_{\alpha\beta})$ $\mathcal{H}_0(\mathbb{F})$-valued cocycle, 190

Chapter 6

$J^\infty(E)$ infinite jet bundle $\varprojlim J^k E$, 211

$\mathcal{H}^i_0(\mathbb{F}_1, \mathbb{F}_2)$ the space $\mathcal{H}^i(\mathbb{F}_1, \mathbb{F}_2) \cap \prod_{j=1}^i \mathcal{L}is(\mathbb{E}^j_1, \mathbb{E}^j_2)$, 216

$(\boldsymbol{P}(E^i), \mathcal{H}^i_0(\mathbb{F}), B, \boldsymbol{p}^i)$ Banach principal bundle with total space $P(E^i) := \bigcup_{x \in B} \mathcal{L}is(\mathbb{F}, E_x)$, 217

$\{\boldsymbol{g}^i_{\alpha\beta}\}_{\alpha\beta \in I}$ transition functions of $P(E^i)$, 218

$\boldsymbol{P}(E)$ abbreviation of $\varprojlim \boldsymbol{P}(E^i)$, 218

Chapter 7

Γ^*_α generalized Christoffel symbol over (U_α, ϕ_α), 229

Chapter 8

$\gamma_1 \approx_x \gamma_2$ equivalence of curves up to acceleration, 249

$\ddot{\gamma}$ acceleration of a curve γ, 249

$T^2_x B$ second order tangent space of B at x, 249

$T^2 M$ second order tangent bundle of B, 249

Subject index

Printed in the United States
by Baker & Taylor Publisher Services